GENETIC MAPS

Locus Maps of Complex Genomes

FIFTH EDITION

Stephen J. O'Brien, Editor

Laboratory of Viral Carcinogenesis, National Cancer Institute

BOOK 5

Human Maps

COLD SPRING HARBOR LABORATORY PRESS 1990

GENETIC MAPS

First edition, March 1980
Second edition, June 1982
Third edition, June 1984
Fourth edition, July 1987
Fifth edition, February 1990

GENETIC MAPS
Locus Maps of Complex Genomes
Fifth Edition
BOOK 5 Human Maps

© 1990 by Cold Spring Harbor Laboratory Press

Printed in the United States of America
Library of Congress Catalog Card Number 84-644938
ISBN 0-87969-346-0
ISSN 0738-5269

Cover design by Leon Bolognese

All Cold Spring Harbor Laboratory Press publications may be ordered directly from Cold Spring Harbor Laboratory, Box 100, Cold Spring Harbor, New York 11724. Phone 1-800-843-4388. In New York (516) 367-8423.

PREFACE

"The map is a sensitive indicator of the changing thought of man, and few of his works seem to be such an excellent mirror of culture and civilization."

Norman J.W. Thrower
Maps and Men, 1972

The early geographic explorers are revered by their countrymen and descendants throughout the world for pioneering the discovery and charting of foreign lands. The precision and attention to detail accorded to their charts and maps are almost always taken for granted today. Astronauts, and then those who saw their photographs from space, have marveled at the resemblance of these real images of the Florida peninsula, the boot of Italy, and the British Isles to the maps we had grown up with. John Noble Wilford, in his fascinating monograph entitled *The Mapmakers,* reminds us that it is only within the last quarter of the twentieth century that it could be said that the earth has been mapped. It is humbling to consider how important these maps have been to our history, our culture, and our way of life today; in this context, the early cartographers were indeed intrepid.

The new explorers of the eukaryote and prokaryote genome will certainly one day be considered with equivalent veneration, since the topography of their charts and maps may prove as daunting as the earth's surface. The animal and plant genomes are products of several hundred million years of biological evolution, and there are clearly far more secrets to be deciphered in the genomes that can be anticipated in future generations. However, genetic detectives have begun to unravel some of the mysteries of genetic organization, and as we enter the final decade of the century, the scientific community (as well as the enlightened public) has begun to grasp the enormous value of generating and expanding genetic maps of man, of agriculturally significant plants and animals, and of model systems that allow us a glimpse of how genes are organized, replicated, and regulated.

The first genetic map was formulated by A.H. Sturtevant in 1913 and consisted of five genes arranged in a linear fashion along the X chromosome of the fruit fly, *Drosophila melanogaster.* In the ensuing decades, gene mapping of numerous species has proceeded deliberately and cumulatively in organisms as diverse as flies, corn, wheat, mink, apes, man, and bacteria. All of these maps, whether based on recombination, restriction, physical or DNA sequence, are predicated on Sturtevant's logical notion that gene order on a chromosome could be displayed as a linear array of genetic markers. The results of these efforts in more than 100 genetically studied organisms are the basis of *GENETIC MAPS: Locus Maps of Complex Genomes.*

During the preparation of the previous edition, it occurred to me that the rate of growth of the gene mapping effort was so rapid that we should prepare to publish future editions in multiple volumes. The fifth edition is a realization of that idea. *Genetic Maps* now consists of six smaller books, each based on an arbitrary subdivision of biological organisms. These are BOOK 1 Viruses; BOOK 2 Bacteria, Algae, and Protozoa; BOOK 3 Lower Eukaryotes; BOOK 4 Nonhuman Vertebrates; BOOK 5 Human Maps; BOOK 6 Plants. Each of these is available in paperback at a modest cost from Cold Spring Harbor Laboratory Press. The entire compendium can be purchased as a hardback volume of 1098 map pages suitable for libraries and research institutions.

Our original intent was to publish in one volume complete, referenced genetic maps of every organism with a substantive group of assigned loci. We intentionally excluded DNA sequences, since these are easily available in several computer databanks (Genbank, EMBL, and others). Text was to be kept at a minimum, and the maps would be both comprehensive and concise. The collection was to be updated every 2–3 years, and each new volume would contain a complete revision, rendering the previous volume obsolete.

The original publication of *Genetic Maps* was an experiment, which I believe now can be judged a success. The execution of such a venture depended heavily on the support and cooperation of thousands of geneticists throughout the worldwide scientific community. This cooperation was graciously extended, and the result was an enormously valuable and accessible collection. On behalf of my colleagues in the many fields that we call genetic biology, I gratefully acknowledge the numerous geneticists, scientists, and readers who have contributed to or corrected these maps. To ensure the continuation of these heroic efforts in the future, readers are encouraged to send to me any suggestions for improvement, particularly suggestions for new maps to be included, as well as constructive criticisms of the present maps. When new organisms are recommended, I would also appreciate names and addresses of prospective authors. In addition, readers are encouraged to supply corrections, reprints, and new mapping data to the appropriate authors who may be included in future editions.

The compilation of each genetic map, like the drafting of the first geographic maps, is an extremely tedious, yet important, assignment. All of the authors deserve special thanks for the large efforts they have expended in contributing their maps. Even in this computer age, we often feel like we are proofreading the telephone book, but I, for one, believe that the final product makes the effort worthwhile. Finally, I acknowledge specifically my editorial assistants, Patricia Johnson and Virginia Frye, who have cheerfully and expeditiously carried the bulk of the editorial activities for the present volume, and Annette Kirk, Lee Martin, and John Inglis of Cold Spring Harbor Laboratory Press for support and advice on the preparation of this edition.

Stephen J. O'Brien, *Editor*

CONTENTS

COMPLETE CONTENTS OF THE FIFTH EDITION

PROTOZOA

Paramecium tetraurelia	K.J. Aufderheide and D. Nyberg
Tetrahymena thermophila	P.J. Bruns

BOOK 3 Lower Eukaryotes

FUNGI

Dictyostelium discoideum	P.C. Newell
Neurospora crassa	
nuclear genes	D.D. Perkins
mitochondrial genes	R.A. Collins
restriction polymorphism	R.L. Metzenberg and J. Grotelueschen
Saccharomyces cerevisiae	
nuclear genes	R.K. Mortimer, D. Schild, C.R. Contopoulou, and J.A. Kans
mitochondrial DNA	L.A. Grivell
Podospora anserina	D. Marcou, M. Picard-Bennoun, and J.-M. Simonet
Sordaria macrospora	G. Leblon, V. Haedens, A.D. Huynh, L. Le Chevanton, P.J.F. Moreau, and D. Zickler
Coprinus cinereus	J. North
Magnaporthe grisea (blast fungus)	D. Z. Skinner, H. Leung, and S.A. Leong
Phycomyces blakesleeanus	M. Orejas, J.M. Diaz-Minguez, M.I. Alvarez, and A.P. Eslava
Schizophyllum commune	C.A. Raper
Ustilago maydis	A.D. Budde and S.A. Leong
Ustilago violacea	A.W. Day and E.D. Garber
Aspergillus nidulans	
nuclear genes	A.J. Clutterbuck
mitochondrial genome	T.A. Brown

NEMATODE

Caenorhabditis elegans	M.L. Edgley and D.L. Riddle

INSECTS

Drosophila melanogaster	
biochemical loci	G.E. Collier
in situ hybridization data	E. Kubli and S. Schwendener
cloned genes	J. Merriam, S. Adams, G. Lee, and D. Krieger
Aedes (Stegomyia) aegypti	L.E. Munstermann
Aedes (Protomacleaya) triseriatus	L.E. Munstermann
Drosophila pseudoobscura	W.W. Anderson
Anopheles albimanus	S.K. Narang and J.A. Seawright
Anopheles quadrimaculatus	S.K. Narang, J.A. Seawright, and S.E. Mitchell
Nasonia vitripennis	G.B. Saul, 2nd

BOOK 4 Nonhuman Vertebrates

RODENTS

Mus musculus (mouse)	
nuclear genes	M.T. Davisson, T.H. Roderick, D.P. Doolittle, A.L. Hillyard, and J.N. Guidi
DNA clones and probes, RFLPs	J.T. Eppig
retroviral and cancer-related genes	C.A. Kozak

Rattus norvegicus (rat)	G. Levan, K. Klinga, C. Szpirer, and J. Szpirer
Cricetulus griseus (Chinese hamster)	
nuclear genes	R.L. Stallings, G.M. Adair, and M.J. Siciliano
CHO cells	G.M. Adair, R.L. Stallings, and M.J. Siciliano
Peromyscus maniculatus (deermouse)	W.D. Dawson
Mesocricetus auratus (Syrian hamster)	R. Robinson
Meriones unquiculatus (Mongolian gerbil)	R. Robinson

OTHER MAMMALS

Felis catus (cat)	S.J. O'Brien
Canis familiaris (dog)	P. Meera Khan, C Brahe, and L.M.M. Wijnen
Equus caballus (horse)	L.R. Weitkamp and K. Sandberg
Sus scrofa domestica L. (pig)	G. Echard
Oryctolagus cuniculus (rabbit)	R.R. Fox
Ovis aries (sheep)	G. Echard
Bos taurus (cow)	J.E.Womack
Mustela vison (American mink)	O.L. Serov and S.D. Pack
Marsupials and Monotremes	J.A. Marshall Graves

PRIMATES

Primate Genetic Maps	N. Creau-Goldberg, C. Cochet, C. Turleau, and J. de Grouchy
Pan troglodytes (chimpanzee)	
Gorilla gorilla (gorilla)	
Pongo pygmaeus (orangutan)	
Hylobates (Nomascus) concolor (gibbon)	
Macaca mulatta (rhesus monkey)	
Papio papio, hamadryas, cynocephalus (baboon)	
Cercopithecus aethiops (African green monkey)	
Cebus capucinus (capuchin monkey)	
Microcebus murinus (mouse lemur)	
Saquinus oedipus (cotton-topped marmoset)	P.A. Lalley
Aotus trivirgatus (owl monkey)	N.S.-F. Ma

FISH

Salmonid fishes *Salvelinus, Salmo, Oncorhynchus*	B. May and K.R. Johnson
Non-Salmonid fishes *Xiphophorus, Poeciliopsis, Fundulus, Lepomis*	D.C. Morizot

AMPHIBIAN

Rana pipiens (leopard frog)	D.A. Wright and C.M. Richards

BIRD

Gallus gallus (chicken)	R.G. Somes, Jr., and J.J. Bitgood

BOOK 5

HUMAN MAPS

The Human Gene Map (<u>Homo sapiens</u>) (2N = 46) as of HGM10

Iva H. Cohen[1], Huey S. Chan[1], Rowena K. Track[1], and Kenneth K. Kidd[1,2]

(1) :   Human Gene Mapping Library, 25 Science Park, New Haven, CT 06511

(2) :   Department of Human Genetics, Yale University School of Medicine, 333 Cedar Street, New Haven, CT 06510

The figures presented here represent the official gene map as compiled at the Tenth International Workshop on Human Gene Mapping (HGM10), and are based on data from the various committee reports of the Workshop proceedings (1). The correct citations for the scientific content of these figures are the committee reports from HGM10 (1). The figures have been produced as a joint effort by the staff of the Yale-HHMI Human Gene Mapping Library (HGML) and the staff of HGM10. Further information about the loci depicted below is available in the proceedings and/or the HGML on-line database.

Bars are used to indicate a region within which a locus or several loci have been mapped. In the figures, an effort has been made to place shorter bars closest to the chromosome, particularly those with a large number of loci mapped to them. A dagger (†) after a locus symbol indicates a tentative assignment based on weak evidence, while a double-dagger (‡) indicates an inconsistent assignment based on conflicting evidence. The symbols of genes mapped within the past year are underlined in the figures. Not included in the figures are those non-polymorphic anonymous DNA segments which have been given a status of D (that is, an assignment based only on the investigator's statement to the DNA Committee, for which no Chromosome Committee evaluation has occurred). An integrated alphabetical listing of loci on all chromosomes follows the figures. Equivalent to underlining in the figures, recently mapped genes are indicated by an asterisk in the table. Though present in the figures, anonymous DNA segments have been omitted from this listing. The DNA Committee Report (2) in the Workshop proceedings contains a sequential listing of all anonymous DNA segments.

The Human Gene Mapping Library is an on-line resource for literature and data relevant to the human genetics research community. The information in the database and its components (LIT, MAP, RFLP, PROBE, and CONTACT) is fully available to members of the scientific community. The HGML does not currently charge for this service nor for its access via Telenet; however, some users may have to pay local telephone or non-U.S. computer network charges. Requests for information from the database, more detailed documentation on its contents and capabilities, and information on the hardware requirements and procedures for obtaining direct interactive access to the Library can be sent by post, by electronic mail, by telephone, or via facsimile to:

> Yale-HHMI Human Gene Mapping Library
> 25 Science Park
> New Haven, CT 06511
> Tel: (203) 786-5515
> FAX: (203) 786-5534
> Email: GENEMAP@YALEVM.BITNET

References

1.   Human Gene Mapping 10 (1989): Tenth International Workshop on Human Gene Mapping. Cytogenet. Cell Genet. 51:1-1147.

2.   Kidd, K. K., Bowcock, A. M., Schmidtke, J., Track, R. K., Ricciuti, F., Hutchings, G., Bale, A., Pearson, P., and Willard, H. F. (1989) Cytogenet. Cell Genet. 51:622-947.

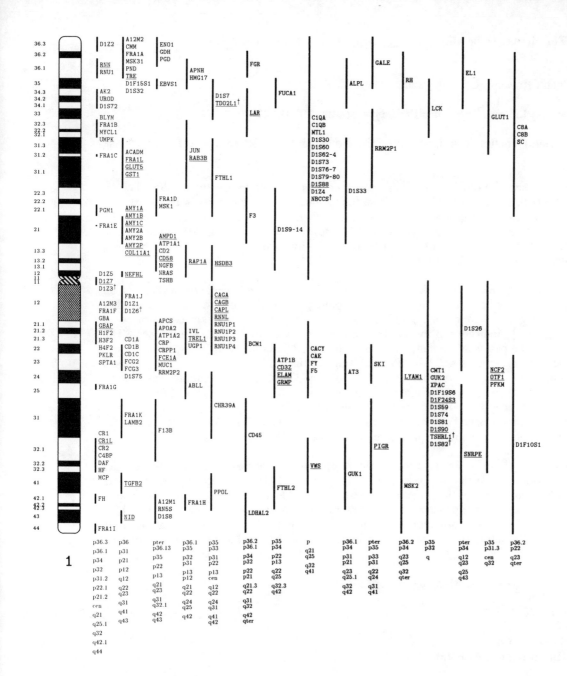

1

Continued next page

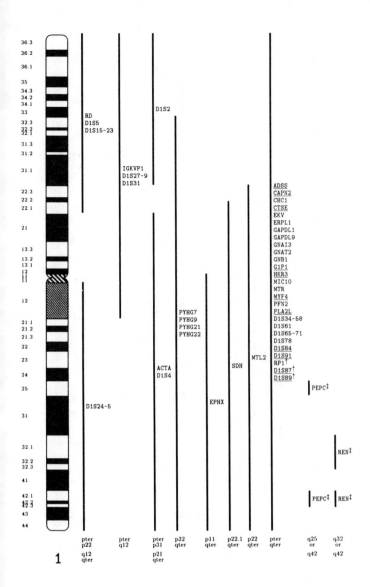

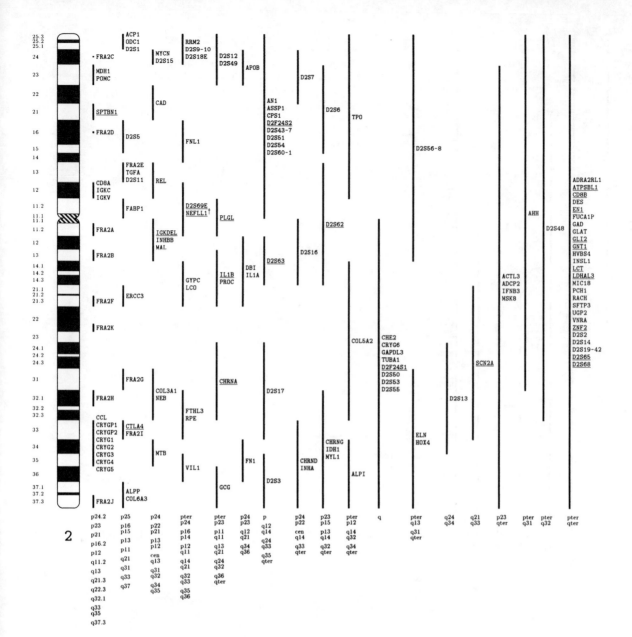

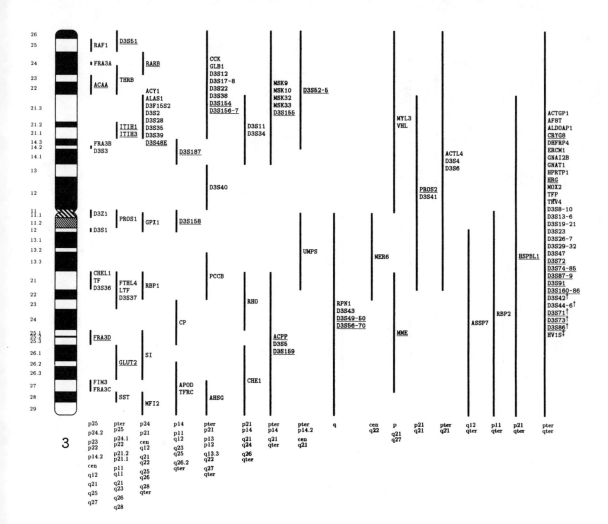

3

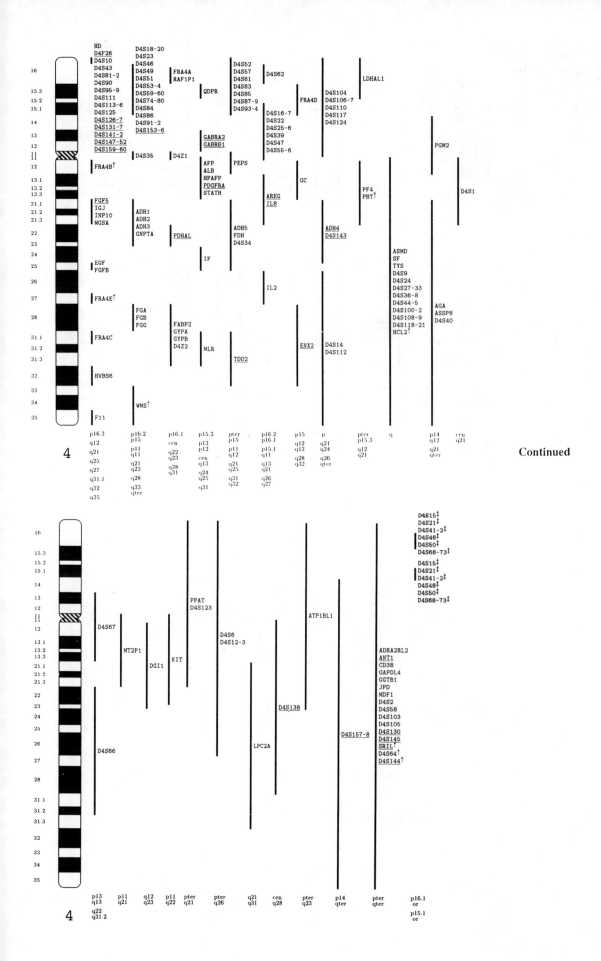

Continued

4

5.8

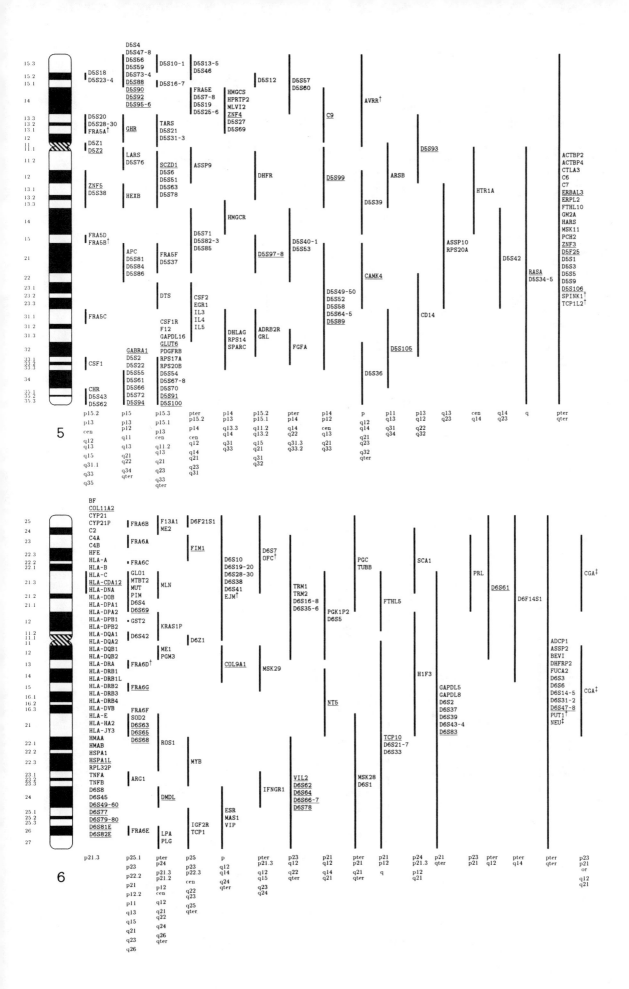

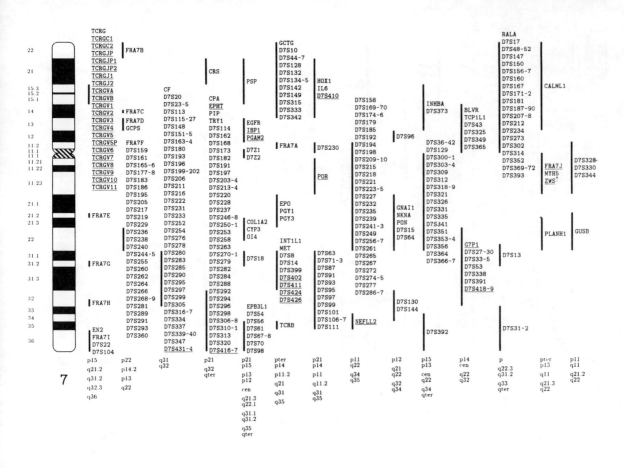

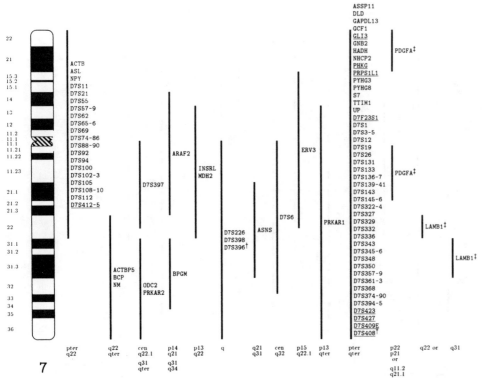

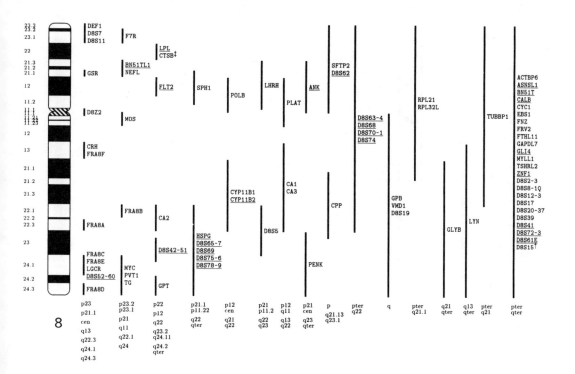

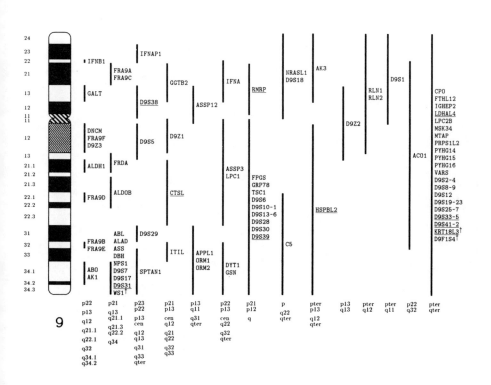

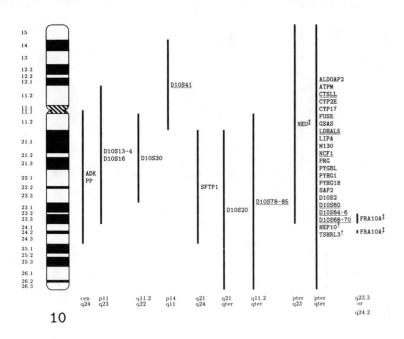

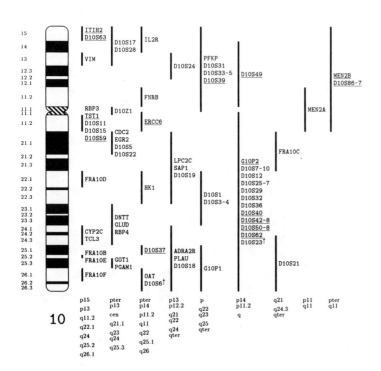

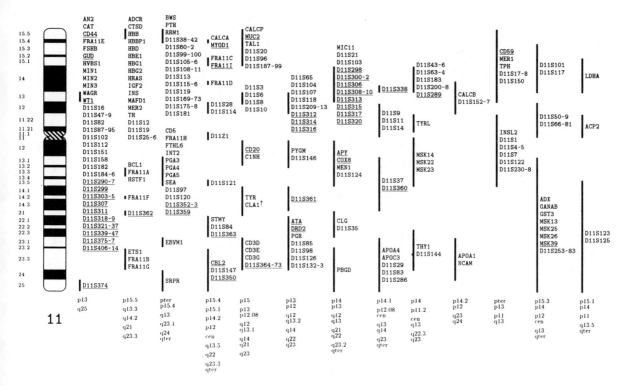

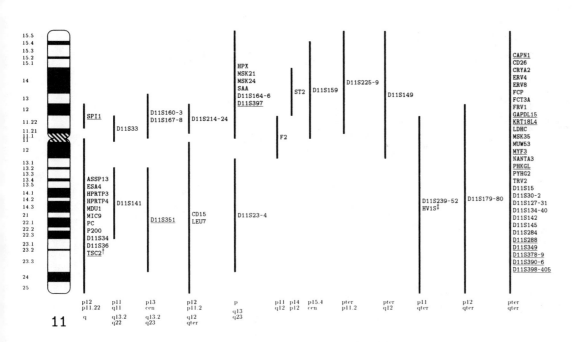

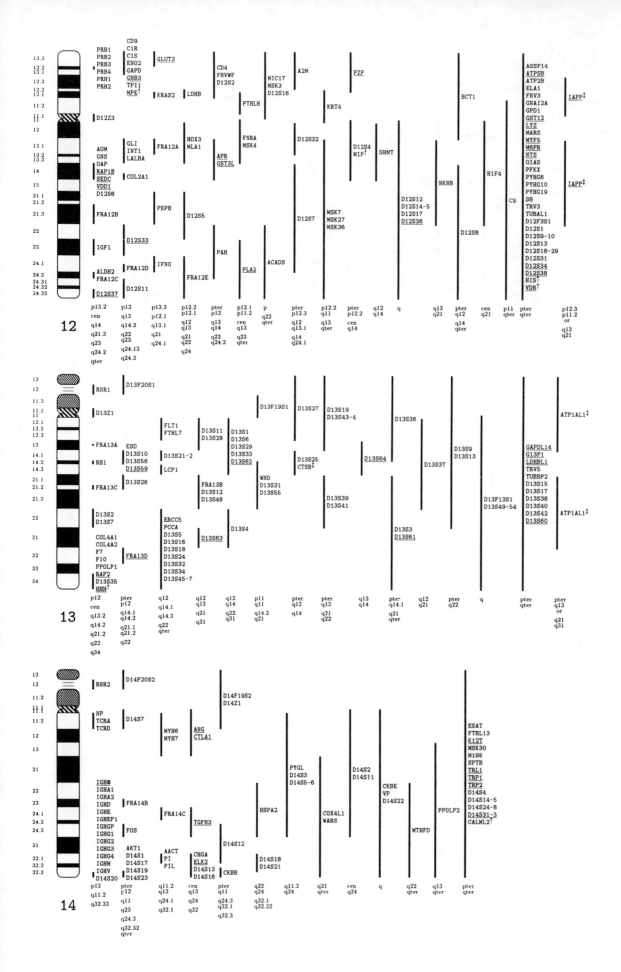

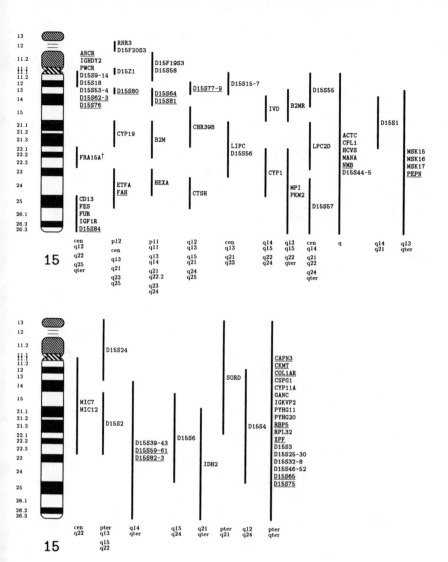

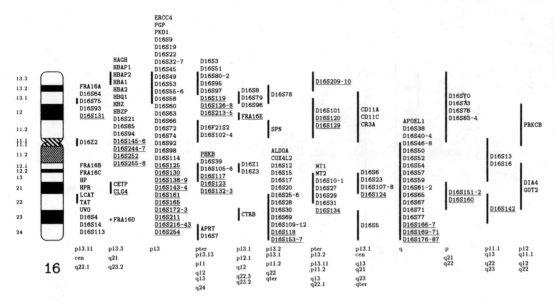

Continued

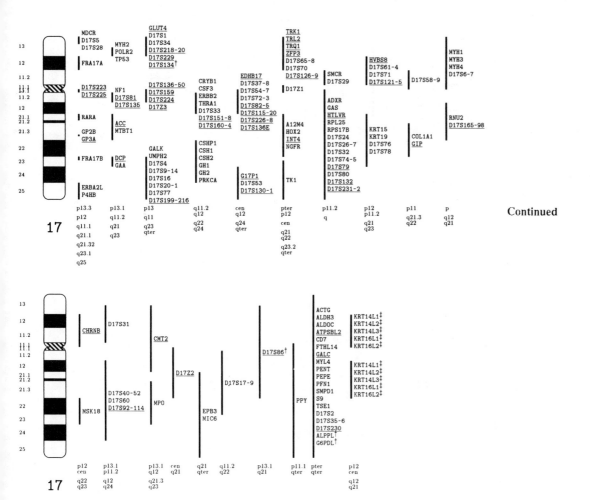

17

Continued

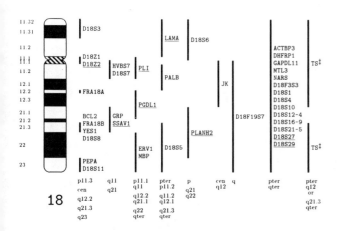

17

18

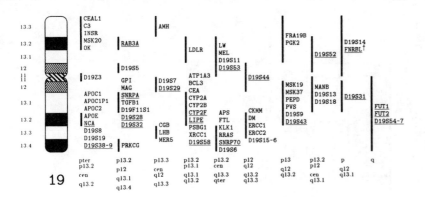

Continued

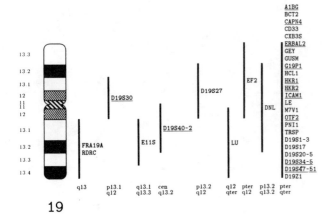

19

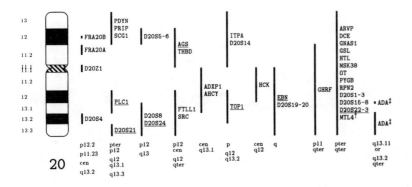

20

5.18

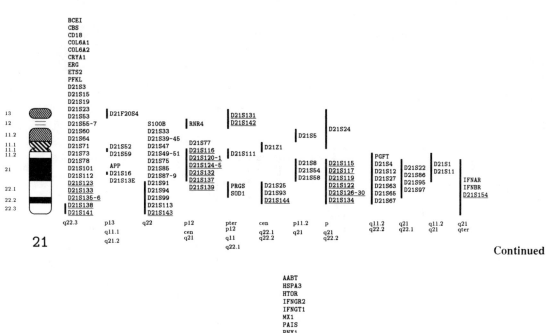

Continued

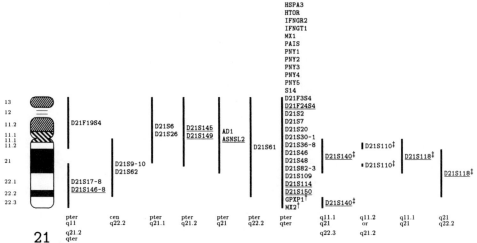

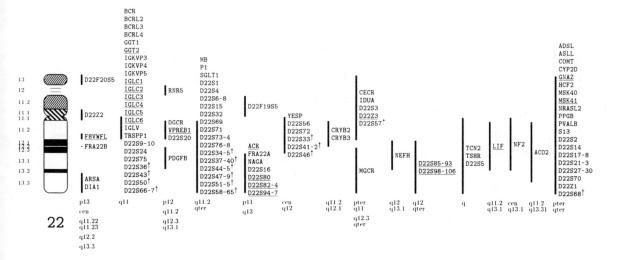

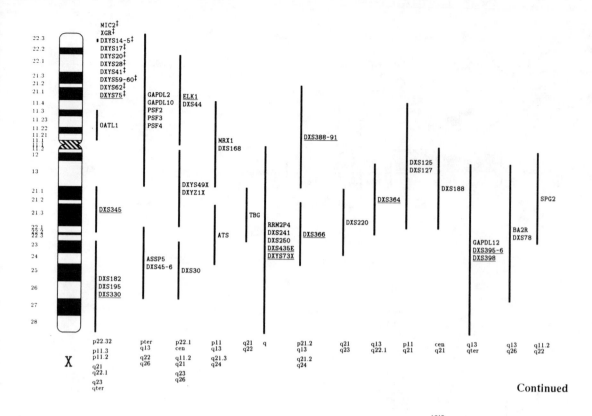

Continued

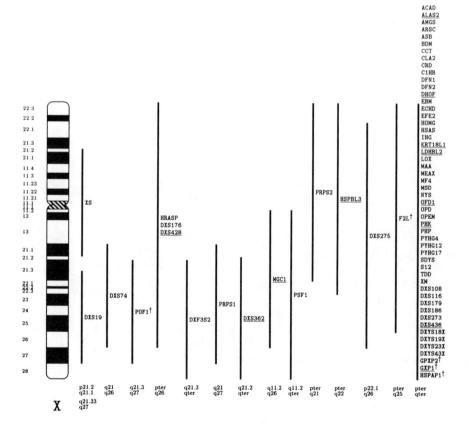

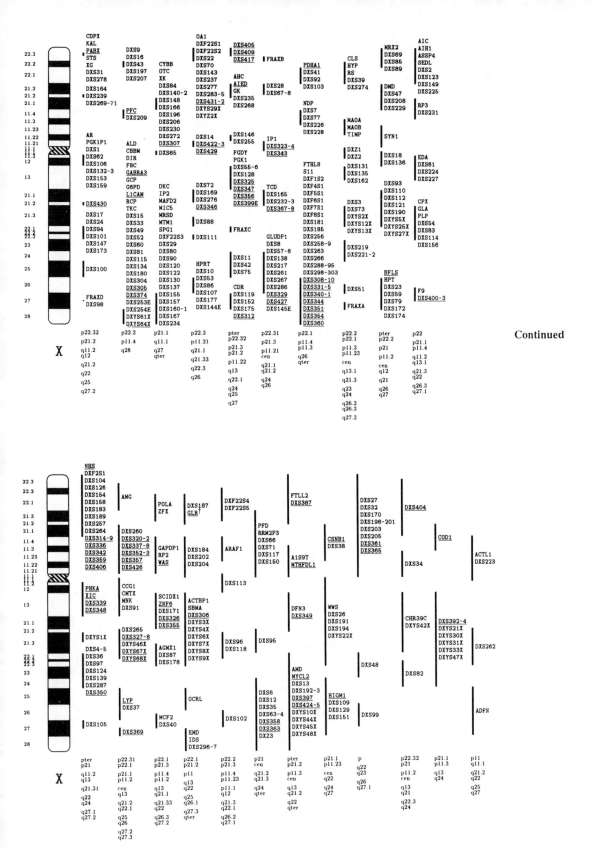

Continued

Continued next page

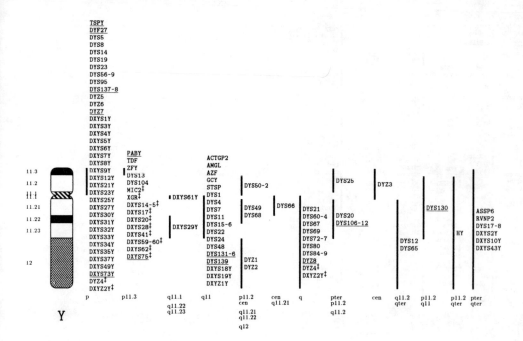

5.22

Symbol	Map location	PIC	Marker name
AABT	21		beta-amino acid transport
AACT	14q32.1	.33	alpha-1-antichymotrypsin
ABL	9q34	.26	Abelson murine leukemia viral (v-abl) oncogene homolog
ABLL	1q24-q25		Abelson murine leukemia viral (v-abl) oncogene homolog-like
ABO	9q34.1-q34.2		ABO blood group
*ACAA	3p23-p22		acetyl-Coenzyme A acyltransferase
ACAD	X		acyl-Coenzyme A dehydrogenase, multiple
ACADM	1p31	.68	acyl-Coenzyme A dehydrogenase, C-4 to C-12 straight-chain
ACADS	12q22-qter		acyl-Coenzyme A dehydrogenase, C-2 to C-3 short chain
*ACC	17q21		acetyl-Coenzyme A carboxylase
ACO1	9p22-q32		aconitase 1, soluble
ACO2	22q11.2-q13.31		aconitase 2, mitochondrial
*ACPP	3q21-qter	.36	acid phosphatase, prostate
ACP1	2p25		acid phosphatase 1, soluble
ACP2	11p11		acid phosphatase 2, lysosomal
*ACR	22q13-qter		acrosin
ACTA	1p21-qter		actin, alpha, skeletal muscle
ACTB	7pter-q22		actin, beta
ACTBP1	Xq13-q22		actin, beta pseudogene 1
ACTBP2	5		actin, beta pseudogene 2
ACTBP3	18		actin, beta pseudogene 3
ACTBP4	5		actin, beta pseudogene 4
ACTBP5	7q22-qter		actin, beta pseudogene 5
ACTBP6	8		actin, beta pseudogene 6
ACTC	15q11-qter	.84	actin, alpha, cardiac muscle
ACTG	17	.38	actin, gamma
ACTGP1	3		actin, gamma pseudogene 1
ACTGP2	Yq11		actin, gamma pseudogene 2
ACTL1	Xp11-q11		actin-like sequence 1
ACTL3	2p23-qter		actin-like sequence 3
ACTL4	3pter-q21		actin-like sequence 4
ACY1	3p21		aminoacylase 1
ADA	20q13.11 or 20q13.2-qter	.27	adenosine deaminase
ADCP1	6		adenosine deaminase complexing protein 1
ADCP2	2p23-qter		adenosine deaminase complexing protein 2
ADCR	11p15.5		adrenocortical carcinoma chromosome region
ADFN	Xq25-q27		albinism-deafness syndrome
ADH1	4q21-q23		alcohol dehydrogenase (class I), alpha polypeptide
ADH2	4q21-q23	.28	alcohol dehydrogenase (class I), beta polypeptide
ADH3	4q21-q23	.59	alcohol dehydrogenase (class I), gamma polypeptide
*ADH4	4q21-q24		alcohol dehydrogenase (class II), pi polypeptide
ADH5	4q21-q25		alcohol dehydrogenase (class III), chi polypeptide
ADK	10cen-q24		adenosine kinase
ADRA2R	10q24-q26	.26	adrenergic, alpha-2-, receptor
ADRA2RL1	2		adrenergic, alpha-2-, receptor-like 1
ADRA2RL2	4		adrenergic, alpha-2-, receptor-like 2
ADRB2R	5q31-q32	.30	adrenergic, beta-2-, receptor, surface
ADSL	22		adenylosuccinate lyase
*ADSS	1		adenylosuccinate synthase
ADX	11q13-qter		adrenodoxin
ADXP1	20cen-q13.1		adrenodoxin pseudogene 1
ADXR	17cen-q25		adrenodoxin reductase
AD1	21pter-q21		Alzheimer disease 1
AFP	4q11-q13	.37	alpha-fetoprotein
AF8T	3		AF8 temperature sensitivity complementing
AGA	4q21-qter		aspartylglucosaminidase
AGMX1	Xq21.33-q22		agammaglobulinemia, X-linked 1 (Bruton)
*AGS	20p12-p11		Alagille syndrome
AHC	Xp21.3-p21.2		adrenal hypoplasia, congenital
AHCY	20cen-q13.1		S-adenosylhomocysteine hydrolase
AHH	2pter-q31		aryl hydrocarbon hydroxylase
AHSG	3q27-q29	.37	alpha-2-HS-glycoprotein
AIC	Xp22		Aicardi syndrome
*AIED	Xp21.3-p21.2		Aland island eye disease (Forsius-Eriksson ocular albinism, ocular albinism type 2)
AIH1	Xp22		amelogenesis imperfecta 1, hypomaturation or hypoplastic (=AMG & AMGS)
AKT1	14q32.32-q32.33	.37	murine thymoma viral (v-akt) oncogene homolog 1
AK1	9q34.1-q34.2	.28	adenylate kinase 1
AK2	1p34		adenylate kinase 2
AK3	9p24-p13		adenylate kinase 3
ALAD	9q34		aminolevulinate, delta-, dehydratase
ALAS1	3p21		aminolevulinate, delta-, synthase 1
*ALAS2	X		aminolevulinate, delta-, synthase 2
ALB	4q11-q13	.59	albumin
ALD	Xq28		adrenoleukodystrophy
ALDH1	9q21.1		aldehyde dehydrogenase 1, soluble
ALDH2	12q24.2	.35	aldehyde dehydrogenase 2, mitochondrial
ALDH3	17		aldehyde dehydrogenase 3
ALDOA	16q22-q24		aldolase A, fructose-bisphosphate
ALDOAP1	3		aldolase A, fructose-bisphosphate, pseudogene 1
ALDOAP2	10		aldolase A, fructose-bisphosphate, pseudogene 2
ALDOB	9q21.3-q22.2	.18	aldolase B, fructose bisphosphate
ALDOC	17		aldolase C, fructose-bisphosphate
ALPI	2q34-q37		alkaline phosphatase, intestinal
ALPL	1p36.1-p34	.34	alkaline phosphatase, liver/bone/kidney
ALPP	2q37	.36	alkaline phosphatase, placental (Regan isozyme)
ALPPL	17		alkaline phosphatase, placental-like

Symbol	Map location	PIC	Marker name
AMD	Xq22-q28		S-adenosylmethionine decarboxylase gene or pseudogene
AMG	Xp22.31-p22.1		amelogenin (=AMGS & AIH)
AMGL	Yq11		amelogenin-like
AMGS	X		amelogenesis (=AIH1 & AMG)
AMH	19p13.3		anti-Mullerian hormone
*AMPD1	1p13		adenosine monophosphate deaminase 1
*AMY1A	1p21	.37	amylase, alpha 1A; salivary
*AMY1B	1p21	.37	amylase, alpha 1B; salivary
*AMY1C	1p21		amylase, alpha 1C; salivary
AMY2A	1p21	.37	amylase, alpha 2A; pancreatic
AMY2B	1p21	.37	amylase, alpha 2B; pancreatic
*AMY2P	1p21		amylase, alpha 2; pancreatic, pseudogene
*ANCR	15q11-q12		Angelman syndrome chromosome region
*ANG	14q11-q13	.33	angiogenin
*ANK	8p21-p11		ankyrin
*ANT1	4		adenine nucleotide translocator 1
AN1	2p		aniridia 1
AN2	11p13		aniridia 2 (without Wilms' tumor, genitourinary abnormalities, and mental retardation)
AOM	12q14		arthroophthalmopathy, progressive (Stickler syndrome)
APC	5q21-q22		adenomatosis polyposis coli
APCS	1q21-q23	.37	amyloid P component, serum
APNH	1p36.1-p35		antiporter, Na+/H+, (amiloride sensitive)
APOA1	11q23-q24	.73	apolipoprotein A-I
APOA2	1q21-q23	.65	apolipoprotein A-II
APOA4	11q23-qter	.55	apolipoprotein A-IV
APOB	2p24-p23	.66	apolipoprotein B (including Ag(x) antigen)
APOC1	19q13.2	.28	apolipoprotein C-I
APOC1P1	19q13.2		apolipoprotein C-I pseudogene 1
APOC2	19q13.2	.79	apolipoprotein C-II
APOC3	11q23-qter	.36	apolipoprotein C-III
APOD	3q26.2-qter	.60	apolipoprotein D
APOE	19q13.2	.36	apolipoprotein E
APOEL1	16q11-q24		apolipoprotein E-like 1
APP	21q21.2	.45	amyloid beta (A4) precursor protein
APPL1	9q31-qter		amyloid beta (A4) precursor protein-like 1
*APR	12q13-q14		apolipoprotein receptor
APRT	16q24	.32	adenine phosphoribosyltransferase
APS	19q13.3-qter		antigen, prostate specific
*APY	11q12-q13		atopy (allergic asthma and rhinitis)
AR	Xq12	.18	androgen receptor (dihydrotestosterone receptor; testicular feminization)
ARAF1	Xp11.4-p11.23		murine sarcoma 3611 viral (v-raf) oncogene homolog 1
ARAF2	7p14-q21		murine sarcoma 3611 viral (v-raf) oncogene homolog 2
*AREG	4q13-q21	.37	amphiregulin
ARG1	6q23	.21	arginase, liver
ARSA	22q13.31-qter		arylsulfatase A
ARSB	5p11-q13		arylsulfatase B
ARSC	X		arylsulfatase C, F isozyme
ARVP	20		arginine vasopressin (neurophysin II)
ASB	X		anemia, sideroblastic/hypochromic
ASL	7pter-q22		argininosuccinate lyase
ASLL	22		argininosuccinate lyase-like
ASMD	4q		anterior segment mesenchymal dysgenesis
ASNS	7q21-q31		asparagine synthetase
*ASNSL1	8pter-q24		asparagine synthetase-like 1
*ASNSL2	21pter-q21		asparagine synthetase-like 2
ASS	9q34-qter	.36	argininosuccinate synthetase
ASSP1	2p25-cen		argininosuccinate synthetase pseudogene 1
ASSP2	6		argininosuccinate synthetase pseudogene 2
ASSP3	9q11-q22	.37	argininosuccinate synthetase pseudogene 3
ASSP4	Xpter-p22		argininosuccinate synthetase pseudogene 4
ASSP5	Xq22-q26		argininosuccinate synthetase pseudogene 5
ASSP6	Y		argininosuccinate synthetase pseudogene 6
ASSP7	3q12-qter		argininosuccinate synthetase pseudogene 7
ASSP8	4q21-qter		argininosuccinate synthetase pseudogene 8
ASSP9	5q11-q12		argininosuccinate synthetase pseudogene 9
ASSP10	5q13-q23		argininosuccinate synthetase pseudogene 10
ASSP11	7	.24	argininosuccinate synthetase pseudogene 11
ASSP12	9p13-q11		argininosuccinate synthetase pseudogene 12
ASSP13	11q		argininosuccinate synthetase pseudogene 13
ASSP14	12		argininosuccinate synthetase pseudogene 14
*ATA	11q22-q23		ataxia telangiectasia (complementation group A)
ATPM	10		ATPase, mitochondrial
*ATPSB	12p13-qter		ATP synthase, H+ transporting, beta
*ATPSBL1	2		ATP synthase, H+ transporting, beta-like 1
*ATPSBL2	17		ATP synthase, H+ transporting, beta-like 2
ATP1AL1	13q21-q31 or 13pter-q13		ATPase, Na+K+, alpha polypeptide-like 1
ATP1A1	1p13		ATPase, Na+K+, alpha 1 polypeptide
ATP1A2	1q21-q23		ATPase, Na+K+, alpha 2 (+) polypeptide
ATP1A3	19q13.1-q13.2	.36	ATPase, Na+K+, alpha 3 polypeptide
ATP1B	1q22-q25	.37	ATPase, Na+K+, beta polypeptide
ATP1BL1	4p16-q23	.28	ATPase, Na+K+, beta polypeptide-like 1
ATP2A	16		ATPase, Ca++ transporting, fast twitch, cardiac muscle

Symbol	Map location	PIC	Marker name
ATP2B	12		ATPase, Ca++ transporting, slow twitch, cardiac muscle
ATS	Xq21.3-q24		Alport syndrome
AT3	1q23-q25.1	.38	antithrombin III
AVR	16		antiviral state regulator
AVRR	5p		antiviral state repressor regulator
AZF	Yq11		azoospermia
*A1BG	19		alpha-1-B glycoprotein
A1S9T	Xp11.3-p11.1		A1S9 temperature sensitivity complementing
A12M1	1q42-q43		adenovirus-12 chromosome modification site 1C
A12M2	1p36		adenovirus-12 chromosome modification site 1A
A12M3	1q21		adenovirus-12 chromosome modification site 1B
A12M4	17q21-q22		adenovirus-12 chromosome modification site 17
A2M	12p13.3-p12.3	.37	alpha-2-macroglobulin
BA2R	Xq13-q26		BALB/c 3T3 ts2 temperature sensitivity complementing
BCEI	21q22.3	.30	breast cancer, estrogen-inducible sequence expressed in
BCL1	11q13.3		B cell CLL/lymphoma 1
BCL2	18q21.3	.36	B cell CLL/lymphoma 2
BCL3	19q13.1-q13.2	.36	B cell CLL/Lymphoma 3
BCM1	1q21.3-q22		B cell activation marker 1
BCP	7q22-qter		blue cone pigment
BCR	22q11	.34	breakpoint cluster region
BCRL2	22q11		breakpoint cluster region-like 2
BCRL3	22q11		breakpoint cluster region-like 3
BCRL4	22q11		breakpoint cluster region-like 4
BCT1	12pter-q12		branched chain aminotransferase 1
BCT2	19		branched chain aminotransferase 2
BDM	X		behavior disorder modifier
BEVI	6		baboon M7 virus integration site
BF	6p21.3	.09	properdin factor B
*BFLS	Xq26-q27		Borjeson-Forssman-Lehmann syndrome
BLVR	7p14-cen		biliverdin reductase
BLYM	1p32		avian lymphoma virus-derived transforming sequence
*BN51T	8pter-q24		BN51 (BHK21) temperature sensitivity complementing
*BN51TL1	8p21		BN51 (BHK21) temperature sensitivity complementing-like 1
BPGM	7q31-q34		2,3-bisphosphoglycerate mutase
*BTS	16		Batten disease (neuronal ceroid-lipofuscinosis; juvenile amaurotic family idiocy)
BWS	11pter-p15.4		Beckwith-Wiedemann syndrome
B2M	15q21-q22.2		beta-2-microglobulin
B2MR	15q13-q15		beta-2-microglobulin regulator
CACY	1q21-q25		calcyclin

Symbol	Map location	PIC	Marker name
CAD	2p22-p21		carbamoyl phosphate synthetase 2, aspartate transcarbamylase, and dihydroorotase
CAE	1q21-q25		cataract, zonular pulverulent (FY-linked)
*CAGA	1q12-q22		calgranulin A
*CAGB	1q12-q22		calgranulin B
*CALB	8	.32	calbindin, 27kd
CALCA	11p15.4	.35	calcitonin/calcitonin-gene-related polypeptide, alpha
CALCB	11p14.2-p12		calcitonin gene-related-polypeptide, beta
CALCP	11pter-p15.1		calcitonin pseudogene
CALML1	7pter-p13	.38	calmodulin-like 1
CALML2	14		calmodulin-like 2
*CAMK4	5q21-q23		Ca++/calmodulin dependent protein kinase IV
*CAPL	1q12-q22		calcium protein, murine placental homolog
*CAPN1	11		calpain, large polypeptide L1
*CAPN2	1		calpain, large polypeptide L2
*CAPN3	15		calpain, large polypeptide L3
*CAPN4	19		calpain, small polypeptide
CAT	11p13	.38	catalase
CA1	8q13-q22		carbonic anhydrase I
CA2	8q22	.38	carbonic anhydrase II
CA3	8q13-q22		carbonic anhydrase III, muscle specific
CA4	16q21-q23		carbonic anhydrase IV
CBBM	Xq28		color blindness, blue monochromatic
*CBL2	11q23.3-qter		Cas-Br-M (murine) ecotropic retroviral transforming sequence
CBS	21q22.3		cystathionine-beta-synthase
CCG1	Xq11-q13		cell cycle, G1 phase defect
CCK	3pter-p21		cholecystokinin
CCL	2q33-q35		cataract, Coppock-like
CCT	X		cataract, congenital, total
CDC2	10q21.1	.36	cell division cycle 2, G1 to S and G2 to M
CDPX	Xp22.32		chondrodysplasia punctata
CDR	Xq27	.38	cerebellar degeneration-related protein
CD1A	1q22-q23		antigen CD1A, a polypeptide
CD1B	1q22-q23		antigen CD1B, b polypeptide
CD1C	1q22-q23		antigen CD1C, c polypeptide
CD2	1p13		antigen CD2 (p50), sheep red blood cell receptor
CD3D	11q23	.11	antigen CD3D, delta polypeptide (TiT3 complex)
CD3E	11q23	.34	antigen CD3E, epsilon polypeptide (TiT3 complex)
CD3G	11q23	.27	antigen CD3G, gamma polypeptide (TiT3 complex)
*CD3Z	1q22-q25		antigen CD3Z, zeta polypeptide (TiT3 complex)
CD4	12pter-p12		antigen CD4 (p55)
CD5	11q13		antigen CD5 (p56-62)
CD7	17		antigen CD7 (p41)
CD8A	2p12	.37	antigen CD8A (p32)
*CD8B	2		antigen CD8B (p37)
CD9	12p13		antigen CD9 (p24)

Symbol	Map location	PIC	Marker name
CD11A	16p13.1-p11		antigen CD11A (p180), lymphocyte function-associated antigen 1; alpha polypeptide
CD11C	16p13.1-p11		antigen CD11C (p150), alpha polypeptide
CD13	15q25-qter		antigen CD13 (p150)
CD14	5q22-q32		antigen CD14
CD15	11q12-qter		antigen CD15
CD18	21q22.3		antigen CD18 (p95), lymphocyte function-associated antigen 1; macrophage antigen 1 (mac-1) beta subunit
*CD20	11q12-q13.1		antigen CD20
CD26	11		antigen CD26 (p250), T cell activation antigen
CD33	19		antigen CD33 (gp67)
CD38	4		antigen CD38 (p45)
*CD44	11p13		antigen CD44 (homing function), includes MDU2, MDU3, MIC4
CD45	1q31-q32		antigen CD45, leukocyte-common antigen/T200 glycoprotein
*CD58	1p13		antigen CD58, (lymphocyte function-associated antigen 3)
*CD59	11pter-p13		antigen CD59 (p18-20)
CEA	19q13.1-q13.2		carcinoembryonic antigen
CEAL1	19p13.3-p13.2		carcinoembryonic antigen-like 1
CECR	22pter-q11		cat eye syndrome chromosome region
CETP	16q21	.53	cholesteryl ester transfer protein, plasma
CF	7q31-q32		cystic fibrosis
CGA	6q12-q21 or 6p23-p21.1	.35	glycoprotein hormones, alpha polypeptide
CGB	19q13.3		chorionic gonadotropin, beta polypeptide
CHC1	1		chromosome condensation 1
CHEL1	3q21		cholinesterase (serum)-like 1
CHEL3	16q11-q23		cholinesterase (serum)-like 3
CHE1	3q26-qter		cholinesterase (serum) 1
CHE2	2q		cholinesterase (serum) 2
CHGA	14q32		chromogranin A, parathyroid secretory protein 1
CHR	5q35		chromate resistance; sulfate transport
*CHRNA	2q24-q32		cholinergic receptor, nicotinic, alpha polypeptide
*CHRNB	17p12-p11		cholinergic receptor, nicotinic, beta polypeptide
CHRND	2q33-qter		cholinergic receptor, nicotinic, delta polypeptide
CHRNG	2q32-qter		cholinergic receptor, nicotinic, gamma polypeptide
CHR39A	1q24-q31		cholesterol repressible protein 39A
CHR39B	15q15-q21		cholesterol repressible protein 39B
CHR39C	Xq13-q21		cholesterol repressible protein 39C
CKBB	14q32.3	.36	creatine kinase, brain
CKBE	14q		creatine kinase, ectopic expression
CKMM	19q13.2-q13.3	.34	creatine kinase, muscle
*CKMT	15		creatine kinase, mitochondrial
CLA1	11q14-q21		cerebellar ataxia 1 (autosomal recessive)
CLA2	X		cerebellar ataxia 2 (X-linked)
CLG	11q21-q22		collagenase, epidermolysis bullosa, dystrophic, (autosomal recessive)
*CLG4	16q21		collagenase IV (basement membrane)
CLS	Xp22.2-p22.1		Coffin-Lowry syndrome
CMM	1p36		cutaneous malignant melanoma/dysplastic nevus
CMTX	Xq11-q13		Charcot-Marie-Tooth neuropathy, X-linked
CMT1	1q		Charcot-Marie-Tooth neuropathy 1
*CMT2	17p13.1-q12		Charcot-Marie-Tooth neuropathy 2
*COD1	Xp21.1-p11.3		cone dystrophy 1 (X-linked)
*COL1AR	15		collagen, type I, alpha, receptor
COL1A1	17q21.3-q22	.43	collagen, type I, alpha 1
COL1A2	7q21.3-q22.1	.60	collagen, type I, alpha 2
COL2A1	12q14.3	.69	collagen, type II, alpha 1
COL3A1	2q31-q32.3	.36	collagen, type III, alpha 1
COL4A1	13q34	.43	collagen, type IV, alpha 1
COL4A2	13q34	.38	collagen, type IV, alpha 2
COL5A2	2q14-q32	.22	collagen, type V, alpha 2
COL6A1	21q22.3		collagen, type VI, alpha 1
COL6A2	21q22.3		collagen, type VI, alpha 2
COL6A3	2q37	.37	collagen, type VI, alpha 3
*COL9A1	6q12-q14		collagen, type IX, alpha 1
*COL11A1	1p21		collagen, type XI, alpha 1
*COL11A2	6p21.3		collagen, type XI, alpha 2
COMT	22		catechol-O-methyltransferase
COX4L1	14q21-q32		cytochrome c oxidase subunit IV-like 1
COX4L2	16q22-q24		cytochrome c oxidase subunit IV-like 2
*COX8	11q12-q13		cytochrome c oxidase subunit VIII
CP	3q23-q25	.36	ceruloplasmin
CPA	7q32-qter		carboxypeptidase A
CPL1	15q11-qter		ceruloplasmin-like 1
CPO	9		coproporphyrinogen oxidase
CPP	8q21.13-q23.1		ceruloplasmin pseudogene
CPS1	2p		carbamoyl phosphate synthetase 1, mitochondrial
CPX	Xq21.3-q22		cleft palate and/or ankyloglossia
CRD	X		choroidoretinal degeneration
CRH	8q13	.11	corticotrophin releasing hormone
CRP	1q21-q23		C-reactive protein
CRPP1	1q21-q23		C-reactive protein pseudogene 1
CRS	7p21		craniosynostosis
CRYA1	21q22.3		crystallin, alpha polypeptide 1
CRYA2	11		crystallin, alpha polypeptide 2
CRYB1	17q11.2-q12	.37	crystallin, beta polypeptide 1
CRYB2	22q11.2-q12.1	.38	crystallin, beta polypeptide 2
CRYB3	22q11.2-q12.1		crystallin, beta polypeptide 3

Symbol	Map location	PIC	Marker name
CRYGP1	2q33-q35		crystallin, gamma polypeptide pseudogene 1
CRYGP2	2q33-q35		crystallin, gamma polypeptide pseudogene 2
CRYG1	2q33-q35	.61	crystallin, gamma polypeptide 1
CRYG2	2q33-q35	.61	crystallin, gamma polypeptide 2
CRYG3	2q33-q35	.61	crystallin, gamma polypeptide 3
CRYG4	2q33-q35	.61	crystallin, gamma polypeptide 4
CRYG5	2q33-q35	.61	crystallin, gamma polypeptide 5
CRYG6	2q		crystallin, gamma polypeptide 6
*CRYG8	3		crystallin, gamma polypeptide 8
CR1	1q32	.37	complement component (3b/4b) receptor 1
*CR1L	1q32		complement component (3b/4b) receptor 1-like
CR2	1q32		complement component (3d/Epstein Barr virus) receptor 2
CR3A	16p13.1-p11		complement component receptor 3, alpha; also known as CD11b (p170), macrophage antigen alpha polypeptide
CS	12p11-qter		citrate synthase
CSF1	5q33		colony-stimulating factor 1 (macrophage)
CSF1R	5q33-q35	.27	colony-stimulating factor 1 receptor, formerly McDonough feline sarcoma viral (v-fms) oncogene homolog
CSF2	5q23-q31		colony-stimulating factor 2 (granulocyte-macrophage)
CSF3	17q11.2-q12		colony-stimulating factor 3 (granulocyte)
CSHP1	17q22-q24	.38	chorionic somatomammotropin pseudogene 1
CSH1	17q22-q24	.38	chorionic somatomammotropin hormone 1
CSH2	17q22-q24	.38	chorionic somatomammotropin hormone 2
*CSNB1	Xp21.1-p11.23		congenital stationary night blindness 1
CSPG1	15	.36	chondroitin sulfate proteoglycan 1
CTH	16		cystathionase
*CTLA1	14q11-q13		cytotoxic T-lymphocyte-associated serine esterase 1
CTLA3	5		cytotoxic T-lymphocyte-associated serine esterase 3
*CTLA4	2q33		cytotoxic T-lymphocyte-associated serine esterase 4
*CTM	16		cataract, Marner
CTRB	16q22.3-q23.2	.59	chymotrypsinogen B
CTSB	8p22 or 13q14		cathepsin B
CTSD	11p15.5		cathepsin D
*CTSE	1		cathepsin E
CTSH	15q24-q25		cathepsin H
*CTSL	9q21-q22		cathepsin L
*CTSLL	10		cathepsin L-like
CXB3S	19pter-q13		coxsackie virus B3 sensitivity
CYBB	Xp21.1	.40	cytochrome b-245, beta polypeptide (chronic granulomatous disease)
CYC1	8		cytochrome c-1
CYP1	15q22-q24	.29	cytochrome P450, subfamily I (aromatic compound-inducible)
CYP2A	19q13.1-q13.2	.37	cytochrome P450, subfamily IIA (phenobarbital-inducible)
CYP2B	19q13.1-q13.2	.37	cytochrome P450, subfamily IIB (phenobarbital-inducible)
CYP2C	10q24.1-q24.3	.36	cytochrome P450, subfamily IIC (mephenytoin 4-hydroxylase)
CYP2D	22		cytochrome P450, subfamily IID (debrisoquine, sparteine, etc., -metabolizing)
CYP2E	10	.16	cytochrome P450, subfamily IIE (ethanol-inducible)
*CYP2F	19q13.1-q13.2		cytochrome p450, subfamily IIF
CYP3	7q21.3-q22.1	.16	cytochrome P450, subfamily III (niphedipine oxidase)
CYP11A	15		cytochrome P450, subfamily XIA (cholesterol side chain cleavage, lipoid adrenal hyperplasia)
CYP11B1	8q21-q22	.36	cytochrome P450, subfamily XIB, polypeptide 1 (steroid 11-beta-hydroxylase)
*CYP11B2	8q21-q22	.36	cytochrome P450, subfamily XIB, polypeptide 2 (steroid 11-beta-hyd hydroxylase)
CYP17	10		cytochrome P450, subfamily XVII (steroid 17-alpha-hydroxylase)
CYP19	15q21		cytochrome P450, subfamily XIX (aromatization of androgens)
CYP21	6p21.3	.38	cytochrome P450, subfamily XXI (steroid 21-hydroxylase, congenital adrenal hyperplasia)
CYP21P	6p21.3	.38	cytochrome P450, subfamily XXI (steroid 21-hydroxylase pseudogene)
C1HR	X		C1AGOH temperature sensitivity complementing
C1NH	11q12-q13.1	.34	complement component 1 inhibitor (angioedema, hereditary)
C1QA	1p		complement component 1, q subcomponent, alpha polypeptide
C1QB	1p		complement component 1, q subcomponent, beta polypeptide
C1R	12p13		complement component 1, r subcomponent
C1S	12p13		complement component 1, s subcomponent
C2	6p21.3	.60	complement component 2
C3	19p13.3-p13.2	.35	complement component 3
C4A	6p21.3	.37	complement component 4A
C4B	6p21.3	.37	complement component 4B

Symbol	Map location	PIC	Marker name
C4BP	1q32		complement component 4 binding protein
C5	9q22-q34		complement component 5
C6	5		complement component 6
C7	5		complement component 7
C8A	1p36.2-p22.1		complement component 8, alpha polypeptide
C8B	1p36.2-p22.1		complement component 8, beta polypeptide
*C9	5p14-p12		complement component 9
DAF	1q32	.54	decay accelerating factor for complement
DBH	9q34		dopamine beta-hydroxylase (dopamine beta-monooxygenase)
DBI	2q12-q21		diazepam binding inhibitor (GABA receptor modulator)
DCE	20		desmosterol-to-cholesterol enzyme
*DCP	17q23		dipeptidyl carboxypeptidase 1 (angiotensin I converting enzyme)
DEF1	8p23		defensin 1
DES	2		desmin
DFN1	X		deafness, progressive
DFN2	X		deafness, perceptive, congenital
DFN3	Xq13-q21.2		deafness, conductive, with fixed stapes
DGCR	22q11.21-q11.23		DiGeorge syndrome chromosome region
DGI1	4q12-q23		dentinogenesis imperfecta 1
DHFR	5q11.2-q13.2	.57	dihydrofolate reductase
DHFRP1	18	.38	dihydrofolate reductase pseudogene 1
DHFRP2	6		dihydrofolate reductase pseudogene 2
DHFRP4	3		dihydrofolate reductase pseudogene 4
DHLAG	5q31-q33		major histocompatibility complex, class II, gamma polypeptide
*DHOF	X		dermal hypoplasia, focal
DIA1	22q13.31-qter		diaphorase (NADH) (cytochrome b-5 reductase)
DIA4	16q12-q22		diaphorase (NADH/NADPH) (cytochrome b-5 reductase)
DIR	Xq28		diabetes insipidus, renal
DKC	Xq27-q28		dyskeratosis congenita
DLD	7		dihydrolipoamide dehydrogenase (E3 component of pyruvate dehydrogenase complex, 2-oxo-glutarate complex, branched chain keto acid dehydrogenase complex)
DM	19q13.2-q13.3		dystrophia myotonia
DMD	Xp21.3-p21.1	.50	muscular dystrophy, Duchenne and Becker types; includes DXS142, DXS164, DXS206, DXS230, DXS239, DXS268, DXS269, DXS270, DXS272.
*DMDL	6q24		dystrophin-like
DNCM	9q12		DNA associated with cytoplasmic membrane
DNL	19p13.2-q13.2		DNase, lysosomal
DNTT	10q23-q24		deoxynucleotidyltransferase, terminal
*DRD2	11q22-q23	.30	dopamine receptor D2

Symbol	Map location	PIC	Marker name
DTS	5q23		diphtheria toxin sensitivity
DYT1	9q32-q34		dystonia, torsion 1 (autosomal dominant)
EBM	X		epidermolysis bullosa, macular type
*EBN	20q		epilepsy, benign neonatal
EBS1	8		epidermolysis bullosa simplex (Ogna)
EBVM1	11q23.1		Epstein Barr virus modification site 1
EBVS1	1p35		Epstein Barr virus insertion site 1
ECHD	X		enoyl-Coenzyme A hydratase: 3-hydroxyacyl Coenzyme A dehydrogenase
EDA	Xq12-q13.1		ectodermal dysplasia, anhidrotic (hypohydrotic)
*EDHB17	17q11-q12		estradiol 17 beta-dehydrogenase
EFE2	X		endocardial fibroelastosis 2
EF2	19pter-q12		elongation factor 2
EGF	4q25	.56	epidermal growth factor
EGFR	7p13-p12	.60	epidermal growth factor receptor, formerly avian erythroblastic leukemia viral (v-erb-b) oncogene homolog
EGR1	5q23-q31		early growth response 1
EGR2	10q21.1	.16	early growth response 2
EJM	6p		epilepsy, juvenile myoclonic
EKV	1		erythrokeratodermia variabilis
*ELAM	1q22-q25		endothelial adhesion molecule
ELA1	12	.25	elastase 1
*ELK1	Xp22.1-p11		ELK1, member of ETS oncogene family
*ELK2	14q32		ELK2, member of ETS oncogene family
ELN	2q31-qter	.27	elastin
EL1	1pter-p34		elliptocytosis 1 (Rh-linked); band 4.1 protein
EMD	Xq27.3-q28		Emery-Dreifuss muscular dystrophy
ENO1	1pter-p36.13		enolase 1, (alpha)
ENO2	12p13		enolase 2, (gamma, neuronal)
*ENX2	4q28-q32		endonexin II
*EN1	2	.37	engrailed homolog 1
EN2	7q36	.21	engrailed homolog 2
EPB3	17q21-qter	.37	erythrocyte surface protein band 3
EPB3L1	7q35-q36		erythrocyte surface protein band 3-like 1
*EPHT	7q32-q36		eph tyrosine kinase/erythropoietin producing hepatoma amplified sequence
EPHX	1p11-qter	.16	epoxide hydroxylase, microsomal (xenobiotic)
EPO	7q21	.31	erythropoietin
*ERBAL2	19		avian erythroblastic leukemia viral (v-erb-a) oncogene homolog-like 2
*ERBAL3	5		avian erythroblastic leukemia viral (v-erb-a) oncogene homolog-like 3

Symbol	Map location	PIC	Marker name	Symbol	Map location	PIC	Marker name
ERBA2L	17q25		avian erythroblastic leukemia viral (v-erb-a) oncogene homolog 2-like	FABP2	4q28-q31		fatty acid binding protein 2, intestinal
ERBB2	17q11.2-q12		avian erythroblastic leukemia viral (v-erb-b2) oncogene homolog 2 (neuro/glioblastoma derived oncogene homolog)	*FAH	15q23-q25		fumarylacetoacetate
				*FCE1A	1q21-q23		Fc fragment of IgE, high affinity receptor for; alpha polypeptide
				FCG2	1q22-q23		Fc fragment of IgG, low affinity II, receptor for (CD32)
ERCC1	19q13.2-q13.3		excision repair cross complementing rodent repair deficiency, complementation group 1 (includes overlapping antisense sequence)	FCG3	1q22-q23		Fc fragment of IgG, low affinity III, receptor for (CD16)
				FCP	11		hemoglobin F cell production
ERCC2	19q13.2-q13.3		excision repair cross complementing rodent repair deficiency, complementation group 2	FCT3A	11		fucosyltransferase, alpha-3- (GDP-fucose: [galactose-beta-1-4] N-acetylglucosamine-alpha -1-3-fucosyltransferase)
ERCC3	2q21		excision repair cross complementing rodent repair deficiency, complementation group 3	FDH	4q21-q25		formaldehyde dehydrogenase
				FES	15q25-qter	.09	feline sarcoma viral (v-fes) oncogene; Fujinami avian sarcoma viral (v-fps) oncogene homolog
ERCC4	16p13.3-p13.11		excision repair cross complementing rodent repair deficiency, complementation group 4	FGA	4q28	.34	fibrinogen, A alpha polypeptide
ERCC5	13q22-q34		excision repair cross complementing rodent repair deficiency, complementation group 5	FGB	4q28	.28	fibrinogen, B beta polypeptide
				FGDY	Xq13		faciogenital dysplasia (Aarskog syndrome)
*ERCC6	10q11		excision repair cross complementing rodent repair deficiency, complementation group 6	FGFA	5q31.3-q33.2		fibroblast growth factor, acidic (endothelial growth factor)
				FGFB	4q25	.27	fibroblast growth factor, basic
ERCM1	3		excision repair complementing defective repair in mouse cells	*FGF5	4q21		fibroblast growth factor-like (fgf.5 oncogene)
ERG	21q22.3		avian erythroblastosis virus E26 (v-ets) oncogene related	FGG	4q28	.25	fibrinogen, gamma polypeptide
ERPL1	1	.34	endogenous retroviral pol gene-like sequence 1 (clone HLM2)	FGR	1p36.2-p36.1	.37	Gardner-Rasheed feline sarcoma viral (v-fgr) oncogene homolog
ERPL2	5		endogenous retroviral pol gene-like sequence 2	FH	1q42.1		fumarate hydratase
				*FIM1	6p23-p22.3		Friend-murine leukemia virus integration site 1 homolog
ERV1	18q22-qter		endogenous retroviral sequence 1				
ERV3	7p15-q22.1	.37	endogenous retroviral sequence 3	FIM3	3q27		Friend-murine leukemia virus integration site 3 homolog
ERV4	11		endogenous retroviral sequence 4				
ERV8	11		endogenous retroviral sequence 8	FLT1	13q12		fms-related tyrosine kinase 1
ESAT	14		esterase activator	*FLT2	8p12		fms-related tyrosine kinase 2
ESA4	11q		esterase A4	FNL1	2p16-p14		fibronectin-like 1
ESB3	16		esterase B3	FNRA	12q11-q13		fibronectin receptor, alpha polypeptide
ESD	13q14.1-q14.2	.56	esterase D/formylglutathione hydrolase	FNRB	10p11.2	.71	fibronectin receptor, beta polypeptide
ESR	6q24-q27	.37	estrogen receptor	*FNRBL	19p		fibronectin receptor, beta polypeptide-like
ETFA	15q23-q25		electron transfer flavoprotein, alpha polypeptide (glutaric aciduria II)	FNZ	8		fibronectin, influences presence on cell surface
				FN1	2q34-q36	.36	fibronectin 1
ETS1	11q23.3	.36	avian erythroblastosis virus E26 (v-ets) oncogene homolog 1	FOS	14q24.3	.12	murine FBJ osteosarcoma viral (v-fos) oncogene homolog
ETS2	21q22.3		avian erythroblastosis virus E26 (v-ets) oncogene homolog 2	FPGS	9cen-q34		folylpolyglutamate synthase
E11S	19q13.1-q13.3		ECHO virus (serotypes 4, 6, 11, 19) sensitivity	FRAXA	Xq27.3		fragile site, folic acid type, rare, fra(X)(q27.3), (macroorchidism, mental retardation)
FABP1	2p11		fatty acid binding protein 1, liver				

Symbol	Map location	PIC	Marker name	Symbol	Map location	PIC	Marker name
FRAXB	Xp22.31		fragile site, aphidicolin type, common, fra(X)(p22.31)	FRA2K	2q22.3		fragile site, folic acid type, rare, fra(2)(q22.3)
FRAXC	Xq22.1		fragile site, aphidicolin type, common, fra(X)(q22.1)	FRA3A	3p24.2		fragile site, aphidicolin type, common, fra(3)(p24.2)
FRAXD	Xq27.2		fragile site, aphidicolin type, common, fra(X)(q27.2)	FRA3B	3p14.2		fragile site, aphidicolin type, common, fra(3)(p14.2)
FRA1A	1p36		fragile site, aphidicolin type, common, fra(1)(p36)	FRA3C	3q27		fragile site, aphidicolin type, common, fra(3)(q27)
FRA1B	1p32		fragile site, aphidicolin type, common, fra(1)(p32)	*FRA3D	3q25		fragile site, aphidicolin type, common, fra(3)(q25)
FRA1C	1p31.2		fragile site, aphidicolin type, common, fra(1)(p31.2)	FRA4A	4p16.1		fragile site, aphidicolin type, common, fra(4)(p16.1)
FRA1D	1p22		fragile site, aphidicolin type, common, fra(1)(p22)	FRA4B	4q12		fragile site, BrdU type, common, fra(4)(q12)
FRA1E	1p21.2		fragile site, aphidicolin type, common, fra(1)(p21.2)	FRA4C	4q31.1		fragile site, aphidicolin type, common, fra(4)(q31.1)
FRA1F	1q21		fragile site, aphidicolin type, common, fra(1)(q21)	FRA4D	4p15		fragile site, aphidicolin type, common, fra(4)(p15)
FRA1G	1q25.1		fragile site, aphidicolin type, common, fra(1)(q25.1)	FRA4E	4q27		fragile site, unclassified, common, fra(4)(q27)
FRA1H	1q42		fragile site, 5-azacytidine type, common, fra(1)(q42)	FRA5A	5p13		fragile site, BrdU type, common, fra(5)(p13)
FRA1I	1q44		fragile site, aphidicolin type, common, fra(1)(q44)	FRA5B	5q15		fragile site, BrdU type, common, fra(5)(q15)
FRA1J	1q12		fragile site, 5-azacytidine type, common, fra(1)(q12)	FRA5C	5q31.1		fragile site, aphidicolin type, common, fra(5)(q31.1)
FRA1K	1q31		fragile site, aphidicolin type, common, fra(1)(q31)	FRA5D	5q15		fragile site, aphidicolin type, common, fra(5)(q15)
*FRA1L	1p31		fragile site, aphidicolin type, common, fra(1)(p31)	FRA5E	5p14		fragile site, aphidicolin type, common, fra(5)(p14)
FRA2A	2q11.2		fragile site, folic acid type, rare, fra(2)(q11.2)	FRA5F	5q21		fragile site, aphidicolin type, common, fra(5)(q21)
FRA2B	2q13		fragile site, folic acid type, rare, fra(2)(q13)	FRA6A	6p23		fragile site, folic acid type, rare, fra(6)(p23)
FRA2C	2p24.2		fragile site, aphidicolin type, common, fra(2)(p24.2)	FRA6B	6p25.1		fragile site, aphidicolin type, common, fra(6)(p25.1)
FRA2D	2p16.2		fragile site, aphidicolin type, common, fra(2)(p16.2)	FRA6C	6p22.2		fragile site, aphidicolin type, common, fra(6)(p22.2)
FRA2E	2p13		fragile site, aphidicolin type, common, fra(2)(p13)	FRA6D	6q13		fragile site, BrdU type, common, fra(6)(q13)
FRA2F	2q21.3		fragile site, aphidicolin type, common, fra(2)(q21.3)	FRA6E	6q26		fragile site, aphidicolin type, common, fra(6)(q26)
FRA2G	2q31		fragile site, aphidicolin type, common, fra(2)(q31)	FRA6F	6q21		fragile site, aphidicolin type, common, fra(6)(q21)
FRA2H	2q32.1		fragile site, aphidicolin type, common, fra(2)(q32.1)	*FRA6G	6q15		fragile site, aphidicolin type, common, fra(6)(q15)
FRA2I	2q33		fragile site, aphidicolin type, common, fra(2)(q33)	FRA7A	7p11.2		fragile site, folic acid type, rare, fra(7)(p11.2)
FRA2J	2q37.3		fragile site, aphidicolin type, common, fra(2)(q37.3)	FRA7B	7p22		fragile site, aphidicolin type, common, fra(7)(p22)
				FRA7C	7p14.2		fragile site, aphidicolin type, common, fra(7)(p14.2)

Symbol	Map location	PIC	Marker name	Symbol	Map location	PIC	Marker name
FRA7D	7p13		fragile site, aphidicolin type, common, fra(7)(p13)	FRA11B	11q23.3		fragile site, folic acid type, rare, fra(11)(q23.3)
FRA7E	7q21.2		fragile site, aphidicolin type, common, fra(7)(q21.2)	FRA11C	11p15.1		fragile site, aphidicolin type, common, fra(11)(p15.1)
FRA7F	7q22		fragile site, aphidicolin type, common, fra(7)(q22)	FRA11D	11p14.2		fragile site, aphidicolin type, common, fra(11)(p14.2)
FRA7G	7q31.2		fragile site, aphidicolin type, common, fra(7)(q31.2)	FRA11E	11p13		fragile site, aphidicolin type, common, fra(11)(p13)
FRA7H	7q32.3		fragile site, aphidicolin type, common, fra(7)(q32.3)	FRA11F	11q14.2		fragile site, aphidicolin type, common, fra(11)(q14.2)
FRA7I	7q36		fragile site, aphidicolin type, common, fra(7)(q36)	FRA11G	11q23.3		fragile site, aphidicolin type, common, fra(11)(q23.3)
*FRA7J	7q11		fragile site, aphidicolin type, common, fra(7)(q11)	FRA11H	11q13		fragile site, aphidicolin type, common, fra(11)(q13)
FRA8A	8q22.3		fragile site, folic acid type, rare, fra(8)(q22.3)	*FRA11I	11p15.1		fragile site, distamycin A type, rare, fra(11)(p15.1)
FRA8B	8q22.1		fragile site, aphidicolin type, common, fra(8)(q22.1)	FRA12A	12q13.1		fragile site, folic acid type, rare, fra(12)(q13.1)
FRA8C	8q24.1		fragile site, aphidicolin type, common, fra(8)(q24.1)	FRA12B	12q21.3		fragile site, aphidicolin type, common, fra(12)(q21.3)
FRA8D	8q24.3		fragile site, aphidicolin type, common, fra(8)(q24.3)	FRA12C	12q24.2		fragile site, BrdU type, rare, fra(12)(q24.2)
FRA8E	8q24.1		fragile site, distamycin A type, rare, fra(8)(q24.1)	FRA12D	12q24.13		fragile site, folic acid type, rare, fra(12)(q24.13)
FRA8F	8q13		fragile site, unclassified, rare, fra(8)(q13)	FRA12E	12q24		fragile site, aphidicolin type, common, fra(12)(q24)
FRA9A	9p21		fragile site, folic acid type, rare, fra(9)(p21)	FRA13A	13q13.2		fragile site, aphidicolin type, common, fra(13)(q13.2)
FRA9B	9q32		fragile site, folic acid type, rare, fra(9)(q32)	FRA13B	13q21		fragile site, BrdU type, common, fra(13)(q21)
FRA9C	9p21		fragile site, BrdU type, common, fra(9)(p21)	FRA13C	13q21.2		fragile site, aphidicolin type, common, fra(13)(q21.2)
FRA9D	9q22.1		fragile site, aphidicolin type, common, fra(9)(q22.1)	*FRA13D	13q32		fragile site, aphidicolin type, common, fra(13)(q32)
FRA9E	9q32		fragile site, aphidicolin type, common, fra(9)(q32)	FRA14B	14q23		fragile site, aphidicolin type, common, fra(14)(q23)
FRA9F	9q12		fragile site, 5-azacytidine type, common, fra(9)(q12)	FRA14C	14q24.1		fragile site, aphidicolin type, common, fra(14)(q24.1)
FRA10A	10q23.3 or 10q24.2		fragile site, folic acid type, rare, fra(10)(q23.3) or fra(10)(q24.2)	FRA15A	15q22		fragile site, aphidicolin type, common, fra(15)(q22)
FRA10B	10q25.2		fragile site, BrdU type, rare, fra(10)(q25.2)	FRA16A	16p13.11		fragile site, folic acid type, rare, fra(16)(p13.11)
FRA10C	10q21		fragile site, BrdU type, common, fra(10)(q21)	FRA16B	16q22.1		fragile site, distamycin A type, rare, fra(16)(q22.1)
FRA10D	10q22.1		fragile site, aphidicolin type, common, fra(10)(q22.1)	FRA16C	16q22.1		fragile site, aphidicolin type, common, fra(16)(q22.1)
FRA10E	10q25.2		fragile site, aphidicolin type, common, fra(10)(q25.2)	FRA16D	16q23.2		fragile site, aphidicolin type, common, fra(16)(q23.2)
FRA10F	10q26.1		fragile site, aphidicolin type, common, fra(10)(q26.1)	*FRA16E	16p12.1		fragile site, distamycin A type, rare, fra(16)(p12.1)
FRA11A	11q13.3		fragile site, folic acid type, rare, fra(11)(q13.3)	FRA17A	17p12		fragile site, distamycin A type, rare, fra(17)(p12)

Symbol	Map location	PIC	Marker name
FRA17B	17q23.1		fragile site, aphidicolin type, common, fra(17)(q23.1)
FRA18A	18q12.2		fragile site, aphidicolin type, common, fra(18)(q12.2)
FRA18B	18q21.3		fragile site, aphidicolin type, common, fra(18)(q21.3)
FRA19A	19q13		fragile site, 5-azacytidine type, common, fra(19)(q13)
FRA19B	19p13		fragile site, folic acid type, rare, fra(19)(p13)
FRA20A	20p11.23		fragile site, folic acid type, rare, fra(20)(p11.23)
FRA20B	20p12.2		fragile site, aphidicolin type, common, fra(20)(p12.2)
FRA22A	22q13		fragile site, folic acid type, rare, fra(22)(q13)
FRA22B	22q12.2		fragile site, aphidicolin type, common, fra(22)(q12.2)
FRDA	9q13-q21.1		Friedreich ataxia
FRV1	11		full length retroviral sequence 1 (band F6, 14.5kb)
FRV2	8		full length retroviral sequence 2 (band F9, 11.2kb)
FRV3	12		full length retroviral sequence 3 (band F21, 3.5kb)
FSHB	11p13	.37	follicle stimulating hormone, beta polypeptide
FTHL1	1p31-p22		ferritin, heavy polypeptide-like 1
FTHL2	1q32.3-q42		ferritin, heavy polypeptide-like 2
FTHL3	2q32-q33		ferritin, heavy polypeptide-like 3
FTHL4	3q21-q23	.34	ferritin, heavy polypeptide-like 4
FTHL5	6p21.3-p12	.36	ferritin, heavy polypeptide-like 5
FTHL6	11q13		ferritin, heavy polypeptide-like 6
FTHL7	13q12		ferritin, heavy polypeptide-like 7
FTHL8	Xq26-q28		ferritin, heavy polypeptide-like 8
FTHL10	5		ferritin, heavy polypeptide-like 10
FTHL11	8		ferritin, heavy polypeptide-like 11
FTHL12	9		ferritin, heavy polypeptide-like 12
FTHL13	14		ferritin, heavy polypeptide-like 13
FTHL14	17		ferritin, heavy polypeptide-like 14
FTL	19q13.3-q13.4		ferritin, light polypeptide
FTLL1	20q12-q13.3		ferritin, light polypeptide-like 1
FTLL2	Xp22.3-p21.2		ferritin, light polypeptide-like 2
FUCA1	1p35-p34	.34	fucosidase, alpha-L- 1, tissue
FUCA1P	2		fucosidase, alpha-L- 1, pseudogene
FUCA2	6		fucosidase, alpha-L- 2, plasma
FUR	15q25-q26		furin, membrane associated receptor protein
FUSE	10		polykaryocytosis promoter
*FUT1	19q		fucosyl transferase 1 (Bombay phenotype included)
*FUT2	19q		fucosyl transferase 2 (secretor status included)
FY	1q21-q25		Duffy blood group
F2	11p11-q12		coagulation factor II (prothrombin)
F2L	Xpter-q25		coagulation factor II (prothrombin)-like
F3	1p22-p21	.33	coagulation factor III
F5	1q21-q25		coagulation factor V
F7	13q34		coagulation factor VII
F7R	8p23.2-p23.1		coagulation factor VII regulator
F8C	Xq28	.56	coagulation factor VIIIc, procoagulant component (hemophilia A)
F8VWF	12pter-p12	.36	coagulation factor VIII VWF (von Willebrand factor)
*F8VWFL	22q11.22-q11.23	.21	coagulation factor VIII VWF (von Willebrand factor)-like
F9	Xq26.3-q27.1	.51	coagulation factor IX (plasma thromboplastic component, 'Christmas disease', hemophilia B)
F10	13q34	.29	coagulation factor X
F11	4q35		coagulation factor XI
F12	5q33-qter		coagulation factor XII (Hageman)
F13A1	6p25-p24	.64	coagulation factor XIII, A1 polypeptide
F13B	1q31-q32.1	.37	coagulation factor XIII, B polypeptide
GAA	17q23		glucosidase, alpha; acid
*GABRA1	5q34-q35		gamma-aminobutyric acid (GABA) A receptor, alpha 1
*GABRA2	4p13-p12		gamma-aminobutyric acid (GABA) A receptor, alpha 2
*GABRA3	Xq28		gamma-aminobutyric acid (GABA) A receptor, alpha 3
*GABRB1	4p13-p12		gamma-aminobutyric acid (GABA) A receptor, beta 1
GAD	2		glutamate decarboxylase
*GALC	17		galactosylceramidase
GALE	1p36-p35		UDP-galactose-4-epimerase
GALK	17q23-q25		galactokinase
GALT	9p13		galactose-1-phosphate uridylyltransferase
GANAB	11q13-qter		glucosidase, alpha; neutral AB
GANC	15		glucosidase, alpha; neutral C
GAPD	12p13		glyceraldehyde-3-phosphate dehydrogenase
GAPDL1	1		glyceraldehyde-3-phosphate dehydrogenase-like 1
GAPDL2	Xpter-q13		glyceraldehyde-3-phosphate dehydrogenase-like 2
GAPDL3	2q		glyceraldehyde-3-phosphate dehydrogenase-like 3
GAPDL4	4		glyceraldehyde-3-phosphate dehydrogenase-like 4
GAPDL5	6p21-qter		glyceraldehyde-3-phosphate dehydrogenase-like 5
GAPDL7	8		glyceraldehyde-3-phosphate dehydrogenase-like 7

Symbol	Map location	PIC	Marker name
GAPDL8	6p21-qter		glyceraldehyde-3-phosphate dehydrogenase-like 8
GAPDL9	1		glyceraldehyde-3-phosphate dehydrogenase-like 9
GAPDL10	Xpter-q13		glyceraldehyde-3-phosphate dehydrogenase-like 10
GAPDL11	18		glyceraldehyde-3-phosphate dehydrogenase-like 11
GAPDL12	Xq13-qter		glyceraldehyde-3-phosphate dehydrogenase-like 12
GAPDL13	7		glyceraldehyde-3-phosphate dehydrogenase-like 13
*GAPDL14	13		glyceraldehyde-3-phosphate dehydrogenase-like 14
*GAPDL15	11		glyceraldehyde-3-phosphate dehydrogenase-like 15
GAPDL16	5q33.1-q35.3		glyceraldehyde-3-phosphate dehydrogenase-like 16
GAPDP1	Xp11.4-p11.21		glyceraldehyde-3-phosphate dehydrogenase pseudogene 1
GAS	17q	.09	gastrin
GBA	1q21		glucosidase, beta; acid
*GBAP	1q21		glucosidase, beta; acid, pseudogene
GC	4q12-q13	.37	group-specific component (vitamin D binding protein)
GCF1	7		growth control factor 1
GCG	2q36-q37		glucagon
GCP	Xq28	.57	green cone pigment (color blindness, deutan)
GCPS	7p13		Greig cephalopolysyndactyly syndrome
GCTG	7pter-p14		gamma-glutamylcyclotransferase
GCY	Yq11		growth control, Y chromosome influenced
GDH	1pter-p36.13		glucose dehydrogenase
GEY	19		green/blue eye color
GGTB1	4		glycoprotein-4-beta-galactosyltransferase 1
GGTB2	9p21-p13		glycoprotein-4-beta-galactosyltransferase 2
GGT1	22q11.1-q11.2	.37	gamma-glutamyltransferase 1
*GGT2	22q11.1-q11.2	.37	gamma-glutamyltransferase 2
*GHR	5p13-p12		growth hormone receptor
GHRF	20p11.23-qter		growth hormone releasing factor
GH1	17q22-q24	.38	growth hormone 1
GH2	17q22-q24	.38	growth hormone 2
*GIP	17q21.3-q22		gastric inhibitory polypeptide
GK	Xp21.3-p21.2		glycerol kinase deficiency
GLA	Xq21.3-q22	.23	galactosidase, alpha
GLAT	2		galactose enzyme activator
GLB1	3pter-p21		galactosidase, beta 1
GLI	12q13		glioma-associated oncogene homolog (zinc finger protein)
*GLI2	2		GLI-Kruppel family member GLI2
*GLI3	7		GLI-Kruppel family member GLI3
*GLI4	8		GLI-Kruppel family member GLI4
GLO1	6p21.3-p21.1		glyoxalase I
*GLR	Xp22.1-p21.2	.49	glycine receptor
GLUD	10q23-q24	.38	glutamate dehydrogenase
GLUDP1	Xq24-q26		glutamate dehydrogenase pseudogene 1
GLUT1	1p35-p31.3	.69	glucose transport 1
*GLUT2	3q26.1-q26.3		glucose transport-like 2
*GLUT3	12p13.3		glucose transport-like 3
*GLUT4	17p13	.36	glucose transport-like 4
*GLUT5	1p31		glucose transport-like 5
*GLUT6	5q33-q35		glucose transport-like 6
GLYB	8q21.1-qter		glycine B complementing
GM2A	5		GM2 ganglioside activator protein
GNAI1	7q21-q22		guanine nucleotide binding protein, alpha inhibiting activity polypeptide 1
GNAI2A	12		guanine nucleotide binding protein, alpha inhibiting activity polypeptide 2A
GNAI2B	3		guanine nucleotide binding protein, alpha inhibiting activity polypeptide 2B
GNAI3	1		guanine nucleotide binding protein, alpha inhibiting activity polypeptide 3
GNAS1	20		guanine nucleotide binding protein, alpha stimulating activity polypeptide 1
GNAT1	3		guanine nucleotide binding protein, alpha transducing activity polypeptide 1
GNAT2	1		guanine nucleotide binding protein, alpha transducing activity polypeptide 2
*GNAZ	22		guanine nucleotide binding protein, alpha z polypeptide
GNB1	1		guanine nucleotide binding protein, beta polypeptide 1
GNB2	7		guanine nucleotide binding protein, beta polypeptide 2
*GNB3	12p13		guanine nucleotide binding protein, beta polypeptide 3
GNPTA	4q21-q23		UDP-N-acetylglucosamine-lysosomal-enzyme N-acetylglucosaminephosphotransferase (mucolipidoses II & III)
GNS	12q14		N-acetylglucosamine-6-sulfatase (Sanfilippo disease IIID)
*GNT1	2		glucuronosyltransferase, phenol-UDP-
GOT1	10q25.3		glutamic-oxaloacetic transaminase 1, soluble
GOT2	16q12-q22		glutamic-oxaloacetic transaminase 2, mitochondrial
GPB	8q		glycerol phosphatase, beta-
GPD1	12		glycerol-3-phosphate dehydrogenase
GPI	19q13.1		glucose phosphate isomerase
GPT	8q24.2-qter		glutamic-pyruvate transaminase
GPXP1	21		glutathione peroxidase pseudogene 1
GPXP2	X		glutathione peroxidase pseudogene 2

5.33

Symbol	Map location	PIC	Marker name
GPX1	3q11-q12	.30	glutathione peroxidase 1
GP2B	17q21.32		glycoprotein IIb (IIb/IIIa complex, platelet, CD41B)
*GP3A	17q21.32	.37	glycoprotein IIIa (platelet, CD61)
GRL	5q31-q32	.37	glucocorticoid receptor
GRLL1	16		glucocorticoid receptor-like 1
*GRMP	1q22-q25		granulocyte membrane protein (140kD)
GRP	18q21	.37	gastrin releasing peptide
GRP78	9q		glucose regulated protein (78kD)
GSAS	10		glutamate gamma-semialdehyde synthetase
GSL	20		galactosialidosis
GSN	9q32-q34	.36	gelsolin
GSR	8p21.1		glutathione reductase
*GST1	1p31	.38	glutathione S-transferase 1, mu (H-b)
*GST12	12		glutathione S-transferase, microsomal
GST2	6p12.2	.38	glutathione S-transferase 2, alpha (H-a)
GST3	11q13-qter		glutathione S-transferase 3, pi
*GST3L	12q13-q14		glutathione S-transferase 3-like
*GUD	11p13		genitourinary dysplasia component of WAGR
GUK1	1q32-q42		guanylate kinase 1
GUK2	1q		guanylate kinase 2
GUSB	7q21.2-q22		glucuronidase, beta
GUSM	19		mouse beta-glucuronidase modifier
*GXP1	X		BN75 temperature sensitivity complementing
GYPA	4q28-q31	.37	glycophorin A (includes MN blood group)
GYPB	4q28-q31	.37	glycophorin B (includes Ss blood group)
GYPC	2q14-q21	.37	glycophorin C (Gerbich blood group)
*G1P1	1		interferon, alpha-inducible protein (clone IFI-4)
G6PD	Xq28	.61	glucose-6-phosphate dehydrogenase
G6PDL	17		glucose-6-phosphate dehydrogenase-like
*G7P1	7q22-q32	.26	kinase-like protein
G10P1	10q25-q26	.07	interferon, alpha-inducible protein (MW 56kD)
*G10P2	10q		interferon, alpha-inducible protein (MW 54kD)
*G13P1	13		interferon, alpha-inducible protein pseudogene of G10P1
*G17P1	17q24-qter		RNA-dependent ATPase, putative RNA helicase
*G19P1	19		protein 80K-H (substrate for protein kinase C)
HADH	7		hydroxyacyl-Coenzyme A dehydrogenase
HAGH	16p13.3		hydroxyacyl glutathione hydrolase
HARS	5		histidyl-tRNA synthetase
HBAP1	16p13.3	.45	hemoglobin, alpha pseudogene 1
HBAP2	16p13.3		hemoglobin, alpha pseudogene 2
HBA1	16p13.3	.45	hemoglobin, alpha 1
HBA2	16p13.3	.45	hemoglobin, alpha 2
HBB	11p15.5	.75	hemoglobin, beta
HBBP1	11p15.5	.37	hemoglobin, beta pseudogene 1
HBD	11p15.5	.38	hemoglobin, delta
HBE1	11p15.5	.37	hemoglobin, epsilon 1
HBG1	11p15.5	.36	hemoglobin, gamma A
HBG2	11p15.5	.37	hemoglobin, gamma G
HBQ1	16p13.3		hemoglobin, theta 1
HBZ	16p13.3	.77	hemoglobin, zeta
HBZP	16p13.3	.77	hemoglobin, zeta pseudogene
HCF2	22	.37	heparin cofactor II
HCK	20q11-q12		hemopoietic cell kinase
HCL1	19		hair color 1 (brown)
HCL2	4q		hair color 2 (red)
HCVS	15q11-qter		human coronavirus sensitivity
HD	4pter-p16.3		Huntington disease
HEP10	10		hepatic protein 10
HEXA	15q23-q24		hexosaminidase A (alpha polypeptide)
HEXB	5q13	.36	hexosaminidase B (beta polypeptide)
HF	1q32		complement component H
HFE	6p21.3		hemochromatosis
*HHH	13q34		hyperornithinemia-hyperammonemia-homocitrullinuria
*HIGM1	Xq24-q27		hyper IgM syndrome
HIS	12		histidase
*HKR1	19		GLI-Kruppel family member HKR1
*HKR2	19		GLI-Kruppel family member HKR2
*HKR3	1		GLI-Kruppel family member HKR3
HK1	10q22		hexokinase 1
HLA-A	6p21.3		major histocompatibility complex, class I
HLA-B	6p21.3		major histocompatibility complex, class I
HLA-C	6p21.3		major histocompatibility complex, class I
*HLA-CDA12	6p21.3		major histocompatibility complex, class I (cosmid cda12)
HLA-DNA	6p21.3		major histocompatibility complex, class II, DN alpha
HLA-DOB	6p21.3		major histocompatibility complex, class II, DO beta
HLA-DPA1	6p21.3		major histocompatibility complex, class II, DP alpha 1
HLA-DPA2	6p21.3		major histocompatibility complex, class II, DP alpha (pseudogene)
HLA-DPB1	6p21.3		major histocompatibility complex, class II, DP beta 1
HLA-DPB2	6p21.3		major histocompatibility complex, class II, DP beta 2 (pseudogene)
HLA-DQA1	6p21.3		major histocompatibility complex, class II, DQ alpha 1
HLA-DQA2	6p21.3		major histocompatibility complex, class II, DQ alpha 2

Symbol	Map location	PIC	Marker name
HLA-DQB1	6p21.3		major histocompatibility complex, class II, DQ beta 1
HLA-DQB2	6p21.3		major histocompatibility complex, class II, DQ beta 2
HLA-DRA	6p21.3		major histocompatibility complex, class II, DR alpha
HLA-DRB1	6p21.3		major histocompatibility complex, class II, DR beta 1
HLA-DRB1L	6p21.3		major histocompatibility complex, class II exon, beta-like
HLA-DRB2	6p21.3		major histocompatibility complex, class II, DR beta 2
HLA-DRB3	6p21.3		major histocompatibility complex, class II, DR beta 3
HLA-DRB4	6p21.3		major histocompatibility complex, class II, DR beta 4
HLA-DVB	6p21.3		major histocompatibility complex, class II, beta, truncated
HLA-E	6p21.3		major histocompatibility complex, class I
HLA-HA2	6p21.3		major histocompatibility complex, class I (T cell surface antigen)
HLA-JY3	6p21.3		major histocompatibility complex, class I (probe JY328)
HMAA	6p21.3		monocyte antigen A
HMAB	6p21.3		monocyte antigen B
HMGCR	5q13.3-q14		3-hydroxy-3-methylglutaryl -Coenzyme A reductase
HMGCS	5p14-p13		3-hydroxy-3-methylglutaryl -Coenzyme A synthase
HMG17	1p36.1-p35	.38	high-mobility group protein 17
HOMG	X		hypomagnesemia, secondary hypocalcemia
HOX1	7p21-p14		homeo box region 1
HOX2	17q21-q22	.22	homeo box region 2
HOX3	12q12-q13	.34	homeo box region 3
HOX4	2q31-q37		homeo box region 4
HP	16q22.1	.53	haptoglobin
HPAFP	4q11-q13		hereditary persistence of alpha-fetoprotein
HPR	16q22.1	.25	haptoglobin-related
HPRT	Xq26	.38	hypoxanthine phosphoribosyltransferase
HPRTP1	3	.07	hypoxanthine phosphoribosyltransferase pseudogene 1
HPRTP2	5p14-p13	.27	hypoxanthine phosphoribosyltransferase pseudogene 2
HPRTP3	11q		hypoxanthine phosphoribosyltransferase pseudogene 3
HPRTP4	11q		hypoxanthine phosphoribosyltransferase pseudogene 4
HPT	Xq26-q27		hypoparathyroidism
HPX	11pter-p11		hemopexin
HRAS	11p15.5	.67	Harvey rat sarcoma viral (v-Ha-ras) oncogene homolog
HRASP	Xpter-q26		Harvey rat sarcoma viral (v-Ha-ras) oncogene homolog pseudogene
*HRG	3		histidine-rich glycoprotein
HSAS	X		hydrocephalus, stenosis of the aqueduct of Sylvius
*HSDB3	1p13-p11		hydroxy-delta 5-steroid dehydrogenase, 3 beta- and steroid delta-i somerase
HSPAP1	X		heat shock 70 kD protein pseudogene 1
HSPA1	6p21.3	.36	heat shock 70 kD protein 1
*HSPA1L	6p21.3		heat shock 70 kD protein-like 1
HSPA2	14q22-q24	.36	heat shock 70 kD protein 2
HSPA3	21	.35	heat shock 70 kD protein 3
*HSPBL1	3p21-qter		heat shock 27 kD protein-like 1
*HSPBL2	9q12-qter		heat shock 27 kD protein-like 2
*HSPBL3	Xpter-q22		heat shock 27 kD protein-like 3
*HSPG	8q22-q24		heparan sulfate proteoglycan, cell surface-associated
HSTF1	11q13.3		heparan secretory transforming protein 1 (Kaposi sarcoma oncogene)
HTL	20		high L-leucine transport
*HTLVR	17q		human T cell leukemia virus (I and II) receptor
HTOR	21		5-hydroxytryptamine oxygenase regulator
HTR1A	5q11-q14		5-hydroxytryptamine receptor 1A
HVBS1	11p13		hepatitis B virus integration site 1
HVBS4	2		hepatitis B virus integration site 4
HVBS6	4q32	.37	hepatitis B virus integration site 6
HVBS7	18q11.1-q11.2		hepatitis B virus integration site 7
*HVBS8	17p12-p11.2		hepatitis B virus integration site 8
HV1S	3 or 11p11-qter		herpes simplex virus type 1 sensitivity
HY	Yp11.2-qter		histocompatibility Y antigen
HYP	Xp22.2-p22.1		hypophosphatemia, vitamin D resistant rickets
H1F2	1q21		H1 histone, family 2
H1F3	6p12-q21		H1 histone, family 3
H1F4	12q11-q21		H1 histone, family 4
H3F2	1q21		H3 histone, family 2
H4F2	1q21		H4 histone, family 2
*IAPP	12p12.3-p11.2 or 12q13-q21		islet amyloid polypeptide
*IBP1	7p13-p12	.32	insulin-like growth factor binding protein
*ICAM1	19		intercellular adhesion molecule 1 (CD54)
IDH1	2q32-qter		isocitrate dehydrogenase 1, soluble
IDH2	15q21-qter		isocitrate dehydrogenase 2, mitochondrial
IDS	Xq27.3-q28		iduronate 2-sulfatase (Hunter syndrome)
IDUA	22pter-q11		iduronidase, alpha-L-
IF	4q24-q25	.36	complement component I
IFNA	9q22-p13	.37	interferon, alpha (leukocyte)
IFNAP1	9p23-p22		interferon, alpha pseudogene 1

Symbol	Map location	PIC	Marker name
IFNAR	21q21-qter		interferon, alpha; receptor
IFNBR	21q21-qter		interferon, beta; receptor
IFNB1	9p22	.32	interferon, beta 1, fibroblast
IFNB3	2p23-qter		interferon, beta 3, fibroblast
IFNG	12q24.1		interferon, gamma
IFNGR1	6q23-q24		interferon, gamma; receptor 1 (binding subunit)
IFNGR2	21		interferon, gamma; receptor 2 (confers anti-viral resistance)
IFNGT1	21		interferon, gamma transducer 1
IFNR	16		interferon production regulator
IGF1	12q23	.53	insulin-like growth factor 1
IGF1R	15q25-qter		insulin-like growth factor 1 receptor
IGF2	11p15.5	.37	insulin-like growth factor 2
IGF2R	6q25-q27		insulin-like growth factor 2 receptor
*IGH@	14q32.33		immunoglobulin heavy chain gene cluster (V,D,J,C)
IGHA1	14q32.33	.46	immunoglobulin alpha 1
IGHA2	14q32.33	.79	immunoglobulin alpha 2 (A2M marker)
IGHD	14q32.33	.37	immunoglobulin delta
IGHDY2	15q11-q12		immunoglobulin heavy chain diversity region 2
IGHE	14q32.33		immunoglobulin epsilon
IGHEP1	14q32.33		immunoglobulin epsilon pseudogene 1
IGHEP2	9		immunoglobulin epsilon pseudogene 2
IGHGP	14q32.33	.39	immunoglobulin gamma pseudogene
IGHG1	14q32.33	.53	immunoglobulin gamma 1 Gm marker)
IGHG2	14q32.33	.49	immunoglobulin gamma 2 Gm marker)
IGHG3	14q32.33	.53	immunoglobulin gamma 3 Gm marker)
IGHG4	14q32.33	.56	immunoglobulin gamma 4 Gm marker)
IGHM	14q32.33	.48	immunoglobulin mu
IGHV	14q32.33	.52	immunoglobulin heavy polypeptide, variable region (many genes)
IGJ	4q21	.23	immunoglobulin J polypeptide, linker protein for immunoglobulin alpha and mu polypeptides
IGKC	2p12	.15	immunoglobulin kappa constant region
*IGKDEL	2q11-q13		immunoglobulin kappa deleting element or like
IGKV	2p12	.22	immunoglobulin kappa variable region
IGKVP1	1pter-q12		immunoglobulin kappa variable region pseudogene 1
IGKVP2	15		immunoglobulin kappa variable region pseudogene 2
IGKVP3	22q11	.37	immunoglobulin kappa variable region pseudogene 3
IGKVP4	22q11	.37	immunoglobulin kappa variable region pseudogene 4
IGKVP5	22q11	.37	immunoglobulin kappa variable region pseudogene 5
*IGLC1	22q11.1-q11.2	.46	immunoglobulin lambda constant region 1 (Mcg marker)
*IGLC2	22q11.1-q11.2	.46	immunoglobulin lambda constant region 2 (Kern-Oz- marker)
*IGLC3	22q11.1-q11.2	.46	immunoglobulin lambda constant region 3 (Kern-Oz+ marker)
*IGLC4	22q11.1-q11.2	.46	immunoglobulin lambda constant region 4 (pseudogene)
*IGLC5	22q11.1-q11.2	.46	immunoglobulin lambda constant region 5 (pseudogene)
*IGLC6	22q11.1-q11.2	.46	immunoglobulin lambda constant region 6 (Kern+ Oz- marker)
IGLV	22q11.1-q11.2	.65	immunoglobulin lambda variable region gene cluster
IHG	X		iris hypoplasia with glaucoma
IL1A	2q12-q21	.32	interleukin 1, alpha
*IL1B	2q13-q21		interleukin 1, beta
IL2	4q26-q27	.07	interleukin 2
IL2R	10p15-p14	.12	interleukin 2 receptor
IL3	5q23-q31	.30	interleukin 3 (colony-stimulating factor, multiple)
IL4	5q23-q31		interleukin 4
IL5	5q23-q31		interleukin 5 (colony-stimulating factor, eosinophil)
IL6	7p21-p14	.79	interleukin 6
*IL8	4q13-q21	.37	interleukin 8
INHA	2q33-qter		inhibin, alpha
INHBA	7p15-p13		inhibin, beta A
INHBB	2cen-q13		inhibin, beta B
INP10	4q21	.37	protein 10 from gamma-interferon induced cell line
INS	11p15.5	.57	insulin
INSL1	2		insulin-like 1
INSL2	11p11-q13		insulin-like 2
INSR	19p13.3-p13.2	.80	insulin receptor
INSRL	7p13-q22		insulin receptor-like
INT1	12q13		murine mammary tumor virus integration site (v-int-1) oncogene homolog
INT1L1	7q31		int-1-like protein 1
INT2	11q13	.36	murine mammary tumor virus integration site (v-int-2) oncogene homolog
*INT4	17q21-q22		murine mammary tumor virus integration site (v-int-4) oncogene homolog
IP1	Xp11.21-cen		incontinentia pigmenti 1 (sporadic, associated with X chromosome rearrangements)
IP2	Xq27-q28		incontinentia pigmenti 2 (familial, male-lethal type)
*ITIH1	3p21.2-p21.1	.37	inter-alpha-trypsin inhibitor (protein HC), H1 polypeptide

Symbol	Map location	PIC	Marker name
*ITIH2	10p15	.33	inter-alpha-trypsin inhibitor (protein HC), H2 polypeptide
*ITIH3	3p21.2-p21.1		inter-alpha-trypsin inhibitor (protein HC), H3 polypeptide
ITIL	9q32-q33	.16	inter-alpha-trypsin inhibitor (protein HC), light polypeptide
ITPA	20p		inosine triphosphatase (nucleoside triphosphate pyrophosphatase)
IVD	15q14-q15		isovaleryl Coenzyme A dehydrogenase
IVL	1q21-q22		involucrin
JK	18q11-q12		Kidd blood group
JPD	4		juvenile periodontitis
JUN	1p32-p31		avian sarcoma virus 17 (v-jun) oncogene homolog
KAL	Xp22.32		Kallmann syndrome
KIT	4p11-q22	.19	Hardy-Zuckerman 4 feline sarcoma viral (v-kit) oncogene homolog
KLK1	19q13.3-qter		kallikrein 1, renal/pancreas/salivary
KRAS1P	6p12-p11	.39	Kirsten rat sarcoma 1 viral (v-Ki-ras1) oncogene homolog, processed pseudogene
KRAS2	12p12.1	.38	Kirsten rat sarcoma 2 viral (v-Ki-ras2) oncogene homolog
KRT4	12p12.2-q11		keratin 4
KRT14L1	17p12-p11 and 17q12-q21		keratin 14-related sequence 1 (20kb)
KRT14L2	17p12-p11 and 17q12-q21		keratin 14-related sequence 2 (10kb)
KRT14L3	17p12-p11 and 17q12-q21		keratin 14-related sequence 3 (6.6kb)
KRT15	17q21-q23		keratin 15
KRT16L1	17p12-p11 and 17q12-q21		keratin 16-related sequence 1 (16kb)
KRT16L2	17p12-p11 and 17q12-q21		keratin 16-related sequence 2 (11kb)
*KRT18L1	X		keratin 18-like 1
*KRT18L3	9		keratin 18-like 3
*KRT18L4	11		keratin 18-like 4
KRT19	17q21-q23		keratin 19
*K12T	14		K12 temperature sensitivity complementing
LALBA	12q13		lactalbumin, alpha-
*LAMA	18p11.32-p11.2		laminin, A polypeptide
LAMB1	7q22 or 7q31	.37	laminin, B1 polypeptide
LAMB2	1q31	.37	laminin, B2 polypeptide
*LAR	1p34-p32		leukocyte antigen related tyrosine phosphatase
LARS	5cen-q11		leucyl-tRNA synthetase
LCAT	16q22.1		lecithin-cholesterol acyltransferase
LCK	1p35-p32		lymphocyte-specific protein tyrosine kinase
LCO	2q14-q21		liver cancer oncogene
LCP1	13q14.3		lymphocyte cytosolic protein 1
*LCT	2	.37	lactase
LDHA	11p15.1-p14	.37	lactate dehydrogenase A
LDHAL1	4pter-p15.3		lactate dehydrogenase A-like 1
LDHAL2	1q42-q44		lactate dehydrogenase A-like 2
*LDHAL3	2		lactate dehydrogenase A-like 3
*LDHAL4	9		lactate dehydrogenase A-like 4
*LDHAL5	10		lactate dehydrogenase A-like 5
LDHB	12p12.2-p12.1		lactate dehydrogenase B
*LDHBL1	13		lactate dehydrogenase B-like 1
*LDHBL2	X		lactate dehydrogenase B-like 2
LDHC	11		lactate dehydrogenase C
LDLR	19p13.2-p13.1	.60	low density lipoprotein receptor (familial hypercholesterolemia)
LE	19		Lewis blood group
LEU7	11q12-qter		leukocyte antigen 7 (CD57)
LGCR	8q24.11-q24.13		Langer-Giedion syndrome chromosome region
LHB	19q13.3		luteinizing hormone beta polypeptide
LHRH	8p21-p11.2		luteinizing-hormone releasing hormone
*LIF	22q11.2-q13.1		leukemia inhibitory factor (differentiation-stimulating factor)
LIPA	10		lipase A, lysosomal acid (Wolman disease)
LIPB	16		lipase B, lysosomal acid
LIPC	15q21-q23	.38	lipase, hepatic
*LIPE	19q13.1-q13.2		lipase, hormone sensitive
LOX	X		lysyl oxidase; cutis laxa-X; Ehlers-Danlos V
LPA	6q26-q27		lipoprotein, Lp(a)
LPC1	9q11-q22		lipocortin I
LPC2A	4q21-q31	.09	lipocortin IIa
LPC2B	9pter-q34		lipocortin IIb
LPC2C	10q21-q22		lipocortin IIc
LPC2D	15q21-q22		lipocortin IId
*LPL	8p22	.57	lipoprotein lipase
LTF	3q21-q23		lactotransferrin
LU	19q12-q13		Lutheran blood group
LW	19p13.2-cen		Landsteiner-Wiener blood group
*LYAM1	1q23-q25		lymphocyte adhesion molecule 1
LYN	8q13-qter		Yamaguchi sarcoma viral (v-yes-1) related oncogene homolog
*LYP	Xq25-q26		lymphoproliferative syndrome
*LYZ	12		lysozyme
*L1CAM	Xq28		L1 cell adhesion molecule (symbol provisional)
MAA	X		microphthalmia or anophthalmia and associated anomalies
MAFD1	11p15.5		major affective disorder 1
MAFD2	Xq27-q28		major affective disorder 2
MAG	19q13.1		myelin associated glycoprotein
MAL	2cen-q13		T cell differentiation protein
MANA	15q11-qter		mannosidase, alpha A, cytoplasmic
MANB	19cen-q13.1		mannosidase, alpha B, lysosomal
MAOA	Xp11.3-p11.23	.46	monoamine oxidase A
MAOB	Xp11.3-p11.23		monoamine oxidase B
MARS	12		methionine tRNA synthetase
MAS1	6q24-q27		MAS1 oncogene
MB	22q11.2-qter	.36	myoglobin
MBP	18q22-qter	.11	myelin basic protein
MCF2	Xq26.3-q27.2		MCF.2 cell line derived transforming sequence
MCP	1q32		membrane cofactor protein
MDCR	17p13.3		Miller-Dieker syndrome chromosome region

Symbol	Map location	PIC	Marker name
MDF1	4		antigen identified by monoclonal antibody A-3A4
MDH1	2p23		malate dehydrogenase, NAD (soluble)
MDH2	7p13-q22		malate dehydrogenase, NAD (mitochondrial)
MDU1	11q		antigen identified by monoclonal antibodies 4F2, TRA1.10, TROP4, and T43
MEAX	X		myopathy with excessive autophagy
MEL	19p13.2-cen	.32	cell line NK14 derived transforming oncogene
MEN1	11q12-q13		multiple endocrine neoplasia I
MEN2A	10p11.2-q11.2		multiple endocrine neoplasia IIA
*MEN2B	10pter-q11.2		multiple endocrine neoplasia IIB
MER1	11pter-p13		antigen identified by monoclonal antibodies W6/34, 5C1, etc. (formerly lethal antigen 1)
MER2	11p15.5		antigen identified by monoclonal antibodies 1D12, 2F7
MER5	19q13.3		antigen identified by monoclonal antibody 2D8
MER6	3cen-q22		antigen identified by monoclonal antibody 1D8
MET	7q31	.81	met proto-oncogene
ME1	6q12		malic enzyme 1, soluble
ME2	6p25-p24		malic enzyme 2, mitochondrial
MFI2	3q28-q29		antigen p97 (melanoma associated) identified by monoclonal antibodies 133.2 and 96.5
MF4	X		metacarpal 4-5 fusion
MGCR	22q12.3-qter		meningioma chromosome region
*MGC1	Xq12-q26		megalocornea 1 (X-linked)
MGSA	4q21	.31	melanoma growth stimulating activity
MIC2	Xp22.32; Yp11.3		antigen identified by monoclonal antibodies 12E7, F21 and O13
MIC5	Xq27-q28		antigen identified by monoclonal antibody R1
MIC6	17q21-qter		antigen identified by monoclonal antibody H207
MIC7	15q11-q22		antigen identified by monoclonal antibody 28.3.7
MIC9	11q		antigen identified by monoclonal antibody 4D12
MIC10	1		antigen identified by monoclonal antibody TRA-2-10
MIC11	11p14-p13		antigen identified by monoclonal antibody 16.3A5
MIC12	15q11-q22		antigen identified by monoclonal antibody 30.2A8
MIC17	12p		antigen identified by monoclonal antibody BB1
MIC18	2		antigen identified by monoclonal antibody AUA1
MIN1	11p13		antigen recognized by monoclonal antibody EJ16
MIN2	11p13		antigen recognized by monoclonal antibody EJ30
MIN3	11p13		antigen recognized by monoclonal antibody EL32
MIP	12cen-q14		major intrinsic protein of lens fiber
MLA1	12q12-q13		melanoma antigen 1(ME491)
MLN	6p21.3-p21.2		motilin
MLR	4q31	.36	mineralocorticoid receptor (aldosterone receptor)
MLVI2	5p14-p13		Moloney murine leukemia virus (MoMuLV) integration site 2 homolog
*MME	3q21-q27		membrane metallo-endopeptidase (neutral endopeptidase, enkephalinase, CALLA, CD10)
MNK	Xcen-q13		Menkes syndrome
MOS	8q11	.11	Moloney murine sarcoma viral (v-mos) oncogene homolog
MOX2	3	.36	antigen identified by monoclonal antibody MRC OX-2
*MPE	12p13		malignant proliferation, esinophil
MPI	15q22-qter		mannose phosphate isomerase
MPO	17q21.3-q23	.25	myeloperoxidase
MRSD	Xq27-q28		mental retardation-skeletal dysplasia
MRX1	Xp11-q13		mental retardation, X-linked 1 (non dysmorphic)
MRX2	Xp22.3-p22.2		mental retardation, X-linked 2
MSD	X		microcephaly with spastic diplegia (Paine syndrome)
MSK1	1p22		antigen identified by monoclonal antibody AJ9
MSK2	1q32-qter		antigen identified by monoclonal antibody T87
MSK3	12p		antigen identified by monoclonal antibody M68
MSK4	12cen-q13		antigen identified by monoclonal antibody A123/A127
MSK7	12q13-qter		antigen identified by monoclonal antibody VI
MSK8	2p23-qter		antigen identified by monoclonal antibody L230
MSK9	3pter-p14		antigen identified by monoclonal antibody K15
MSK10	3pter-p14		antigen identified by monoclonal antibody AJ425
MSK11	5		antigen identified by monoclonal antibody SR84
MSK13	11q13-qter		antigen identified by monoclonal antibody Q14
MSK14	11cen-q13		antigen identified by monoclonal antibody JF23
MSK15	15q13-qter		antigen identified by monoclonal antibody SV13
MSK16	15q13-qter		antigen identified by monoclonal antibody A0122
MSK17	15q13-qter		antigen identified by monoclonal antibody F23
MSK18	17q22-q23		antigen identified by monoclonal antibody J143

Symbol	Map location	PIC	Marker name
MSK19	19q12-q13.2		antigen identified by monoclonal antibody F8 (gylcoprotein MW 95kD)
MSK20	19pter-p13.2		antigen identified by monoclonal antibody F10 (glycoprotein MW 50kD)
MSK21	11pter-cen		antigen identified by monoclonal antibody G344
MSK22	11cen-q13		antigen identified by monoclonal antibody A124
MSK23	11cen-q13		antigen identified by monoclonal antibody T43
MSK24	11pter-cen		antigen identified by monoclonal antibody NP13
MSK25	11q13-qter		antigen identified by monoclonal antibody MC139
MSK26	11q13-qter		antigen identified by monoclonal antibody K117
MSK27	12q13-qter		antigen identified by monoclonal antibody MG6
MSK28	6q21-qter		antigen identified by monoclonal antibody MG2
MSK29	6q12-q15		antigen identified by monoclonal antibody A42
MSK30	14		antigen identified by monoclonal antibody A42
MSK31	1p36		antigen identified by monoclonal antibody SR75
MSK32	3pter-p14		antigen identified by monoclonal antibody K66
MSK33	3pter-p14		antigen identified by monoclonal antibody SR3
MSK34	9		antigen identified by monoclonal antibody CNT/6
MSK35	11		antigen identified by monoclonal antibody SR27
MSK36	12q13-qter		antigen identified by monoclonal antibody CNT/11
MSK37	19q12-q13.2		antigen identified by monoclonal antibody S7
MSK38	20		antigen identified by monoclonal antibody O5
*MSK39	11q13-qter		antigen recognized by monoclonal antibody 5.1H11
MSK40	22		antigen identified by monoclonal antibody F35/9
*MSK41	22		antigen identified by monoclonal antibody E3
MTAP	9		methylthioadenosine phosphorylase
MTB	2q34-q35		microtubule (beta) associated protein
MTBT1	17q21	.16	microtubule (beta) associated protein tau 1
MTBT2	6p21		microtubule (beta) associated protein tau-like
MTHFD	14q22-q32		5,10-methylenetetrahydrofolate dehydrogenase, 5,10-methylenetetrahydrofolatecyclohydrolase, 10-formyltetrahydrofolate synthetase
*MTHFDL1	Xp11.3-p11.1		5,10-methylenetetrahydrofolate dehydrogenase, 5,10- methylenetetra hydrofolatecyclohydrolase, 10-formyltetrahydrofolate synthetase-like 1
MTL1	1p		metallothionein-like 1
MTL2	1p22-qter		metallothionein-like 2
MTL3	18		metallothionein-like 3
MTL4	20		metallothionein-like 4
MTM1	Xq27-q28		myotubular myopathy 1
MTR	1		tetrahydropteroylglutamate methyltransferase
MT1	16q13-q22.1		metallothionein 1
MT2	16q13-q22.1	.22	metallothionein 2
MT2P1	4p11-q21	.38	metallothionein 2 pseudogene 1 (processed)
MUC1	1q21-q23		mucin 1, urinary, peanut lectin binding
*MUC2	11p15		mucin 2, intestinal
MUT	6p21	.37	methylmalonyl Coenzyme A mutase
MUW53	11		antigen identified by monoclonal antibody 53.6
MX1	21		myxovirus (influenza) resistance 1, homolog of murine
MX2	21		myxovirus (influenza) resistance 2, homolog of murine
MYB	6q22-q23	.37	avian myeloblastosis viral (v-myb) oncogene homolog
MYC	8q24	.49	avian myelocytomatosis viral (v-myc) oncogene homolog
MYCL1	1p32	.37	avian myelocytomatosis viral (v-myc) oncogene homolog 1, lung carcinoma derived
*MYCL2	Xq22-q28		avian myelocytomatosis viral (v-myc) oncogene homolog 2
MYCN	2p24	.59	avian myelocytomatosis viral (v-myc) related oncogene, neuroblastoma derived
*MYF3	11		myogenic factor 3
*MYF4	1		myogenic factor 4
*MYF5	12		myogenic factor 5
MYH1	17pter-p11		myosin, heavy polypeptide 1, skeletal muscle, adult
MYH2	17p13.1	.43	myosin, heavy polypeptide 2, skeletal muscle, adult
MYH3	17pter-p11	.37	myosin, heavy polypeptide 3, skeletal muscle, embryonic
MYH4	17pter-p11		myosin, heavy polypeptide 4, skeletal muscle
MYH5	7cen-q11.2		myosin, heavy polypeptide 5, skeletal muscle, adult
MYH6	14q11.2-q13	.34	myosin, heavy polypeptide 6, cardiac muscle, alpha
MYH7	14q11.2-q13		myosin, heavy polypeptide 7, cardiac muscle, beta
MYLL1	8		myosin, light polypeptide, cardiac muscle-like 1
MYL1	2q32.1-qter		myosin, light polypeptide 1, alkali; skeletal, fast
MYL3	3p		myosin, light polypeptide 3, alkali; ventricular, skeletal, slow
MYL4	17	.19	myosin, light polypeptide 4, alkali; atrial, embryonic
*MYOD1	11p15.4		myogenic differentiation 1
*M6PR	12		mannose-6-phosphate receptor (cation dependent)
M7V1	19		baboon M7 virus receptor

Symbol	Map location	PIC	Marker name
M130	10		external membrane protein (MW 130kD)
M195	14		external membrane protein (MW 195kD)
NAGA	22q13-qter		acetylgalactosaminidase, alpha-N-
NANTA3	11		N-acetylneuraminyltransferase, alpha-3-
NARS	18		asparaginyl-tRNA synthetase
NBCCS	1p		nevoid basal cell carcinoma syndrome
*NCA	19q13.2		non-specific cross reacting antigen
NCAM	11q23-q24	.29	cell adhesion molecule, neural (CD56)
*NCF1	10		neutrophil cytosolic factor 1 (47 kD, chronic granulomatous disease, autosomal 1)
*NCF2	1cen-q32		neutrophil cytosolic factor 2 (65 kD, chronic granulomatous disease, autosomal 2)
NDP	Xp11.4-p11.3		Norrie disease(pseudoglioma)
NEB	2q31-q32		nebulin
NEFH	22q12.1-q13.1		neurofilament, heavy polypeptide
*NEFHL	1p12		neurofilament, heavy polypeptide-like
NEFL	8p21	.36	neurofilament, light polypeptide
*NEFLL1	2p12-q11		neurofilament, light polypeptide-like 1
*NEFLL2	7q34-q35		neurofilament, light polypeptide-like 2
NEU	10pter-q23 or 6		neuraminidase
NF1	17q11.2		neurofibromatosis 1 (von Recklinghausen disease, Watson disease)
NF2	22q11-q13.1		neurofibromatosis 2 (bilateral acoustic neuroma)
NGFB	1p13	.37	nerve growth factor, beta polypeptide
NGFR	17q21-q22	.37	nerve growth factor receptor
NHCP1	16		non-histone chromosome protein 1
NHCP2	7		non-histone chromosome protein 2
*NHS	Xp22.3-p21.1		Nance-Horan syndrome (congenital cataracts and dental anomalies)
*NID	1q43		nidogen
NKNA	7q21-q22	.31	neurokinin A/substance P
NKNB	12q13-q21		neurokinin B
NM	7q22-qter		neutrophil migration
*NMB	15q11-qter		neuromedin B
*NMOR1	16		NAD(P)H menadione oxidoreductase 1, dioxin-inducible
NP	14q11.2		nucleoside phosphorylase
NPS1	9q34		nail patella syndrome 1
NPY	7pter-q22	.22	neuropeptide Y
NRAS	1p13	.37	neuroblastoma RAS viral (v-ras) oncogene homolog
NRASL1	9p	.37	neuroblastoma RAS viral (v-ras) oncogene homolog-like 1
NRASL2	22		neuroblastoma RAS viral (v-ras) oncogene homolog-like 2
*NTS	12		neurotensin
*NT5	6q14-q21		5' nucleotidase (CD73)
NYS	X		nystagmus
OAP	12q14		osteoarthrosis, precocious
OAT	10q26	.34	ornithine aminotransferase
OATL1	Xp11.3-p11.21	.48	ornithine aminotransferase-like 1
OA1	Xp22.3		ocular albinism 1 (Nettleship-Falls)
OCRL	Xq25-q26.1		oculocerebrorenal syndrome of Lowe
ODC1	2p25	.32	ornithine decarboxylase 1
ODC2	7q31-qter		ornithine decarboxylase 2
OFC	6pter-p21.3		orofacial cleft
*OFD1	X		oral-facial-digital syndrome I
OIAS	12		2',5'-oligoisoadenylate synthetase
OI4	7q21.3-q22.1		osteogenesis imperfecta type IV
OK	19pter-p13.2		OK blood group
OPD	X		otopalatodigital syndrome
OPEM	X		ophthalmoplegia, external, with myopia
ORM1	9q31-qter		orosomucoid 1
ORM2	9q31-qter		orosomucoid 2
OT	20		prepro-oxytocin (neurophysin I)
OTC	Xp21.1	.68	ornithine carbamoyltransferase
*OTF1	1cen-q32		octamer-binding transcription factor 1
*OTF2	19		octamer-binding transcription factor 2
*PABX	Xp22.32	.50	pseudoautosomal boundary region, X-linked
*PABY	Yp11.3		pseudoautosomal boundary region, Y-linked
PAH	12q22-q24.2	.82	phenylalanine hydroxylase
PAIS	21		phosphoribosylaminoimidazole synthetase
PALB	18q11.2-q12.1	.38	prealbumin
PBGD	11q23.2-qter	.37	porphobilinogen deaminase
PBT	4q12-q21		piebald trait
PC	11q		pyruvate carboxylase
PCCA	13q22-q34	.36	propionyl Coenzyme A carboxylase, alpha polypeptide
PCCB	3q13.3-q22	.37	propionyl Coenzyme A carboxylase, beta polypeptide
PCH1	2		protein spot in 2-D gels (MW 250kD; pI 7.0)
PCH2	5		protein spot in 2-D gels (MW 250kD; pI 8.3)
PDGFA	7p22-p21 or 7q11.2-q21.1		platelet-derived growth factor alpha polypeptide
PDGFB	22q12.3-q13.1	.37	platelet-derived growth factor beta polypeptide (simian sarcoma viral (v-sis) oncogene homolog)
*PDGFRA	4q11-q13		platelet-derived growth factor receptor alpha polypeptide
PDGFRB	5q33-q35		platelet-derived growth factor receptor, beta polypeptide
*PDHAL	4q22-q23		pyruvate dehydrogenase, E1 alpha polypeptide-like
*PDHA1	Xp22.1		pyruvate dehydrogenase, E1 alpha polypeptide
PDYN	20pter-p12	.33	prodynorphin
PENK	8q23-q24	.43	proenkephalin
PENT	17	.34	phenylethanolamine N-methyltransferase
PEPA	18q23		peptidase A

Symbol	Map location	PIC	Marker name
PEPB	12q21		peptidase B
PEPC	1q25 or 1q42		peptidase C
PEPD	19q12-q13.2		peptidase D
PEPE	17		peptidase E
*PEPN	15q13-qter	.30	peptidase N, amino-
PEPS	4p11-q12		peptidase S
*PFC	Xp11.4		properdin P factor, complement
PFD	Xp21-p11		properdin P factor, complement deficiency
PFKL	21q22.3	.35	phosphofructokinase, liver type
PFKM	1cen-q32		phosphofructokinase, muscle type
PFKP	10pter-p11.1		phosphofructokinase, platelet type
PFKX	12		phosphofructokinase, polypeptide X
PFN1	17		profilin 1
PFN2	1		profilin 2
PF4	4q12-q21	.37	platelet factor 4
PGAM1	10q25.3		phosphoglycerate mutase 1 (brain)
*PGAM2	7p13-p12		phosphoglycerate mutase 2 (muscle)
PGA3	11q13		pepsinogen 3, group I (pepsinogen A)
PGA4	11q13		pepsinogen 4, group I (pepsinogen A)
PGA5	11q13		pepsinogen 5, group I (pepsinogen A)
PGC	6pter-p21.1	.24	progastricsin (pepsinogen C)
PGD	1p36.3-p36.13	.30	phosphogluconate dehydrogenase
*PGDL1	18q12.2-q21.1		phosphogluconate dehydrogenase-like 1
PGFT	21q11.2-q22.2		phosphoribosylglycinamide formyltransferase
PGK1	Xq13	.50	phosphoglycerate kinase 1
PGK1P1	Xq12		phosphoglycerate kinase 1, pseudogene 1
PGK1P2	6p21-q12	.33	phosphoglycerate kinase 1, pseudogene 2
PGK2	19pter-p13.1		phosphoglycerate kinase 2
PGM1	1p22.1		phosphoglucomutase 1
PGM2	4p14-q12		phosphoglucomutase 2
PGM3	6q12		phosphoglucomutase 3
PGP	16p13		phosphoglycolate phosphatase
PGR	11q22-q23	.28	progesterone receptor
PGY1	7q21	.37	P glycoprotein 1/multiple drug resistance 1
PGY3	7q21	.25	P glycoprotein 3/multiple drug resistance 3
*PHK	X		phosphorylase kinase deficiency, liver (glycogen storage disease type VIII)
*PHKA	Xq12-q13		phosphorylase kinase, alpha
*PHKB	16q12-q13		phosphorylase kinase, beta
*PHKG	7		phosphorylase kinase, gamma (muscle)
*PHKGL	11		phosphorylase kinase, gamma-like
PHP	X		panhypopituitarism
PI	14q32.1	.81	alpha-1-antitrypsin (protease inhibitor)
*PIGR	1q31-q41		polymeric immunoglobulin receptor
PIL	14q32.1	.54	alpha-1-antitrypsin-like
PIM	6p21	.32	pim oncogene homolog
PIP	7q32-qter	.30	prolactin-induced protein
PKD1	16p13		polycystic kidney disease 1 (autosomal dominant)
PKLR	1q21		pyruvate kinase, liver and RBC
PKM2	15q22-qter		pyruvate kinase, muscle
PLANH1	7q21.3-q22	.37	plasminogen activator inhibitor, type I
*PLANH2	18q21-q22		plasminogen activator inhibitor, type II (arginine-serpin)
PLAT	8p12-q11.2	.37	plasminogen activator, tissue
PLAU	10q24-qter	.33	plasminogen activator, urokinase
*PLA2	12q23-qter		phospholipase A2, pancreatic
*PLA2L	1		phospholipase A2, pancreatic-like
*PLC1	20q12-q13.1		phospholipase C, subtype 148
PLG	6q26-q27	.37	plasminogen
*PLGL	2p11-q11		plasminogen-like
*PLI	18p11.1-q11.2		alpha-2-plasmin inhibitor
PLP	Xq21.3-q22	.15	proteolipid protein (Pelizaeus-Merzbacher disease)
PND	1p36	.37	pronatriodilatin
PNI1	19		protein spot NC22 in 2-D gels (MW 50kD, pI 5.00)
PNI2	16		protein spot in 2-D gels (MW 33kD)
PNY1	21		protein spot in 2-D gels (MW 82kD)
PNY2	21		protein spot in 2-D gels (MW 65kD)
PNY3	21		protein spot in 2-D gels (MW 33kD)
PNY4	21		protein spot in 2-D gels (MW 72kD)
PNY5	21		protein spot in 2-D gels (MW 40kD)
POF1	Xq21.3-q27		premature ovarian fail 1
POLA	Xp22.1-p21.3		polymerase (DNA directed), alpha
POLB	8p12-p11		polymerase (DNA directed), beta
POLR2	17p13.1	.28	polymerase (RNA) II (DNA directed) large polypeptide
POMC	2p23	.36	proopiomelanocortin (adrenocorticotropin/beta-lipotropin)
PON	7q21-q22		paraoxonase
*POR	7q11.2		P-450 (cytochrome) oxidoreductase
PP	10q11.1-q24		pyrophosphatase (inorganic)
PPAT	4pter-q21		phosphoribosyl pyrophosphate amidotransferase
PPGB	22		protective protein for beta-galactosidase
PPOL	1q41-q42	.37	poly (ADP-ribose) polymerase (NAD + ADP-ribosyltransferase)
PPOLP1	13q34	.19	poly (ADP-ribose) polymerase pseudogene 1
PPOLP2	14q13-q32		poly (ADP-ribose) polymerase pseudogene 2
PPY	17p11.1-qter	.37	pancreatic polypeptide
PRB1	12p13.2	.60	proline-rich protein BstNI subfamily 1
PRB2	12p13.2	.60	proline-rich protein BstNI subfamily 2
PRB3	12p13.2	.60	proline-rich protein BstNI subfamily 3

Symbol	Map location	PIC	Marker name
PRB4	12p13.2	.60	proline-rich protein BstNI subfamily 4
PRG	10		proteoglycan, secretory granule
PRGS	21q22.1		phosphoribosylglycinamide synthetase
PRH1	12p13.2		proline-rich protein HaeIII subfamily 1
PRH2	12p13.2		proline-rich protein HaeIII subfamily 2
PRIP	20pter-p12	.16	prion protein
PRKAR1	7p13-qter		protein kinase, cAMP-dependant, type I, regulatory polypeptide
PRKAR2	7q31-qter	.28	protein kinase, cAMP-dependant, type II, regulatory polypeptide
PRKCA	17q22-q24		protein kinase C, alpha polypeptide
PRKCB	16p12-q11.1		protein kinase C, beta polypeptide
PRKCG	19q13.4	.55	protein kinase C, gamma polypeptide
PRL	6p23-p21.1		prolactin
PRM1	16		protamine 1
PROC	2q13-q21	.34	protein C (inactivator of coagulation factors Va and VIIIa)
PROS1	3p11-q11.2		protein S, alpha
*PROS2	3p21-q21		protein S, beta
PRPS1	Xq21-q27		phosphoribosyl pyrophosphate synthetase 1
*PRPS1L1	7p22-qter		phosphoribosyl pyrophosphate synthetase 1-like 1
PRPS1L2	9		phosphoribosyl pyrophosphate synthetase 1-like 2
PRPS2	Xpter-q21		phosphoribosyl pyrophosphate synthetase 2
PSBG1	19q13.1-q13.2		pregnancy specific beta-1-glycoprotein 1
PSF1	Xq12-q28		protein spot in 2-D gels (MW 24kD)
PSF2	Xpter-q13		protein spot in 2-D gels (MW 27kD)
PSF3	Xpter-q13		protein spot in 2-D gels (MW 37kD)
PSF4	Xpter-q13		protein spot in 2-D gels (MW 40kD)
PSP	7p21-p15		phosphoserine phosphatase
PTH	11pter-p15.4	.37	parathyroid hormone
PTHLH	12p12.1-p11.2		parathyroid hormone-like hormone
PUT1	6		polypeptide m44p in lymphocyte membrane
PVALB	22		parvalbumin
PVS	19q12-q13.2		poliovirus sensitivity
PVT1	8q24		mouse pvt-1 oncogene homolog, MYC activator
PWCR	15q11-q12		Prader-Willi syndrome chromosome region
PYGB	20		phosphorylase, glycogen; brain
PYGBL	10		phosphorylase, glycogen; brain-like A (ras related)
PYGL	14q11.2-q24.3	.37	phosphorylase, glycogen; liver (Hers disease, glycogen storage disease type VI)
PYGM	11q12-q13.2	.46	phosphorylase, glycogen; muscle (McArdle syndrome, glycogen storage disease type V)
PYHG1	10		protein spot in 2-D gels (MW 218kD)
PYHG2	11		protein spot in 2-D gels (MW 178KD)
PYHG3	7		protein spot in 2-D gels (MW 106kD)
PYHG4	X		protein spot in 2-D gels (MW 104kD)
PYHG6	12		protein spot in 2-D gels (MW 85kD)
PYHG7	1p32-qter		protein spot in 2-D gels (MW 82kD)
PYHG8	7		protein spot in 2-D gels (MW 80kD)
PYHG9	1p32-qter		protein spot in 2-D gels (MW 79kD)
PYHG10	12		protein spot in 2-D gels (MW 77kD)
PYHG11	15		protein spot in 2-D gels (MW 74kD)
PYHG12	X		protein spot in 2-D gels (MW 55kD)
PYHG14	9		protein spot in 2-D gels (MW 37kD)
PYHG15	9		protein spot in 2-D gels (MW 35kD)
PYHG16	9		protein spot in 2-D gels (MW 38kD)
PYHG17	X		protein spot in 2-D gels (MW 32kD)
PYHG18	10		protein spot in 2-D gels (MW 31kD)
PYHG19	12		protein spot in 2-D gels (MW 12kD)
PYHG20	15		protein spot in 2-D gels (MW 10kD)
PYHG21	1p32-qter		protein spot in 2-D gels (MW 10kD)
PYHG22	1p32-qter		protein spot in 2-D gels (MW <10kD)
*PZP	12p13-p12.2		pregnancy zone protein
P1	22q11.2-qter		P blood group (P one antigen)
P4HB	17q25		procollagen-proline, 2-oxoglutarate 4 dioxygenase, beta polypeptide (protein disulfide isomerase; thyroid hormone binding protein p55)
P200	11q		cell surface protein
QDPR	4p15.3	.58	quinoid dihydropteridine reductase
*RAB3A	19p13.2		RAB3A, member RAS oncogene family
*RAB3B	1p32-p31		RAB3B, member RAS oncogene family
RACH	2		acetylcholinesterase derepressor
RAF1	3p25	.64	murine leukemia viral (v-raf-1) oncogene homolog 1
RAF1P1	4p16.1	.36	murine leukemia viral (v-raf-1) oncogene pseudogene 1
RALA	7p		simian leukemia viral (v-ral) oncogene homolog
*RAP1A	1p13-p12		RAP1A, member of RAS oncogene family (K-rev)
*RAP1B	12q14		RAP1B, member of RAS oncogene family (K-rev)

Symbol	Map location	PIC	Marker name
*RAP2	13q34		RAP2, member of RAS oncogene family (K-rev)
RARA	17q21.1	.33	retinoic acid receptor, alpha
*RARB	3p24		retinoic acid receptor, beta
*RASA	5q		RAS p21 protein activator (GTPase activating protein)
RBP1	3q21-q22	.37	retinol-binding protein 1, cellular
RBP2	3p11-qter		retinol-binding protein 2, cellular
RBP3	10q11.2	.44	retinol-binding protein 3, interstitial
RBP4	10q23-q24		retinol-binding protein 4, interstitial
*RBP5	15		retinoic acid-binding protein 5
RB1	13q14.2	.80	retinoblastoma 1 (including osteosarcoma)
RCP	Xq28	.57	red cone pigment (color blindness, protan)
RD	1pter-p22.1		Radin blood group
RDRC	19q13.1-qter		RD114 virus receptor
REL	2p13-p12		avian reticuloendotheliosis viral (v-rel) oncogene homolog
REN	1q32 or 1q42	.37	renin
RH	1p36.2-p34		Rhesus blood group
RHO	3q21-q24	.31	rhodopsin
RLN1	9pter-q12		relaxin 1 (H1)
RLN2	9pter-q12		relaxin 2 (H2)
*RMRP	9p21-p12		RNA component of mitochondrial RNA processing endoribonuclease
*RNN	1p36.1		tRNA asparagine
*RNNL	1q12-q22		tRNA asparagine-like
RNR1	13p12		RNA, ribosomal 1
RNR2	14p12		RNA, ribosomal 2
RNR3	15p12		RNA, ribosomal 3
RNR4	21p12		RNA, ribosomal 4
RNR5	22p12		RNA, ribosomal 5
RNU1	1p36.1		RNA, U1 small nuclear
RNU1P1	1q12-q22		RNA, U1 small nuclear pseudogene 1
RNU1P2	1q12-q22		RNA, U1 small nuclear pseudogene 2
RNU1P3	1q12-q22		RNA, U1 small nuclear pseudogene 3
RNU1P4	1q12-q22		RNA, U1 small nuclear pseudogene 4
RNU2	17q12-q21		RNA, U2 small nuclear
RN5S	1q42-q43		RNA, 5S
ROS1	6q21-q22		avian UR2 sarcoma virus oncogene (v-ros) homolog
RPE	2q32-q33.3		ribulose-5-phosphate-3-epimerase
RPL21	8pter-q21.1		ribosomal protein L21 (gene or pseudogene)
RPL25	17q		ribosomal protein L25 (gene or pseudogene)
RPL32	15		ribosomal protein L32
RPL32L	8pter-q21.1		ribosomal protein L32-like
RPL32P	6p21.3		ribosomal protein L32 pseudogene
RPN1	3q		ribophorin I
RPN2	20		ribophorin II
RPS14	5q31-q33		ribosomal protein S14
RPS17A	5q33-qter		ribosomal protein S17A (gene or pseudogene)
RPS17B	17q		ribosomal protein S17B (gene or pseudogene)
RPS20A	5q13-q23		ribosomal protein S20A (gene or pseudogene)
RPS20B	5q33-qter		ribosomal protein S20B (gene or pseudogene)
RP1	1		retinitis pigmentosa 1
RP2	Xp11.4-p11.2		retinitis pigmentosa 2
RP3	Xp21.1-p11.4		retinitis pigmentosa 3
RRAS	19q13.3-qter		related RAS viral (r-ras) oncogene homolog
RRM1	11p15.5-p15.4	.37	ribonucleotide reductase M1 polypeptide
RRM2	2p25-p24		ribonucleotide reductase M2 polypeptide
RRM2P1	1p33-p31		ribonucleotide reductase M2 polypeptide pseudogene 1
RRM2P2	1q21-q23		ribonucleotide reductase M2 polypeptide pseudogene 2
RRM2P3	Xp21-p11		ribonucleotide reductase M2 polypeptide pseudogene 3
RRM2P4	Xq		ribonucleodide reductase M2 polypeptide pseudogene 4
RS	Xp22.2-p22.1		retinoschisis
RVNP2	Y		retroviral sequences NP2
SAA	11pter-p11	.36	serum amyloid A
SAP1	10q21-q22		sphingolipid activator protein 1
SAP2	10		sphingolipid activator protein 2
SBMA	Xq13-q22		spinal and bulbar muscular atrophy (Kennedy disease)
SC	1p36.2-p22.1		Scianna blood group
SCA1	6p24-p21.3		spinal cerebellar ataxia (olivopontocerebellar ataxia)
SCG1	20pter-p12		secretogranin 1
SCIDX1	Xq13-q21.1		severe combined immunodeficiency, X-linked 1
*SCN2A	2q21-q33	.31	sodium channel, type II, alpha polypeptide
*SCZD1	5q11.2-q13.3		schizophrenia disorder 1
SDH	1p22.1-qter		succinate dehydrogenase 1 (1 of 2 polypeptides)
SDYS	X		Simpson dysmorphia syndrome
SEA	11q13		S13 avian erythroblastosis oncogene homolog
*SEDC	12q14		spondyloepiphyseal dysplasia
SEDL	Xp22		spondyloepiphyseal dysplasia, late
SF	4q		Stoltzfus blood group
SFTP1	10q21-q24	.36	surfactant-associated protein 1
SFTP2	8p	.30	surfactant-associated protein 2
SFTP3	2	.16	surfactant-associated protein 3
SGLT1	22q11.2-qter		sodium/glucose transporter 1
SHMT	12q12-q14		serine hydroxymethyltransferase
SI	3q25-q26		sucrase-isomaltase
SKI	1q22-q24		avian sarcoma viral (v-ski) oncogene homolog
SMCR	17p11.2		Smith-Magenis syndrome chromosome region
SMPD1	17		sphingomyelin phosphodiesterase 1, acid lysosomal

Symbol	Map location	PIC	Marker name
*SNRPA	19q13.1		small nuclear ribonucleoprotein polypeptide A
*SNRPE	1q25-q43	.16	small nuclear ribonucleoprotein polypeptide E
*SNRP70	19q13.3-qter		small nuclear ribonucleoprotein 70kD polypeptide (RNP antigen)
SOD1	21q22.1	.37	superoxide dismutase 1, soluble
SOD2	6q21	.36	superoxide dismutase 2, mitochondrial
SORD	15pter-q21		sorbitol dehydrogenase
SPARC	5q31-q33	.64	secreted protein, acidic, cysteine-rich (osteonectin)
SPG1	Xq27-q28		spastic paraplegia, complicated
SPG2	Xq12-q22		spastic paraplegia, uncomplicated
SPH1	8p21.1-p11.22		spherocytosis 1 (clinical type II)
SPINK1	5		serine protease inhibitor, Kazal type 1
*SPI1	11p12-p11.22		spleen focus forming virus (SFFV) proviral integration oncogene spi1
SPN	16p11.2		sialophorin (gpL115, leukosialin, CD43)
SPTAN1	9q33-q34		spectrin, alpha, non-erythrocytic 1 (alpha-fodrin)
SPTA1	1q21	.38	spectrin, alpha, erythrocytic 1
SPTB	14	.36	spectrin, beta, erythrocytic
*SPTBN1	2p21	.27	spectrin, beta, non-erythrocytic 1
SRC	20q12-q13		avian sarcoma viral (v-src) oncogene homolog
*SRIL	4		sorcin-like
SRPR	11q24-q25		signal recognition particle receptor, 'docking protein'
*SSAV1	18q21	.37	simian sarcoma-associated virus 1/gibbon ape leukemia virus-related endogenous retroviral element
SST	3q28	.46	somatostatin
STATH	4q11-q13		statherin
STMY	11q22	.37	stromelysin
STS	Xp22.32	.48	steroid sulfatase (microsomal)
STSP	Yq11		steroid sulfatase (microsomal) pseudogene
ST2	11p14.3-p12		suppression of tumorigenicity 2
SYN1	Xp11.2		synapsin I
S7	7		surface antigen (chromosome 7) 2
S8	12		surface antigen (chromosome 12) 1
S9	17		surface antigen (chromosome 17) 1
S11	Xq26-q28		surface antigen (X-linked) 2
S12	X		surface antigen (X-linked) 3
S13	22		surface antigen (chromosome 22)
S14	21		surface antigen (chromosome 21)

Symbol	Map location	PIC	Marker name
S100B	21q22		S100 protein, beta polypeptide (neural)
TAL1	11p15		T cell acute lymphoblastic leukemia 1
TARS	5p13-cen		threonyl-tRNA synthetase
TAT	16q22.1	.44	tyrosine aminotransferase
TBG	Xq21-q22		thyroxin binding globulin
TCD	Xq21.1-q21.2		tapeto-choroidal dystrophy, progressive choroidemia
TCL3	10q24		T cell lymphoma 3 associated breakpoint
TCN2	22q		transcobalamin II; macrocytic anemia
TCP1	6q25-q27	.29	t-complex 1
TCP1L1	7p14-cen		t-complex-like 1
TCP1L2	5		t-complex-like 2
*TCP10	6q	.53	t-complex 10 (a murine tcp homolog)
TCRA	14q11.2	.41	T cell receptor, alpha (V,D,J,C)
TCRB	7q35	.75	T cell receptor, beta cluster
TCRD	14q11.2	.46	T cell receptor, delta (V,D,J,C)
TCRG	7p15	.57	T cell receptor, gamma cluster
*TCRGC1	7p15		T cell receptor, gamma, constant region C1
*TCRGC2	7p15		T cell receptor, gamma, constant region C2
*TCRGJP	7p15		T cell receptor, gamma, joining segment JP
*TCRGJP1	7p15		T cell receptor, gamma, joining segment JP1
*TCRGJP2	7p15		T cell receptor, gamma, joining segment JP2
*TCRGJ1	7p15		T cell receptor, gamma, joining segment J1
*TCRGJ2	7p15		T cell receptor, gamma, joining segment J2
*TCRGVA	7p15		T cell receptor, gamma, variable region VA (pseudogene)
*TCRGVB	7p15		T cell receptor, gamma, variable region VB (pseudogene)
*TCRGV1	7p15		T cell receptor, gamma, variable region V1 (pseudogene)
*TCRGV2	7p15		T cell receptor, gamma, variable region V2
*TCRGV3	7p15		T cell receptor, gamma, variable region V3
*TCRGV4	7p15		T cell receptor, gamma, variable region V4
*TCRGV5	7p15		T cell receptor, gamma, variable region V5
*TCRGV5P	7p15		T cell receptor, gamma, variable region V5P (pseudogene)
*TCRGV6	7p15		T cell receptor, gamma, variable region V6 (pseudogene)
*TCRGV7	7p15		T cell receptor, gamma, variable region V7 (pseudogene)
*TCRGV8	7p15		T cell receptor, gamma, variable region V8
*TCRGV9	7p15		T cell receptor, gamma, variable region V9
*TCRGV10	7p15		T cell receptor, gamma, variable region V10
*TCRGV11	7p15		T cell receptor, gamma, variable region V11

Symbol	Map location	PIC	Marker name
TDD	X		testicular 17,20-desmolase deficiency
TDF	Yp11.3		testis determining factor
*TDO2	4q31-q32		tryptophan 2,3-dioxygenase
*TDO2L1	1p35-p33		tryptophan 2,3-dioxygenase-like 1
TF	3q21	.07	transferrin
TFP	3		transferrin pseudogene
TFRC	3q26.2-qter	.15	transferrin receptor (p90)
TG	8q24	.38	thyroglobulin
TGFA	2p13	.33	transforming growth factor, alpha
TGFB1	19q13.1	.36	transforming growth factor, beta 1
*TGFB2	1q41		transforming growth factor, beta 2
*TGFB3	14q24		transforming growth factor, beta 3
TH	11p15.5	.41	tyrosine hydroxylase
THBD	20p12-cen		thrombomodulin
THRA1	17q11.2-q12	.18	thyroid hormone receptor, alpha 1 (avian erythroblastic leukemia viral (v-erb-a) oncogene homolog 1, formerly ERBA1)
THRB	3p24.1-p22	.37	thyroid hormone receptor, beta (avian erythroblastic leukemia vira 1 (v-erb-a) oncogene homolog 2)
THY1	11q22.3-q23	.33	Thy-1 cell surface antigen
TIMP	Xp11.3-p11.23	.45	tissue inhibitor of metalloproteinase (erythroid potentiating activity)
TKC	Xq28-qter		torticollis, keloids, cryptorchidism and renal dysplasia
TK1	17q23.2-q25.3	.38	thymidine kinase, soluble
TK2	16		thymidine kinase, mitochondrial
TNFA	6p21.3		tumor necrosis factor, alpha-
TNFB	6p21.3	.11	tumor necrosis factor, beta-
*TOP1	20q12-q13.2		topoisomerase (DNA) I
TPH	11p15-p13		tryptophan hydroxylase (tryptophan 5-monooxygenase)
TPI1	12p13		triosephosphate isomerase 1
TPO	2pter-p12	.37	thyroid peroxidase
TP53	17p13.1	.20	tumor protein p53
*TRE	1p36		tRNA glutamic acid
*TREL1	1q21-q22		tRNA glutamic acid-like 1
*TRK1	17pter-p12		tRNA lysine 1
*TRL1	14		tRNA leucine 1
*TRL2	17pter-p12		tRNA leucine 2
TRM1	6p23-q12		tRNA methionine initiator 1
TRM2	6p23-q12		tRNA methionine initiator 2
*TRP1	14		tRNA proline 1
*TRP2	14		tRNA proline 2
*TRQ1	17pter-p12		tRNA glutamine 1
TRSP	19	.23	tRNA, phosphoserine (opal suppressor)
TRSPP1	22q11		tRNA, phosphoserine (opal suppressor) pseudogene 1
TRV2	11		truncated retroviral sequence 2 (band T2, 21.0kb)
TRV3	12		truncated retroviral sequence 3 (band T3, 17.8kb)
TRV4	3		truncated retroviral sequence 4 (band T4; 14.9kb)
TRV5	13		truncated retroviral sequence 5
TRY1	7q32-qter		trypsin 1
TS	18pter-q12 or 18q21.3-qter		thymidylate synthetase
TSC1	9q		tuberous sclerosis 1
*TSC2	11q		tuberous sclerosis 2
TSE1	17		tissue-specific extinguisher 1
TSHB	1p13	.61	thyroid stimulating hormone, beta
TSHR	22q11-q13		thyroid stimulating hormone receptor
TSHRL1	1q		thyroid stimulating hormone receptor-like 1
TSHRL2	8		thyroid stimulating hormone receptor-like 2
TSHRL3	10		thyroid stimulating hormone receptor-like 3
*TSPY	Yp		testis specific protein Y-linked
*TST1	10q11.2	.33	transforming sequence, thyroid 1
TTIM1	7		T cell tumor invasion & metastasis 1
TUBAL1	12	.36	tubulin, alpha-like 1
TUBA1	2q		tubulin, alpha 1 (testis specific)
TUBB	6pter-p21		tubulin, beta polypeptide
TUBBP1	8pter-q21		tubulin, beta polypeptide pseudogene 1
TUBBP2	13		tubulin, beta polypeptide pseudogene 2
TYR	11q14-q21	.37	tyrosinase
TYRL	11p11.2		tyrosinase-like
TYS	4q		sclerotylosis
UGP1	1q21-q22		UDP glucose pyrophosphorylase 1
UGP2	2		UDP glucose pyrophosphorylase 2
UMPH2	17q23-q25		uridine 5'-monophosphate phosphohydrolase 2
UMPK	1p32		uridine monophosphate kinase
UMPS	3cen-q21		uridine monophosphate synthetase (orotate phosphoribosyl transferase and orotidine-5'-decarboxylase)
UP	7		uridine phosphorylase
UROD	1p34		uroporphyrinogen decarboxylase
UVO	16q22.1		uvomorulin
VARS	9		valyl-tRNA sythetase
*VDD1	12q14		vitamin D dependency 1
VDI	16		vesicular stomatitis virus defective interfering particle suppression
*VDR	12	.37	vitamin D receptor
VHL	3p		von-Hippel Lindau syndrome
VIL1	2q35-q36		villin 1
*VIL2	6q22-q27		villin 2
VIM	10p13		vimentin
VIP	6q24-q27		vasoactive intestinal peptide
VMD1	8q		vitelliform macular dystrophy, atypicall
VNRA	2		vitronectin receptor, alpha polypeptide
VP	14q		variegate porphyria (protoporphyrinogen oxidase)

Symbol	Map location	PIC	Marker name
*VPREB1	22q11.2		pre-B lymphocyte gene-1 (non-standard provisional symbol)
*VWS	1q32-q41		Van der Woude syndrome
WAGR	11p13		Wilms tumor, aniridia, genitourinary abnormalities, and mental retardation triad
WARS	14q21-q32		tryptophanyl-tRNA synthetase
*WAS	Xp11.4-p11.21		Wiskott-Aldrich syndrome
WMS	4q33-qter		William syndrome
WND	13q14.2-q21		Wilson disease
WS1	9q34		Waardenburg syndrome, type 1
*WT1	11p13		Wilms tumor 1
WWS	Xq11-q22		Wieacker-Wolff syndrome
XG	Xp22.32		Xg blood group
XGR	Xp22.32; Yp11.3		expression of XG and MIC2 on erythrocytes
*XIC	Xq12-q13		X chromosome inactivation center
XK	Xp21.1		Kell blood group precursor (McLeod phenotype)
XM	X		Xm(a) antigen
XPAC	1q		fast kinetic complementation DNA repair in xeroderma pigmentosum, group A
*XPF	15		xeroderma pigmentosum, complementation group F
XRCC1	19q13.1-q13.2		X-ray repair complementing defective repair in Chinese hamster cells 1
XS	Xp21.2-q21.1		suppressor of LU antigens
YESP	22q11-q12		Yamaguchi sarcoma viral (v-yes-1) oncogene homolog pseudogene
YES1	18q21.3		Yamaguchi sarcoma viral (v-yes-1) oncogene homolog 1
*ZFP3	17pter-p12		zinc finger protein 3
ZFX	Xp22.1-p21.3	.50	zinc finger protein, X-linked
ZFY	Yp11.3		zinc finger protein, Y-linked
*ZNF1	8		zinc finger protein 1 (A11-500)
*ZNF2	2		zinc finger protein 2 (A1-5)
*ZNF3	5		zinc finger protein 3 (A8-51)
*ZNF4	5p14-p13		zinc finger protein 4 (5CMP1)
*ZNF5	5q12-q13		zinc finger protein 5 (5CMP1)
*ZNF6	Xq13-q21.1		zinc finger protein 6 (CMPX1)
*ZWS	7q11		Zellweger syndrome

Please call my attention
to errors of omission
or commission.
V.A.M.

Victor A. McKusick, M.D.
The Johns Hopkins Hospital
Baltimore, MD 21205
September 15, 1989

THE HUMAN GENE MAP

The following information, updated continuously since 1973, is based in large part on nine International Workshops on Human Gene Mapping. The first was organized by Dr. Frank Ruddle and held in New Haven in June 1973. The second, known as the Rotterdam Conference, was organized by Dr. Dirk Bootsma and held in The Netherlands in July 1974. The third, organized by Dr. Victor McKusick, was held in Baltimore in October 1975. The fourth, organized by Dr. John Hamerton, was held in Winnipeg in August 1977. The fifth, organized by Dr. John Evans, was held in Edinburgh in July 1979. The sixth, organized by Dr. Kåre Berg, was held in Oslo in June-July 1981. The seventh, organized by Dr. Robert Sparkes, was held in Los Angeles in August 1983. The eighth, organized by Dr. Albert de la Chapelle, was held in Helsinki in August 1985. The ninth, organized by Dr. Jean Frézal, was held in Paris in September 1987. The tenth, organized by Drs. Kenneth Kidd and Frank Ruddle, was held in New Haven in June 1989. (An interim workshop, HGM9.5, was held in New Haven, August 28-31, 1988.) The first six were sponsored exclusively by the National Foundation-March of Dimes (now March of Dimes Birth Defects Foundation), which published the proceedings of the first eight workshops as part of its *Birth Defects: Original Article Series* (BD:OAS). The proceedings also appear in *Cytogenetics and Cell Genetics* (Table 1). The number of loci mapped by the time of each workshop and updated counts are given in Table 2. The numbers of mapped loci can be compared with the counts of total loci in *Mendelian Inheritance in Man* (MIM), as given in Table 3. This human gene mapping information is available online through OMIM (Online MIM).

TABLE 1
Publication of Proceedings of Human Gene Mapping Workshops

Workshop	Location	BD:OAS	Cytogenetics and Cell Genetics
HGM1	New Haven (1973)	X(3):1-216, 1974	13:1-216, 1974
HGM2	Rotterdam (1974)	XI(3):1-310, 1975	14:162-480, 1975
HGM3	Baltimore (1975)	XII(7):1-452, 1976	16:1-452, 1976
HGM4	Winnipeg (1977)	XIV(4):1-730, 1978	22:1-730, 1978
HGM5	Edinburgh (1979)	XV(11):1-236, 1979	25:2-236, 1979
HGM6	Oslo (1981)	XVIII(2):1-343, 1982	32:1-343, 1982
HGM7	Los Angeles (1983)	XX(2):1-666, 1984	37:1-666, 1984
HGM8	Helsinki (1985)	XXI(4):1-832, 1985	40:1-832, 1985
HGM9	Paris (1987)	------------	46:1-762, 1987
HGM9.5	New Haven (1988)	------------	49:2-58, 1988

Table of Contents

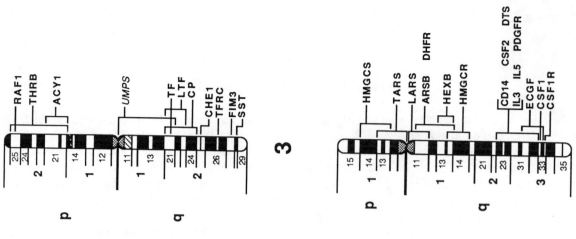

THE HUMAN GENE MAP
(selected 'anchor' loci)

A confirmed assignment ENO1
A provisional assignment ....*DHPR*
Gene cluster ....**MHC**

SCALE
(in megabases)

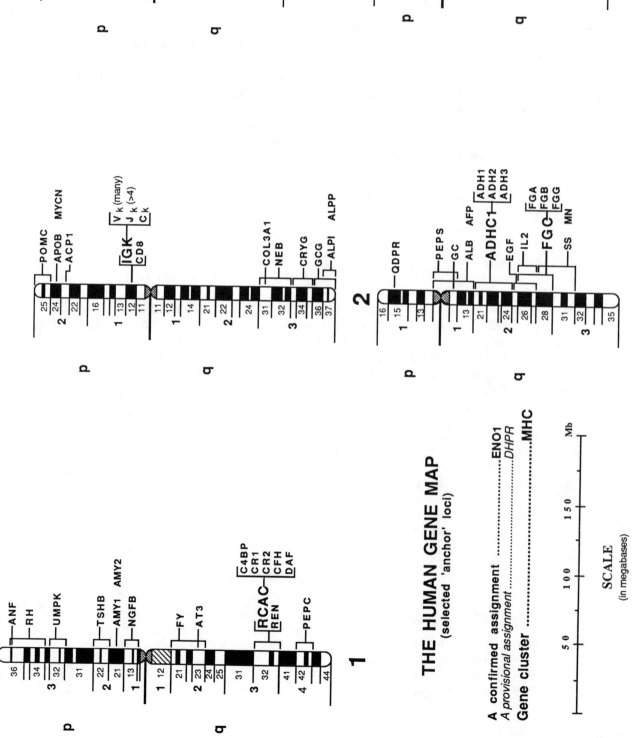

Figure 1a

5.48

Figure 1b

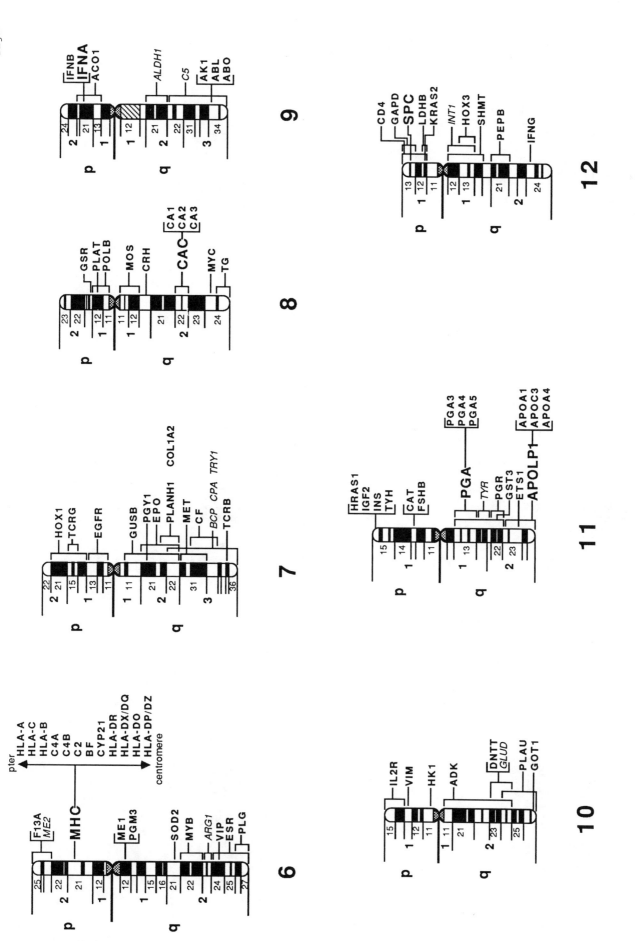

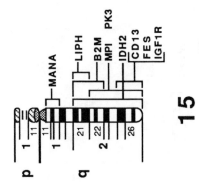

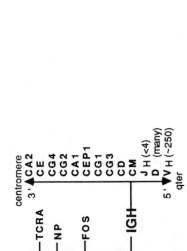

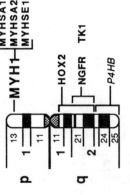

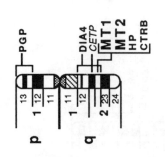

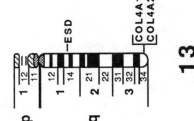

Figure 1c

5.50

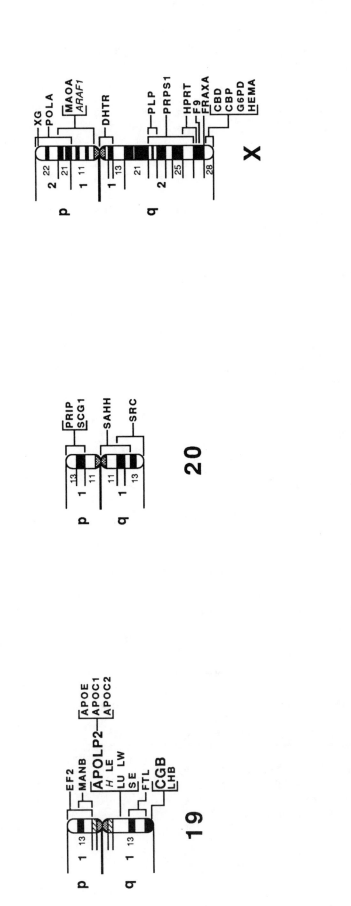

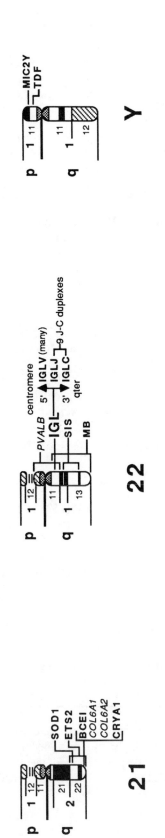

TABLE 2
Number of Loci Assigned at Each of the Ten Human Gene Mapping Workshops

Conference	Number of autosomal assignments				X chromosome assignments	
	Confirmed	Provisional	Tentative†	Total	Confirmed	Tentative
New Haven (1973)	31	28	5	64	88	67
Rotterdam (1974)	48	32	6	86	91	70
Baltimore (1975)	72	46	7	125	95	80
Winnipeg (1977)	83	82	11	176	102	96
Edinburgh (1979)	123	87	20	230	112	101
Oslo (1981)	180	120	45	345	116	118
Los Angeles (1983)	247	161	80	488	118	136
Helsinki (1985)	351	216	66	633	123	157
Paris (1987)	461	392	75	928	140	168
New Haven (1989)	669	669	77	1415*	158	170

*Plus 62 antigens, 25 O'Farrell protein spots, and 90 'like' genes (an unknown portion of which are pseudogenes), to a **grand total, for autosomes, of 1592**. Also, many pseudogenes and anonymous DNA segments have been mapped to specific autosomes.

†Includes 'Inconsistent' and 'Limbo' entries.

TABLE 3
Number of loci identified mainly by mendelizing phenotypes*†
From McKusick: *Mendelian Inheritance in Man*

MIM	Autosomal dominant	Autosomal recessive	X-linked	Total	Grand Total
1966 (1st ed.)	269(+568)	237(+294)	68(+51)	574(+913)	1487
1968 (2nd ed.)	344(+449)	280(+349)	68(+55)	692(+853)	1545
1971 (3rd ed.)	415(+528)	365(+418)	86(+64)	866(+1010)	1876
1975 (4th ed.)	583(+635)	466(+481)	93(+78)	1142(+1194)	2336
1978 (5th ed.)	736(+753)	521(+596)	107(+98)	1364(+1447)	2811
1983 (6th ed.)	934(+893)	588(+710)	115(+128)	1637(+1731)	3368
1986 (7th ed.)	1172(+1029)	610(+810)	124(+162)	1906(+2001)	3907
1988 (8th ed.)‡	1442(+1117)	626(+851)	139(+171)	2207(+2139)	4346
September 15, 1989	1798(+1162)	639(+901)	159(+172)	2596(+2235)	4831

* Numbers in parentheses refer to loci not fully identified or validated.

† Increasingly loci identified by molecular and cellular genetic methods are included despite the lack of mendelian variation in phenotype.

‡ 8th edition closed March 1, 1988, distributed July 1, 1988.

The methods for mapping genes (see Table 4 for counts) are symbolized as follows:

1. **A** = *in situ* DNA-RNA or DNA-DNA annealing ('hybridization'); e.g., ribosomal RNA genes to acrocentric chromosomes; kappa light chain genes to chromosome 2.

2. **AAS** = deductions from the amino acid sequence of proteins; e.g., linkage of delta and beta hemoglobin loci from study of hemoglobin Lepore. (Includes deductions of hybrid protein structure by monoclonal antibodies; e.g., close linkage of MN and SS from study of Lepore-like MNSs blood group antigen.)

3. **C** = chromosome mediated gene transfer (CMGT); e.g., cotransfer of galactokinase and thymidine kinase. (In conjunction with this approach fluorescence-activated flow sorting can be used for transfer of specific chromosomes.)

4. **Ch** = chromosomal change associated with particular phenotype and not proved to represent linkage (Fc), deletion (D), or virus effect (V); e.g., loss of 13q14 band in some cases of retinoblastoma. ('Fragile sites,' observed in cultured cells with or without folate-deficient medium or BrdU treatment, fall into this class of method; e.g., fragile site at Xq27.3 in one form of X-linked mental retardation. Fragile sites are useful as markers in family linkage studies; e.g., FS16q22 and haptoglobin.)

5. **D** = deletion mapping (concurrence of chromosomal deletion and phenotypic evidence of hemizygosity), trisomy mapping (presence of three alleles in the case of a highly polymorphic locus), or gene dosage effects (correlation of trisomic state of part or all of a chromosome with 50% more gene product). Includes "loss of heterozygosity" (loss of alleles) in malignancies. Examples: acid phosphatase-1 to chromosome 2; glutathione reductase to chromosome 8. Includes DNA dosage; e.g., fibrinogen loci to 4q2.

6. **EM** = exclusion mapping, i.e., narrowing the possible location of loci by exclusion of parts of the map by deletion mapping, extended to include negative lod scores from families with marker chromosomes and negative lod scores with other assigned loci; e.g., support for assignment of MNSs to 4q.

7. **F** = linkage study in families; e.g., linkage of ABO blood group and nail-patella syndrome. (When a chromosomal heteromorphism or rearrangement is one trait, **Fc** is used; e.g., Duffy blood group locus on chromosome 1. When 1 or both of the linked loci are identified by a DNA polymorphism, **Fd** is used; e.g., Huntington disease on chromosome 4. F = L in the HGM workshops.)

8. **H** = based on presumed homology; e.g., proposed assignment of TF to 3q. Includes Ohno's law of evolutionary conservatism of X chromosome in mammals. Mainly heuristic or confirmatory.

9. **HS** = DNA/cDNA molecular hybridization in solution ('Cot analysis'); e.g., assignment of Hb beta to chromosome 11 in derivative hybrid cells.

10. **L** = lyonization; e.g., OTC to X chromosome. (L = family linkage study in the HGM workshops.)

11. **LD** = linkage disequilibrium; e.g., beta and delta globin genes (HBB, HBD).

12. **M** = Microcell mediated gene transfer (MMGT); e.g., a collagen gene (COL1A1) to chromosome 17.

13. **OT** = ovarian teratoma (centromere mapping); e.g., PGM3 and centromere of chromosome 6.

14. **R** = irradiation of cells followed by 'rescue' through fusion with nonirradiated (nonhuman) cells (Goss-Harris method of radiation-induced gene segregation); e.g., order of genes on Xq. (Also called cotransference. The complement of cotransference = recombination. Cotransference can be observed also when mitotic chromosomes are used for activated-oncogene-mediated cellular transformation; e.g., Porteous et al., HGM8.)

15. **RE** = Restriction endonuclease techniques; e.g., fine structure map of the beta-globin cluster (HBBC) on 11p; physical linkage of 3 fibrinogen genes (on 4q) and APOA1 and APOC3 (on 11p).
    REa = combined with somatic cell hybridization; e.g., NAG (HBBC) to 11p.
    REb = combined with chromosome sorting; e.g., insulin to 11p. Includes Lebo's adaptation (dual laser chromosome sorting and spot blot DNA analysis); e.g., MGP to 11q. (For this method, using flow sorted chromosomes, W is the symbol adopted by the HGM workshops.)
    REn = neighbor analysis in restriction fragments.
    REc = hybridization of cDNA to genomic fragment, e.g., A-11 on Xq.

16. **S** = 'segregation' (cosegregation) of human cellular traits and human chromosomes (or segments of chromosomes) in particular clones from interspecies somatic cell hybrids; e.g., thymidine kinase to chromosome 17. When with restriction enzyme, REa; with hybridization in solution, HS.

17. **V** = induction of microscopically evident chromo-
    somal change by a virus; e.g., adenovirus 12 changes
    on chromosomes 1 and 17.

18. **X/A** = X-autosome translocation in female with X-
    linked recessive disorder; e.g., assignment of Duch-
    enne muscular dystrophy to Xp21.

The certainty with which assignment of loci to chro-
mosomes or the linkage between two loci has been esta-
blished has been graded into the following classes:
C = confirmed - observed in at least two laboratories or in
    several families.
P = provisional - based on evidence from one laboratory
    or one family.
I = inconsistent - results of different laboratories disagree.
L = limbo - evidence not as strong as that provisional, but
    included for heuristic reasons. (Same as 'tentative'.)

TABLE 4
Number of Autosomal Loci Mapped by the Several Methods
September 15, 1989

Method	*No. of loci mapped*
Somatic cell hybridization (S, REa, HS, M)	1017
In situ hybridization (A)	568
Family linkage study (F, Fc, Fd)	424
Dosage effect (D)	141
Restriction enzyme fine analysis (RE, REc, REn)	131
Chromosome aberrations (Ch)	108
Homology of synteny (H)	85
Radiation induced gene segregation (R)	18
Others (LD, OT, EM, V, REb, C, AAS)	133
Total (many mapped by 2 or more methods)	2609

Gene Map of the Autosomes

As indicated in Table 3, more than 2300 loci are known with confidence to exist on autosomes, on the basis mainly of characteristic mendelian patterns of inheritance of alternative forms of particular traits. (Another 2000 loci have been less securely identified on autosomes.) As indicated by the following data, some mapping information is available concerning almost three-fourths of the well confirmed autosomal loci. In addition to the loci listed here, anonymous DNA segments (some expressed), antigens defined by monoclonal antibodies, surface antigens, some 'like' genes, and pseudogenes, and function-unknown electrophoretic (O'Farrell) protein spots have been assigned to individual autosomes. The number in parentheses after the name of each item in the disorder field indicates whether the mutation was positioned by mapping the 'wildtype' gene (1), by the mapping disease phenotype itself (2), or by both approaches (3). Multiple allelic disorders in the disorder column are separated by semicolons. Brackets indicate 'nondisease' and braces indicate susceptibilities.

Chromosome No. 1

Location	Symbol	Status	Title	MIM #	Method	Comments	Disorder	Mouse
1pter-p36.13	A12M2	P	Adenovirus-12 chromosome modification site-1p	10292	V			
1pter-p36.13	ENO1, PPH	C	Enolase-1	17243	S, F, R		Enolase deficiency (1)	4(Eno-1)
1pter-p36.13	GDH	C	Glucose dehydrogenase	13809	S, F			
1pter-p36	HLM2	C	Oncogene HLM2	13119	REa, F			
1pter-p32	GALE	C	UDP galactose-4-epimerase	23035	S, LD		Galactose epimerase deficiency (1)	
1pter-p22.1	TFS1	P	Transformation suppressor-1	19019	S, H			
1pter-p21	GLUT5	P	Glucose transporter, kidney	13823	REa, A			
1p36.3	RNU1	C	RNA, U1 small nuclear	18068	REa, A	?same as A12M2		
1p36.3-p36.1	NB, NBS	C	Neuroblastoma (neuroblastoma suppressor)	25670	Ch, D		Neuroblastoma (2)	
1p36.2	ANP, ANF, PND	C	Atrial natriuretic peptide; pronatriodilatin	10878	REa, A, H			4(Pnd)
1p36.2-p36.13	PGD	C	6-Phosphogluconate dehydrogenase	17220	F, S			4(Pgd)
1p36.2-p36.1	FGR, SRC2	P	Oncogene FGR	16494	A, REb, REa, Fd	same as SRC2		
1p36.2-p34	ALPL, HOPS	C	Alkaline phosphatase, liver/bone/kidney type	17176	S, H, Fd, F, A		Hypophosphatasia, infantile 24150 (3); ?Hypophosphatasia, adult 14630 (1)	?4(Akp-2); ?1(Akp-1)
1p36.2-p34	EL1	C	Elliptocytosis-1 (protein 4.1)	13050	F, REb	linked to RH	Elliptocytosis-1 (3)	
1p36.2-p34	EKV	C	Erythrokeratodermia variabilis	13320	F	theta = 0.03 with RH	Erythrokeratodermia variabilis (2)	
1p36.2-p34	RD	C	Radin blood group	11162	F			
1p36.2-p34	RH	C	RHESUS BLOOD GROUP CLUSTER		F, D, Fd	Order: 1pter--D-C-E--cen	Erythroblastosis fetalis (1); ?Rh-null hemolytic anemia (1)	
1p36.2-p34	RHC	C	Rhesus system c polypeptide	11170	F, D, Fd			
1p36.2-p34	RHD	C	Rhesus system D polypeptide	11168	F, D, Fd			
1p36.2-p34	RHE	C	Rhesus system E polypeptide	11169	F, D, Fd			
1p36.2-p34	SC	C	Scianna blood group	11175	F			
1p36.1-p35	APNH	P	Antiporter, sodium-potassium ion, amiloride-sensitive	10731	A			
1p36.1	TRN	P	tRNA asparagine	18988	REa, RE			
1p36	BCDS2	P	Breast cancer, ductal	21142	Ch, F, D		Breast cancer, ductal (2)	
1p36	CMM, HCMM	P	Malignant melanoma, cutaneous	15560	F, Fd		Malignant melanoma, cutaneous (2)	
1p36	TRE, TRNE	P	tRNA glutamic acid	18064	A			
1p35-p32	LCK	P	Lymphocyte-specific tyrosine kinase	15339	A, H			4(lck)
1p35-p31.3	GLUT1	C	Glucose transporter protein-1	13814	REa, A, Fd			
1p34	AK2	C	Adenylate kinase-2, mitochondrial	10302	S, F, R	probably in 1p33		4(Ak-2)
1p34	FUCA1, FUCA	C	Alpha-L-fucosidase-1	23000	S, F, R, A, REa		Fucosidosis (1)	4(Fuca)

Location		Symbol	Title	MIM#	Method	Comments	Disorder	Mouse
1p34	L	FUCT	Alpha-L-fucosidase regulator	13683	LD	?very close to FUCA1		
1p34	C	UROD	Uroporphyrinogen decarboxylase	17610	S, A, REa		Porphyria cutanea tarda (1); Porphyria, hepatoerythropoietic (1)	
1p34-p12	P	HMG17	Nonhistone chromosomal protein HMG-17	16391	REa			
1p33	P	SCL	Stem-cell leukemia	18475	Ch, RE			
1p33-p32	P	LAR	Leukocyte antigen related tyrosine phosphorylase	17959	A			
1p32	P	BLYM1, BLYM	Oncogene BLYM1, chicken bursal lymphoma	16483	A			
1p32	P	MYCL1, LMYC	Oncogene MYC, lung carcinoma-derived	16485	REa, A, F	?same as SCL; proximal to MYCL1		Y(L-myc)
1p32	P	TCL5	T-cell leukemia/lymphoma-5	18704	Ch, RE			
1p32	C	UMPK	Uridine monophosphate kinase	19171	S, R, F			
1p32-p31	P	RAB3B	Brain antigen RAB3B	17951	REb, A			
1p32-p31	P	JUN	Oncogene JUN	16516	REa, A			
1p31	P	ACADM, MCAD	Acyl-CoA dehydrogenase, medium chain	20145	REa,A		Acyl-CoA dehydrogenase, medium chain, deficiency of (1)	
1p31	C	GST1	Glutathione-S-transferase-1	13835	A			
1p22.1-qter	P	SDH	Succinate dehydrogenase	18547	S	1 of 2 polypeptides		
1p22.1	C	PGM1	Phosphoglucomutase-1	17190	F, S, R			4(Pgm-2)
1p22.1-q21.1	P	CD3Z, TCRZ	Antigen CD3Z, zeta polypeptide	18678	REa, A			1(Cd-3z)
1p22	P	C8A	C8, alpha polypeptide	12095	F	close to PGM1; alpha and gamma coded by separate genes		
1p22	P	C8B	C8, beta polypeptide	12096	F	close to PGM1	C8 deficiency, type II (2)	
1p22	C	NGFB	Nerve growth factor, beta	16203	REa, H, A, Fd, RE	same 310kb fragment as TSHB		3(Ngfb)
1p22	C	TSHB	Thyroid stimulating hormone, beta subunit	18854	REa, RE, Fd	centromeric to NGFB	Hypothyroidism, nongoitrous (1)	3(Tshb)
1p22	C	NRAS1	Oncogene NRAS1	16479	REa, A	1p12-p11 by A; ?same as NGF		3(Nras)
1p22-p21	C	F3, TFA	Clotting factor III	13439	S, REb, Fd, A			
1p22-q23	C	CMT1	Charcot-Marie-Tooth disease, slow nerve conduction type	11820	F, Fd	ca. 15cM from FY	Charcot-Marie-Tooth disease, slow nerve conduction type (2)	
1p22-qter	P	EPHX, EPOX	Epoxide hydroxylase, microsomal (epoxide hydrolase)	13281	REa		?Fetal hydantoin syndrome (1)	1(Eph-1)
1p21	C	AMY2A	Amylase, pancreatic, alpha 2A	10465	F, A, REa			3(Amy-2)
1p21	C	AMY2B	Amylase, pancreatic, alpha 2B	10466	F, A, REa			
1p21	C	AMY1	Amylase, salivary	10470	F, A, REa	multiple amylase genes		3(Amy-1)
1p21	P	COL11A1	Collagen, type XI, alpha 1	12028	REb, A			
1p21-p13	P	AMPD1	Adenosine monophosphate deaminase-1	10277	REa, A			
1p21-q23	C	APOA2	Apolipoprotein A-II	10767	REa, A, Fd			1(Alp-2, Apoa-2)
1p21-qter	C	ACTA, ASMA	Actin, skeletal muscle alpha chains	10261	REa, H	?near centromere suggested by mouse data		3(Acts)
1p13	L	ASG	Aspermiogenesis factor	10842	Ch	1q25 = conflicting localization		
1p13	C	CD2	T-cell surface CD2 antigen	18699	REa, A			3(Cd-2)
1p13	P	HSDB	3-beta-hydroxysteroid dehydrogenase/isomerase	20181	A		Adrenal hyperplasia II (1)	

Note: C8 deficiency, type I (2) is associated with C8A.

Location	Symbol(s)		MIM	Title	Disorder	Comments	Methods	Mouse
1p13	CD58, LFA3	C	15342	Antigen CD58, lymphocyte function-associated antigen 3		?same as MSK1; gene cloned	S, REa, A	
1p13-p12	RAP1A, KREV1	P	17952	RAS-related protein RAP1A			A	
1p13-p11	ATP1A1	P	18231	ATPase, sodium-potassium, alpha-1 polypeptide			REa, REb, A	3(Atpa-1)
1p	C1QA	P	12055	Complement component, C1q, A chain	?C1q deficiency (1)		REa	
1p	C1QB	C	12057	Complement component C1q, B chain	?C1q deficiency (1)		REa	
1p	NBCCS, BCNS, NBCS	L	10940	Nevoid basal cell carcinoma syndrome	?Nevoid basal cell carcinoma syndrome (2)		F, Fd	
1cen-q32	ATP1A2	P	18234	Sodium-potassium-ATPase, alpha-2 polypeptide			REa, A	1(Atpa-3)
1cen-q32	BCM1, BLAST1	P	10953	B-cell activation marker			REa, A	
1cen-q32	OTF1	P	15325	Lymphoid octamer-binding transcription factor-1			REa	
1cen-q32	NCF2	P	23371	Neutrophil cytosolic factor-2	Chronic granulomatous disease due to deficiency of NCF-2 (1)		REa	
1cen-q32	PFKM	P	23280	Phosphofructokinase, muscle type	Glycogen storage disease, type VII (1)		S	
1q	ATP1B	P	18233	Sodium-potassium ATPase, beta polypeptide		1qh	REa, A	1(Atpb)
1q11	D1Z1	C	12637	Satellite DNA III			A	
1q12-q21	CAGRA, CFAG, CFA	P	21971	Cystic fibrosis antigen (calgranulin A)		probable cluster with CAGRB, CACY, CAPL; on mouse 3	S, REa	
1q12-q22	TRNL	P	18989	tRNA asparagine-like			REa, RE	
1q12-q23	CRP	C	12326	C-reactive protein		probably close to CRP	REa, A	
1q12-q23	APCS, SAP	C	10477	Amyloid P component, serum	[?Amyloidosis, secondary, susceptibility to] (2)		REa, A, Fd	1(Sap)
1q2	CAE	C	11620	Cataract, zonular pulverulent	Cataract, zonular pulverulent (2)	close to FY	F	
1q2	EL2	L	13060	Elliptocytosis-2	?Elliptocytosis-2 (2)	?linked to FY; ?same locus as alpha-spectrin	F	
1q21	A12M3	P	10294	Adenovirus-12 chromosome modification site-1q2		class 1, U2 snRNA pseudogenes, 18069, at this site	V	
1q21	FCE1A	P	14714	Fc IgE receptor, alpha polypeptide			?REa, A	1(Fce-1a)
1q21	FLGR	P	13594	Filaggrin	?Ichthyosis vulgaris, 14670 (1)		REa, A	
1q21	GBA	C	23080	Acid beta-glucosidase (glucocerebrosidase)	Gaucher disease (1)		S, A, D	3(Gba)
1q21	H1F2	P	14271	H1 histone, family 2			REa, A	
1q21	H3F2	P	14278	H3 histone, family 2			REa, A	
1q21	H4F2, HCB	P	14275	H4 histone, family 2		100-200 histone genes; some on chromosome 6 and 12, as well as perhaps 7	REa, A	
1q21	IVL	P	14736	Involucrin			REa, A	
1q21	MUC1, PUM	C	15834	Mucin, urinary, peanut lectin binding		5cM proximal to SPTA1	REa, A, Fd	
1q21	SPTA1	C	18286	Spectrin, alpha, erythrocytic-1	Elliptocytosis-2 (2); Pyropoikilocytosis (1); Spherocytosis, recessive (1)	?same locus as EL2	REa, A, Fd	1(Spna-1)
1q21-q22	ADPRT	P	16101	ADP ribosyltransferase			A	
1q21-q22	FY	C	11070	Duffy blood group	[Vivax malaria, susceptibility to] (1)	distal to 1qh; 20cM from AMY1	F, Fc, Fd	
1q21-q22	PKLR, PK1, PKR	C	26620	Pyruvate kinase, liver and RBC type	PK deficiency hemolytic anemia (1)		REa, A	

Location	Symbol	C/P/L	Name	MIM	Method	Comments	Disorder	Mouse
1q21-q23	UGP1	C	Uridyl diphosphate glucose pyrophosphorylase-1	19175	S, R			
1q21-q25	CACY	P	Calcyclin	11411	A			3(Cacy)
1q22-q23	CD1A	P	Thymocyte antigen CD1A	18837	A, REa	genes A, B, C, D in cluster		
1q22-q23	CD1B	P	Thymocyte antigen CD1B	18836	A, REa			
1q22-q23	CD1C	P	Thymocyte antigen CD1C	18834	A, REa			
1q22-q24	SKI, SK, D1S3	P	Oncogene Sloan-Kettering, chicken virus	16478	REa, A			
1q22-q25	LAM1	P	Lymphocyte-adhesion molecule-1	15324	A			
1q23	F5	C	Clotting factor V	22740	REa, A, Fd		Factor V deficiency (1)	
1q23	MS336	P	Minisatellite 33.6	15756	REa, A			
1q23-q24	FCG2, IGFR2, CD32	C	Fc fragment of IgG, low affinity II, receptor for (CD32) (immunoglobulin G Fc receptor II)	14679	REb, REn	FCG2 and FCG3 within 250kb		1(Ly-17, Cd32)
1q23-q24	FCG3, CD16, IGFR3	C	Fc fragment of IgG, low affinity III, receptor for (CD16) (immunoglobulin G Fc receptor III)	14674	REb, REn	FCG2 and FCG3 within 250kb		
1q23.1-q23.9	AT3	C	Antithrombin III	10730	F, D, A, REa	ca. 10cM from FY	Antithrombin III deficiency (3)	1(At-3)
1q24-q25	ABLL, ARG	P	Oncogene ARG (or ABL-like)	16469	REa, A			1(Abll)
1q24-q31	CHR39A	P	Cholesterol repressible protein 39A	11846	REa			
1q31	LAMB2	C	Laminin B2	15029	REa, Fd			1(Lamb-2)
1q31-q32	CD45, LCA, T200	C	Leukocyte common antigen (T200)	15146	A, S			1(Ly-5)
1q31-q32.1	F13B	C	Clotting factor XIII, B subunit	13458	Fd, A			
1q31-q32.1	MCT	L	Microcephaly, true	25120	Ch		?Microcephaly, true (2)	
1q31-q41	PIGR	P	Polymeric immunoglobulin receptor	17388	REa, A			
1q31-q41	TRK	P	Oncogene TRK	16497	REa, A	NA2 protein in TRK oncogene		
1q31-q41	TPM	P	Tropomyosin, nonmuscle	19103	REa, A			
1q32	RCAC	C	REGULATOR OF COMPLEMENT ACTIVATION CLUSTER			MCP, CR1, CR2, DAF, C4BP in 750kb segment		
1q32	C4BP, C4BR	C	Complement component-4 binding protein	12083	F, REa, A, RE			1(C4bp)
1q32	CR1, C3BR	C	Complement component-3b, C3b, receptor	12062	F, REa, A, RE		CR1 deficiency (1); ?SLE (1)	
1q32	CR2, C3DR	C	Complement component-3d, C3d, receptor	12065	F, REa, A, RE			
1q32	DAF	C	Decay-accelerating factor of complement	12524	REa, A			
1q32	HF, CFH	C	Complement factor H	13437	F, REa, RE, H			1(Cfh)
1q32	LPS, PIT, VDWS	C	Lip pit syndrome (van der Woude syndrome)	11930	Ch, Fd	del (1q32-q41); linked to REN	van der Woude syndrome (2)	
1q32	MCP	P	Membrane cofactor protein	12092	REa, A, REn			
1q32	REN	C	Renin	17982	REa, A, D, Fd	q32.3-q42.3 excluded by D; q42 = conflicting assignment		1(Ren-1)
1q32	RNPE	P	E protein of small nuclear ribonucleoprotein complexes	12826	REa, A			
1q32.1-q42	GUK1	C	Guanylate kinase-1	13927	S, D			

Location	Symbol	Type	Title	MIM #	Status	Comments	Disorder	Mouse
1q32.1-q42	GUK2	C	Guanylate kinase-2	13928	S, D	genetic independence of GUK1 and GUK2 unproved		
1q32.3-q42.3	DH	L	Diaphragmatic hernia	22240	Ch		?Diaphragmatic hernia (2)	
1q41	TGFB2	P	Transforming growth factor beta-2	19022	REa, A			1(Tgf-b2)
1q41-q42	PPOL, PARP	C	Poly-ADP-ribose polymerase (NAD(+) ADP-ribosyltransferase)	17387	REa, A	?processed pseudogenes on 13, 14q22	?Fanconi anemia (1); ?Xeroderma pigmentosum (1)	
1q42	PEPC	C	Peptidase C	17000	S, R, Fd	1q25, 1q32 = conflicting localizations		1(Pep-3)
1q42-q43	RN5S	C	5S ribosomal RNA genes	18042	A			
1q42-q43	A12M1	P	Adenovirus-12 chromosome modification site-1q1	10293	V	same site as A12M1		
1q42-qter	XPAC, XPA	P	Fast kinetic complementation DNA repair in xeroderma pigmentosum, group A	27870	S, H	should be on 1p in light of mouse location	?Xeroderma pigmentosum A (1)	4(Xpa)
1q42.1	FH	C	Fumarate hydratase	13685	S, R, D		?Fumarase deficiency (1)	
1q43	NID	P	Nidogen	13139	A			
Chr.1	ADSS, ADEH	P	Adenylosuccinate synthetase (Ade(-)H-complementing)	10306	REa			
Chr.1	CAGRB	P	Calgranulin B	21972	REa			
Chr.1	CAPL	P	Calcium placental protein	11421	REa			3(Capl)
Chr.1	GNAI3	P	Guanine nucleotide binding protein, alpha inhibiting, polypeptide 3	13937	REa			
Chr.1	GNAT2	P	Guanine nucleotide binding protein, alpha transducing, polypeptide 2	13934	REa			17(Gnat-2)
Chr.1	GNB1	P	Guanine nucleotide binding protein, beta polypeptide 1	13938	REa			19(Gnb-1)
Chr.1	HKR3	P	Oncogene HKR3	16527	REa			
Chr.1	KRL3	P	Kruppel-like zinc finger protein-3	14943	REa			
Chr.1	MTR	P	5-Methyltetrahydrofolate:L-homocysteine S-methyltransferase; tetrahydropteroyl-glutamate methyltransferase	15657	S			
Chr.1	MYF4	P	Myogenic factor-4	15998	REa			
Chr.1	CHC1, RCC1	P	Regulator of chromosome condensation	17971	REb			
Chr.1	CTSE	P	Cathepsin E	11689	REa			
Chr.1	PFN2, PFL	P	Profilin-2	17659	REa			
Chr.1	PLA2L	P	Pancreatic phospholipase A2-like	see 17241	REa			
Chr.1	TPR	P	Tumor potentiating region (translocated promoter region)	18994	REa	fused with MET in chemically induced tumor		

In addition: 2 surface antigens defined by monoclonals, 4 O'Farrell protein spots, 13 'like' genes, 9 pseudogenes, and 10 fragile sites (HGM10). The order of closely linked loci (of ENO1 and 6PGD; of EL1, RH, and FUCA; of UMPK and SC; and of FY and CAE) is uncertain. 'However, the following order of loosely linked segments seems established': 6PGD--RH--UMPK--PGM1--AMY--1qh12--FY--PEPC. (From Rao et al, Am. J. Hum. Genet. 31: 680-696, 1979.)

Chromosome No. 2

Location	Symbol	Status	Title	MIM #	Method	Comments	Disorder	Mouse
2pter-p25.1	COI	L	Coloboma of iris	12020	Ch		?Coloboma of iris (2)	
2pter-p12	TPO, TPX	C	Thyroid peroxidase	27450	REa		Thyroid iodine peroxidase deficiency (1)	
2p25	ACP1	C	Acid phosphatase-1	17150	D, S			12(Acp-1)
2p25	AN1	P	Aniridia-1	10620	F	linked to ACP1 see 14q24	Aniridia-1 (2)	
2p25	CAP	L	Cataract, anterior polar	11565	Fc		?Cataract, anterior polar (2)	
2p25	ODC1	C	Ornithine decarboxylase-1	16564	REa, A			12(Odc-1)
2p25	POMC	C	Proopiomelanocortin	17683	REa	?close to ACP1	ACTH deficiency (1)	12(Pomc-1)
2p25-p24	RRM2	P	Ribonucleotide reductase, M2 subunit	18039	REa, A	pseudogenes on 1p, 1q, Xp		4,7,12,13(Rrm-2)
2p24	APOB	C	Apolipoprotein B	10773	REa, A	1 gene for liver apo-B100 and gut apo-B48; Ag linked	Hypobetalipoproteinemia (1); ?Abetalipoproteinemia (1); Hyperbetalipoproteinemia (1); Apolipoprotein B-100, defective (1)	12(Apob)
2p24	MYCN, NMYC	C	Oncogene NMYC	16484	REa, A	proximal to APOB		12(N-myc)
2p23	MDH1	C	Malate dehydrogenase, soluble	15420	S	proximal to APOB		
2p23-qter	ERCC3	P	Excision-repair, complementing defective, in Chinese hamster, number 3	13351	S			
2p23-qter	IFNB3	C	Beta-3-interferon	14786	S, REa			
2p22-p21	CAD	P	CAD trifunctional protein of pyrimidine biosynthesis	11401	REa, A, S			
2p22-p11	GLAT	P	Galactose enzyme activator	13703	S			
2p13	TGFA	C	Transforming (or tumor) growth factor, alpha type	19017	REa, A			
2p13-cen	REL	P	Oncogene REL, avian reticuloendotheliosis	16491	REa			11(Rel)
2p12	IGK, KM	C	IMMUNOGLOBULIN KAPPA LIGHT CHAIN GENE CLUSTER		A, REa	2p11.2 by high resolution in situ mapping; order: pter-C-J-V-cen		6(Igk)
2p12	IGKV	C	Variable region of kappa light chain	14698	REa, A	25+ genes in 4 classes		
2p12	IGKJ	C	J region of kappa light chain	14697	REa, A	5 genes		
2p12	IGKC	C	Constant region of kappa light chain	14720	REa, A	1 gene		
2p12	IGKDE	P	Immunoglobulin kappa polypeptide deleting element	14678	RE			
2p12	CD8, T8, LEU2	C	Leu-2 T-cell antigen (T8 lymphocyte antigen)	18691	REa, A	distal to IGK		6(Lyt-3; ?Ly-2)
2p11	FABP1, FABPL	C	Fatty acid binding protein, liver	13465	REa, A			6(Fabp-l; ?lvp-1)
2p11-q11	PLGL	P	Plasminogen-like	17334	REa, A			
2p	AKE	L	Acrokeratoelastoidosis	10185	F	?linked to ACP1, JK, IGKC	?Acrokeratoelastoidosis (2)	
2p	CPS1	P	Carbamoylphosphate synthetase I (mitochondrial CPS)	23730	REa	urea cycle enzyme	Carbamoylphosphate synthetase I deficiency (1)	
2p	OAK	L	Optic atrophy, Kjer type	16550	F	?linked to JK; lod = 2.15 at theta 0.14 male, 0.27 female	?Optic atrophy, Kjer type (2)	
2cen-q13	INHBB	P	Inhibin, beta-2	14739	REa			1(Inhbb)

Location	Symbol		Title	MIM #	Method	Comments	Disorder	Mouse
2cen-q13	MAL	P	T-lymphocyte maturation-associated protein	18886	REa			
2q	TUBA1	P	Tubulin, alpha, testis-specific	19111	REa			
2q12-q21	DBI	P	Diazepam binding inhibitor	12595	A			
2q13-q14	PROC	C	Protein C	17686	H, REa, A	tight linkage to IL1B in mouse	Protein C deficiency (1)	
2q13-q21	IL1A	C	Interleukin-1, alpha	14776	H, REa, A			2(Il-1a)
2q13-q21	IL1B	P	Interleukin-1, beta form	14772	REa, A, H			2(Il-1b)
2q14-q21	GYPC, GE, GPC	P	Blood group Gerbich (glycophorin C)	11075	A			
2q14-q21	LCO, LCA	P	Liver cancer oncogene (Oncogene LCA)	16532	REa, REb, A			
2q31	COL3A1	C	Collagen III, alpha-1 chain	12018	REa, A		Ehlers-Danlos syndrome, type IV (3)	
2q31	COL5A2	C	Collagen V, alpha-2 chain	12019	REa, A	very close to COL3A1		
2q31-q32	NEB	C	Nebulin	16165	REa, Ch, A			
2q31-q37	HOX4	P	Homeo box-4	14298	REa, H, A	linked to Km in mice		6(Hox-4)
2q31-qter	ELN	P	Elastin	13016	A	?Hox5		
2q32-q33.3	RPE	C	Ribulose 5-phosphate 3-epimerase	18048	S, D			
2q32-q34	CHRND, ACHRD	P	Cholinergic receptor, nicotinic, delta polypeptide	10072	H, REa, A, LD	linked to Idh-1 in mouse		1(Achr-d)
2q32-q34	CHRNG, ACHRG	P	Cholinergic receptor, nicotinic, gamma polypeptide	10073	H, REa, LD	tightly linked to CHRND?		1(Achr-g)
2q32.1-qter	MYL1	C	Myosin light chain, skeletal fast	16078	REa, H			1(Myl-1)
2q33	CTLA4	P	Cytotoxic T-lymphocyte-associated serine esterase-IV	12389	A			
2q33-q35	CHE2	I	Cholinesterase, serum, 2	17750	F	see chr.16		
2q33-q35	CRYG, CCL	C	CRYSTALLIN, GAMMA POLYPEPTIDE CLUSTER		REa, A		Cataract, Coppock-like (3)	1(Cryg)
2q33-q35	CRYG1	C	Crystallin, gamma polypeptide 1	12366	REa, A			
2q33-q35	CRYG2	C	Crystallin, gamma polypeptide 2	12367	REa, A			
2q33-q35	CRYG3	C	Crystallin, gamma polypeptide 3	12368	REa, A			
2q33-q35	CRYG4	C	Crystallin, gamma polypeptide 4	12369	REa, A			
2q33-q35	CRYG5	C	Crystallin, gamma polypeptide 5	12371	REa, A			
2q33-q35	CRYG6	P	Crystallin, gamma polypeptide 6	12372	REa, A			
2q33-qter	INHA	P	Inhibin, alpha	14738	REa			1(Inha)
2q33.3	IDH1	C	Isocitrate dehydrogenase, soluble	14770	S, D			1(Idh-1)
2q34	TCL4	P	T-cell leukemia/lymphoma-4	18686	Ch, RE			
2q34-q35	MAP2	C	Microtubule-associated protein-2	15713	REa, A			
2q34-q36	FN1	C	Fibronectin-1	13560	S, REa, A	structural gene; see chr. 8, 11	?Ehlers-Danlos syndrome, type X (3)	1(Fn-1)
2q35	WS1	L	Waardenburg syndrome, type I	19350	Ch	paracentric inversion; 2q37.3 = alternative location	?Waardenburg syndrome, type I (2)	
2q35-q36	VIL	P	Villin	19304	A			1(Vil)
2q36-q37	GCG	C	Glucagon	13803	REa, A, RE		[?Hyperproglucagonemia] (1)	2(Gcg)
2q37	ALPI	C	Alkaline phosphatase, adult intestinal	17174	A, REa, RE			2(Alp-3)
2q37	ALPP, PLAP	C	Alkaline phosphatase, placental	17180	REb, REa, A, RE	close to ALPP		4(Akp-2)
2q37	ALPPL	P	Alkaline phosphatase, placental-like	17181	REa, A			
2q37	COL6A3	P	Collagen VI, alpha-3 chain	12025	REa, A	close to CRBP1		9(Crbp-2)
2q37	RMSA	C	Rhabdomyosarcoma, aveolar	26822	Ch	see 11p	Rhabdomyosarcoma, aveolar (2)	
Chr.2	ADCP2	C	Adenosine deaminase complexing protein-2	10272	S			

5.62

Chr.2	ADRA2CR	P	Alpha-2C-adrenergic receptor	10426	REa	
Chr.2	CD8B	P	Antigen CD8B	18673	REa	6(Ly-3)
Chr.2	DES	P	Desmin	12566	REa	
Chr.2	EN1	P	Engrailed-1	13129	REa	
Chr.2	GAD	P	Glutamate decarboxylase	26610	REa	?Pyridoxine dependency with seizures (1)
Chr.2	GLI2	P	Oncogene GLI2	16523	REa	
Chr.2	GNT1	P	Phenol UDP-glucuronosyltransferase	19174	REa	
Chr.2	LTS, LAC, LPH	C	Lactase (lactase-phlorizin hydrolase)	22300	REa, Fd	?Lactase deficiency, congenital (1); ?Lactase deficiency, adult, 22310 (1)
Chr.2	SCN2A, NAC2	P	Sodium channel, neuronal (type II)	18239	REa	4 related genes on chr.2
Chr.2	RACH	P	Acetylcholinesterase regulator, or derepressor	10068	D	
Chr.2	SFTP3	C	Pulmonary surfactant-associated protein-3, 18kD	17864	REa	SFTP2 in previous listing
Chr.2	SPTB2, SPTBN1	P	Nonerythroid spectrin, beta type	18279	REa	
Chr.2	UGP2	P	Uridyl diphosphate glucose pyrophosphorylase-2	19176	S	
Chr.2	UV24	P	Ultraviolet damage, repair of, in UV24	19207	S	
Chr.2	VNRA	P	Vibronectin receptor, alpha polypeptide	19321	REa	
Chr.2	ZNF2	P	Zinc finger protein 2	19450	REa	

In addition: 1 surface antigen, defined by monoclonals, 1 O'Farrell protein spot, 6 'like' genes, 8 pseudogenes, and 8 fragile sites (HGM10).

Chromosome No. 3

Location	Symbol	Status	Title	MIM #	Method	Comments	Disorder	Mouse
3pter-p21	CCK	C	Cholecystokinin	11844	REb, REa			6(Raf-1)
3p25	RAF1	C	Oncogene RAF1	16476	REa			
3p25-p24	TSP1	P	Testis-specific protein-1	18742	REa, A	homologous sequences on 10q24-q25 linked to RAF1		
3p25-p24	VHL	C	von Hippel-Lindau syndrome	19330	Fd, D		von Hippel-Lindau syndrome (2)	
3p24.1-p22	THR, THRB, THR1, ERBA2	C	Thyroid hormone receptor, beta (ERBA2)	19016	REa, A		Thyroid hormone resistance, 27430, 18857 (3)	
3p24	RARB, RAR2, HAP	C	Retinoic acid receptor, beta polypeptide	18022	REa, A	= HAP = HBV-activated protein		
3p23-p22	ACAA	P	Peroxisomal 3-oxoacyl-CoA thiolase	26151	REa, A		Pseudo-Zellweger syndrome (1)	
3p23-p21	SCLC1, SCCL	C	Small-cell cancer of lung	18228	Ch, D	centromeric to ERBA2	Small-cell cancer of lung (2)	
3p21.2-p21.1	ITIH1	P	Inter-alpha-trypsin inhibitor, heavy chain-1	14727	A			
3p21.2-p21.1	ITIH3	P	Inter-alpha-trypsin inhibitor, heavy chain-3	14665	A			
3p21	ACY1	C	Aminoacylase-1	10462	S			9(Acy)
3p21	ALAS1	C	Delta-aminolevulinate synthase	12529	REa, A			
3p21-p14.2	GLB1	C	Beta-galactosidase-1	23050	S, EM	3p14.2-p11 excluded	GM1-gangliosidosis (1); Mucopolysaccharidosis IVB (1)	9(Bgl)
3p14.2	RCC1, RCC	C	Renal cell carcinoma	14470	Fc, Ch	at site of FRA3B	Renal cell carcinoma (2)	
3p14.2-qter	APOD	P	Apolipoprotein D	10774	REb, A			
3p13-q12	GPX1	C	Glutathione peroxidase-1	13832	S, REa		Hemolytic anemia due to glutathione peroxidase deficiency (1)	
3p11.1-q11.2	PSA, PROS	P	Protein S, alpha polypeptide	17688	REa	2 protein S genes in primates	Protein S deficiency (1)	
3p11.2-q11.2	PSB, PROS2	P	Protein S, beta polypeptide	17703	REa			
3p	MYL3	P	Myosin light chain alkali, ventricular and skeletal slow	16079	REa, H			9(Myl-3)
3cen-q22	MER6, RHN	P	Rh-null, regulator type	26815	S	monoclonal ID8	Rh-null disease (1)	
3q12-q13	MOX2	C	MRC OX-2 antigen	15597	REa, A			
3q13	UMPS, OPRT	C	Orotate phosphoribosyltransferase/OMP decarboxylase (UMP synthase)	25890	S, A		Oroticaciduria (1)	
3q13.3-q22	PCCB	C	Propionyl CoA carboxylase, beta polypeptide	23205	REa, A	pccB complementation group	Propionicacidemia, type II or pccB type (1)	
3q21	TF	C	Transferrin	19000	S, H, REa, D, A		Atransferrinemia (1)	9(Trf)
3q21-q22	RBP1, CRBP1	C	Cellular retinol binding protein I	18026	REa, A			9(Crbp-1)
3q21-q23	LTF	C	Lactotransferrin	15021	REa, A			9(Ltf)
3q21-q23	RP1	P	Retinitis pigmentosa-1	18010	Fd		Retinitis pigmentosa-1 (2)	
3q21-q24	CP	C	Ceruloplasmin	11770	F, H, REa, A	ca. 15cM from TF	[Hypoceruloplasminemia, hereditary] (1)	9(Cp)
3q21-q24	RHO	C	Rhodopsin	18038	REa, A			
3q21-q27	MME, CD10, CALLA	P	Membrane metallo-endopeptidase (common acute lymphocytic leukemia antigen)	12052	REa, A			
3q21-qter	ACPP	P	Acid phosphatase, prostate-specific	17179	REa			
3q21-qter	RBP2, CRBP2	P	Cellular retinol binding protein II	18028	REa	close to CRBP1		9(Crbp-2)
3q25-q26	SI	P	Sucrase-isomaltase	22290	REa, A	coamplified with CHE1	Sucrose intolerance (1)	
3q25.2	ACHE	L	Acetylcholinesterase	10074	F			

Location	Symbol	Status	Name	MIM	Method	Comments	Disorder	Mouse
3q25.2	CHE1	C	Pseudocholinesterase-1	17740	F, D, A	distal to CP, TF	Postanesthetic apnea (1)	
3q26.1-q26.3	GLUT2	P	Glucose transporter protein-2	13816	REa, A			
3q26.2	TFRC	C	Transferrin receptor	19001	S, H, REa, A			
3q27	FIM3	P	Friend murine leukemia virus integration site 3, homolog of	13677	REa, A			
3q27-q29	AHSG	C	Alpha-2HS-glycoprotein	13868	F, S, D, REa, A	linked to TF, CHE1; ?order = cen-TF-CHE1-AHSG		
3q28	SST	C	Somatostatin	18245	REa			16(Smst)
3q29	MAP97, MFJ1, MFI2	P	Melanoma-associated antigen p97	15575	S, A	identical to TFRC		
Chr.3	AF8T	P	Temperature sensitive, tsAF8, complement	11695	S			
Chr.3	CRYG8	P	Crystallin, gamma polypeptide 8	12373	REa			
Chr.3	DHFRP4	C	Dihydrofolate reductase pseudogene-4	see 12606	REa			
Chr.3	GAP43	P	Neuron growth-associated protein 43	16206	REa			16(Gap43)
Chr.3	GNAI2B	P	Guanine nucleotide binding protein, alpha inhibiting, polypeptide 2	13936	REa			9(Gnai-2)
Chr.3	GNAT1	P	Guanine nucleotide binding protein, alpha transducing, polypeptide 1	13933	REa			9(Gnat-1)
Chr.3	GSTM	P	Glutathione-S-transferase, mu class	13838	REa			
Chr.3	HRG	P	Histidine-rich glycoprotein	14264	REa		Thrombophilia due to elevated HRG	
Chr.3	HV1S	I	Herpes virus sensitivity	14245	S	see chr. 11		
Chr.3	RPN1	P	Ribophorin I	18047	REa			
Chr.3	TRV4	P	Truncated endogenous retroviral sequence-4	19097	REa			6(Rpm-1)

In addition: 7 surface antigens, most defined by monoclonals, 4 'like' genes, 4 pseudogenes, and 4 fragile sites (HGM10) and 32.

Chromosome No. 4

Location	Symbol	Status	Title	MIM #	Method	Comments	Disorder	Mouse
4pter-p15	RAF2	P	Oncogene RAF2	see 16476	REa	processed pseudogene		6(Raf-2)
4pter-q21	PPAT	P	Phosphoribosylpyrophosphate amidotransferase	17245	S, H			
4p16.3	HD	C	Huntington disease	14310	Fd	distal to D4S10	Huntington disease (2)	
4p16.1	D4S10	C	G8 DNA segment	see 14310	REa, A, D	theta .03-.05, vs. HD		
4p15.3	QDPR, DHPR	C	Quinoid dihydropteridine reductase	26163	S, A, REa		Phenylketonuria due to dihydropteridine reductase deficiency (1)	5(Qdpr)
4p14-q12	PGM2	C	Phosphoglucomutase-2	17200	S			5(Pgm-1)
4p13-p12	GABRA2	P	Gamma-aminobutyric acid receptor, alpha-2 polypeptide	13714	A			
4p13-p12	GABRB1	P	Gamma-aminobutyric acid receptor, beta-1 polypeptide	13719	A			
4p1-q12	PEPS	C	Peptidase S	17025	S, EM			5(Pep-7)
4cen-q21	MT2P1	C	Metallothionein II processed pseudogene	see 15636	REa, A, F			
4q11-q12	KIT	C	Oncogene KIT	16492	REa, A, H			5(Kit)
4q11-q13	ALB	C	Albumin	10360	F, A, REa	linked to GC	Analbuminemia (1); [Dysalbuminemic hyperthyroxinemia] (1); [Dysalbuminemic hyperzincemia] (1)	5(Alb-1)
4q11-q13	AFP	C	Alpha-fetoprotein	10415	H, A	order: 5'-ALB-3'--5'-AFP-3'		5(Afp)
4q11-q13	HPAFP	P	Hereditary persistence of alpha-fetoprotein	10414	Fd, F	?same locus as AFP	[Hereditary persistence of alpha-fetoprotein] (3)	
4q11-q13	JPD, JP	P	Periodontitis, juvenile	17065	F	linked to GC, which is probably between DGI1 and JP	Periodontitis, juvenile (2)	
4q11-q13	STATH, STR	P	Statherin	18447	REa			
4q12	GC, DBP	C	Group-specific component (vitamin D binding protein)	13920	F, Fc, H, D, Ch, REa, A	4q13-q21.1 by in situ hybridization		
4q2	PBT	L	Piebald trait	17280	Ch, H		?Piebaldism (2)	?5(W)
4q2	PDGFRA, PDGFR2	C	Platelet-derived growth factor receptor alpha polypeptide	17349	A, REa			
4q12-q21	PF4	P	Platelet factor 4	17346	REa, A			
4q13-q21	AREG	P	Amphiregulin	10464	REa, A			
4q13-q21	DGI1	C	Dentinogenesis imperfecta-1	12549	F, Fd	ca. 11cM from GC	Dentinogenesis imperfecta-1 (2)	
4q13-q21	GRO, MGSA	P	Melanoma growth stimulatory activity	15573	REa, A			
4q13-q21	IL8	P	Interleukin-8	14693	REa, A			
4q21	INP10, IP10	C	Interferon-inducible cytokine IP-10	14731	A, REa, F	?involved in monocytic leukemia with t(4;11)(q21;q23)		
4q21	IGCJ, JCH	P	J region of immunoglobulin heavy chain	14779	REa, A		?Leukemia, acute lymphocytic, with 4/11 translocation (3)	5(Igj)
4q21-q23	GNPTA	P	N-acetyl-alpha-glucosaminylphosphotransferase	25250	F, S, D		Mucolipidosis II (1); Mucolipidosis III (1)	

Location		Symbol	Title	MIM	Method	Comments	Disorder	Mouse
4q21-q24	P	ADH4	Alcohol dehydrogenase, class II, pi polypeptide	10374	REa, A			
4q21-q24	C	FDH	Formaldehyde dehydrogenase	13649	S			
4q21-q25	P	ADHX, ADH5	Alcohol dehydrogenase, class III	10371	S			
4q21-q31	P	LPC2A	Lipocortin IIa	15171	S,F			
4q21-qter	C	AGA	Aspartylglucosaminidase	20840	REa, REb, A	4q24-qter (M. Smith)	Aspartylglucosaminuria (1)	
4q22	C	ADHC1	ALCOHOL DEHYDROGENASE, CLASS I, CLUSTER		REa	ADH1,ADH2,ADH3 loci for alpha, beta, and gamma chains		3(Adh-1,3)
4q22	C	ADH1	Alcohol dehydrogenase, alpha polypeptide	10370	REa			
4q22	C	ADH2	Alcohol dehydrogenase, beta polypeptide	10372	REa			
4q22	C	ADH3	Alcohol dehydrogenase, gamma polypeptide	10373	REa			
4q23-q27	P	RGS	Rieger syndrome	18050	Ch, Fd	chr.21 and others implicated in some cases; ?not in 4q26	Rieger syndrome (2)	
4q25	C	EGF	Epidermal growth factor	13153	REa, H, F, RE	linked to ADH3; cen-ADH3-EGF-IL2-qter		3(Egf)
4q25	C	IF	Complement component I (C3b inactivator)	21703	REa, Fd, A, RE	40kb distal to EGF	C3b inactivator deficiency (1)	
4q26-q27	C	IL2, TCGF	T-cell growth factor (interleukin-2)	14768	REa, A, F			3(Il-2)
4q26-q28	C	FGC	FIBRINOGEN GENE CLUSTER		RE, REa, D, LD	likely order: gamma-alpha-beta		
4q26-q28	C	FGB	Fibrinogen, beta chain	13483	RE, REa, D, LD, A	4q31 by A; proximal to GYPB/GYPA	Dysfibrinogenemia, beta types (1)	
4q26-q28	C	FGA	Fibrinogen, alpha chain	13482	RE, REa, H, D, LD, A		Dysfibrinogenemia, alpha types (1)	
4q26-q28	C	FGG	Fibrinogen, gamma chain	13485	F, REa, H, RE, D, LD, A	linked to MN	Dysfibrinogenemia, gamma types (1); Hypofibrinogenemia, gamma types (1)	
4q28-q31	P	ASMD	Anterior segment mesenchymal dysgenesis	10725	F	linked to MN	Anterior segment mesenchymal dysgenesis (2)	
4q28-q31	P	ENX2	Endonexin II	13123	REa, A			
4q28-q31	P	FABP2	Fatty acid binding protein, intestinal	13464	REa, A			3(Fabp-i)
4q28-q31	C	SS, GYPB	Ss blood group (glycophorin B)	11174	F, Fc, AAS, EM			
4q28-q31	C	MN, GYPA	MN blood group (glycophorin A)	11130	F, Fc, AAS, EM, A, D, Fc	male lod = 3.79 at theta 0.32 vs. GC		
4q28-q31	C	TYS	Sclerotylosis	18160	F	tightly linked to MN	Sclerotylosis (2)	
4q28-q31	P	HCL2, RHC	Red hair color	26630	F	see unassigned linkage groups, Ib4		
4q28-q31	C	SF	Stoltzfus blood group	11180	F	ca. 25cM from MNSs		
4q31-q32	C	TPH2, TRPO	Tryptophan oxygenase	19107	REa, A			
4q31.1	C	MLR, MCR, MR	Mineralocorticoid receptor	26435	REa, M, A		Pseudohypoaldosteronism (1)	
4q32.1	P	HVBS6, HCC2	Hepatitis B virus integration site 6 (hepatocellular carcinoma-2)	14238	REa, A		Hepatocellular carcinoma (2)	
Chr.4	P	ADRA2BR	Alpha-2B-adrenergic receptor (renal type)	10425	REa			
Chr.4	P	ANT1	Adenine nucleotide translocator 1 (ADP/ATP translocator of skeletal muscle)	10322	REa, REb			
Chr.4	P	ATPBL1	Sodium-potassium-ATPase, beta-polypeptide-like	18237	REa			

Chr.4	CD38	P	Antigen CD38 of acute lymphoblastic leukemia	10727	S			
Chr.4	F11	P	Factor XI	26490	A			Factor XI deficiency (1)
Chr.4	FGFB	P	Fibroblast growth factor, basic	13492	REa, A	many alternate names		
Chr.4	GTB	I	Galactosyltransferase, 4-beta	13706	S	see chr.9		
Chr.4	LAG5	P	Leukocyte antigen group five	15145	S			
Chr.4	MNB	L	Mannosidase, beta-	24851	H	linked in mouse to Adl-3	3(Bmn)	
Chr.4	PDE1A	P	Phosphodiesterase-1A	17189	S			
Chr.4	TS13	P	Temperature sensitivity complementation, ts13	18732	S			

In addition: 1 antigen, 4 'like' genes, 1 pseudogene, and 4 fragile sites (HGM10). Possible order: CEN--GC--DGI1--SS--MN--FGG. FGB--FGA--FGG in this order in 50kb segment.

Chromosome No. 5

Location	Symbol	Status	Title	MIM #	Method	Comments	Disorder	Mouse
5pter-q11	RARS	P	Arginyl-tRNA synthetase	10782	S	very close to LARS		
5p14	MLVI2	P	Maloney leukemia virus integration site-2	15433	REa, A			
5p14-p13	HMGCS	C	3-hydroxy-3-methylglutaryl coenzyme A synthase; HMG CoA synthase	14294	S, REa, A	like HMGCR, regulated transcriptionally by steroid; ?2 genes closely situated		
5p14-p13	ZNF4	P	Zinc finger protein-4	19452	REa, A			
5p13	C9	P	Complement component-9	12094	REa, A		C9 deficiency (1)	
5p13-p12	GHR	P	Growth hormone receptor	13917	REa, A		Laron dwarfism, 26250 (1)	
5p13-cen	TARS	P	Threonyl-tRNA synthetase	18779	S	linked to LARS		
5cen-q11	LARS, RNTLS	C	Leucyl-tRNA synthetase	15135	S, H			
5q	GAP	P	Guanine triphosphatase (GTPase) activating protein	13915	REa	excluded from 5q13-q15		13(Gap)
5q11	MFD1	L	Treacher Collins mandibulofacial dysostosis	15450	Ch	t(5;13)(q11;p11)	?Treacher Collins mandibulofacial dysostosis (2)	
5q11-q13	ARSB	C	Arylsulfatase B	25320	S		Maroteaux-Lamy syndrome (1)	13(As-1)
5q11.1-q13.2	DHFR	C	Dihydrofolate reductase	12606	S, REa, H, D	5q23 = conflicting localization; to other chrs. with amplification	?Megaloblastic anemia (1)	13(Dhfr)
5q11.2-q13	HTR1A	P	5-hydroxytryptamine-1A receptor	10976	REa, A			
5q11.2-q13.3	SCZD1	P	Schizophrenia-1	18151	Ch, Fd	cosegregation with partial trisomy	Schizophrenia (2)	
5q12-q13	ZNF5	P	Zinc finger protein-5	19453	REa, A			
5q12-q32	MAR	P	Macrocytic anemia, refractory	15355	Ch	resulting from 5q-	Macrocytic anemia of 5q- syndrome, refractory (2)	
5q13	HEXB	C	Beta-hexosaminidase, beta chain	26880	S, Ch, D		Sandhoff disease (1)	13(Hex-2)
5q13.3-q14	HMGCR	C	3-hydroxy-3-methylglutaryl coenzyme A reductase; HMG CoA reductase	14291	REa, A			
5q21-q23	CAMK4	P	Ca(2+)-calmodulin-dependent protein kinase type IV of brain	11408	REa, A			
5q22-q23	APC, GS, FPC	C	Adenomatous polyposis of the colon (Gardner syndrome; familial polyposis coli)	17510	D, Fd		Gardner syndrome (2); Polyposis coli, familial (2);?Familial colorectal cancer (2)	
5q23	DTS	C	Diphtheria toxin sensitivity	12615	S, M		[Diphtheria, susceptibility to) (1)	
5q23-q31	CD14	P	Monocyte differentiation antigen CD14	15812	REa, A			
5q23-q31	EGR1	C	Early growth response-1	12899	REa, A			18(Egr-1)
5q23-q31	IL3	P	Interleukin-3	14774	REa, A, H, RE	9kb from CSF2		11(Il-3)
5q23-q32	CSF2, GMCSF	C	Granulocyte-macrophage colony-stimulating factor	13896	REa, A, RE	order: cen-CSF2-CSF1-FMS-qter		11(Csfgm)
5q23.3-q31.1	IL5	C	Interleukin 5	14785	Ch, A			11(Il-5)
5q31	GRL	C	Glucocorticoid receptor, lymphocyte	13804	S, REa, Fd, H, A		Cortisol resistance (1)	18(Grl-1)
5q31	IL4	C	Interleukin-4	14778	REa, A			11(Il-4)
5q31-q32	ADRB2R, BAR2	C	Beta-2-adrenergic receptor	10969	S, REa, A			

Location	Symbol		Title	MIM	Method	Comments	Disorder	Mouse
5q31-q32	PDGFRB, PDGFR, PDGFR1	P	Platelet-derived growth factor receptor-1	17341	REa, A	between GMCSF and FMS		18(Pdgfr)
5q31-q33	EMTB, RPS14	C	Emetine resistance (ribosomal protein S14)	13062	S	see 18046		11(Sparc)
5q31-q33	SPARC, ON	C	Secreted protein, acidic, cysteine-rich (osteonectin)	18212	REa, A, ?Fd			
5q31.3-q33.2	ECGF	C	Endothelial cell growth factor	13122	REa, A			18(Ii)
5q32	HLADG, DHLAG	C	Histocompatibility class II antigens, gamma chain	14279	S, REb, REa, A			
5q32-q34	CBP68	P	Calcium-binding protein p68	11407	REa, A			
5q33-qter	F12, HAF	C	Clotting factor XII (Hageman factor)	23400	REa, A		Factor XII deficiency (1)	
5q33.1?	CMD1	L	Campomelic dysplasia with sex reversal	21197	Ch	?or 8q21.4; balanced translocation	?Campomelic dysplasia with sex reversal (2)	
5q33.1	CSF1, MCSF	P	Macrophage colony stimulating factor	12042	A, REa			3(Csfm)
5q33.2-q33.3	CSF1R, FMS	C	Oncogene FMS (McDonough feline sarcoma)	16477	REa, A	= receptor for CSF1; FMS2 is 5' end		18(Fms)
5q34	GLUT6	P	Glucose transporter-6	13824	REa, A			
5q34-q35	GABRA1	P	Gamma-aminobutyric acid receptor, alpha-1 polypeptide	13716	A			
5q35	CHR	C	Chromate resistance (sulfate transport)	11884	S			
Chr.5	C6	P	Complement component-6	21705	A, H, RE	linked to C7 in dog, marmoset	C6 deficiency (1); Combined C6/C7 deficiency (1)	
Chr.5	C7	P	Complement component-7	21707	A, H, RE	linked to C6 in dog, marmoset	C7 deficiency (1)	
Chr.5	ERBAL3, EAR3	P	ERBA-related gene-3	13289	REb			
Chr.5	FGFA	P	Fibroblast growth factor, acidic	13491	REa	same locus as ECGF		
Chr.5	GM2A	P	GM2-activator protein	27275	S		GM2-gangliosidosis, AB variant (1)	
Chr.5	HARS	P	Histidyl-tRNA synthetase	14281	S			
Chr.5	HFSP	P	Hanukah factor serine protease	14005	REa			
Chr.5	SPINK1, PSTI	P	Serine protease inhibitor, Kazal type I (pancreatic secretory trypsin inhibitor)	16779	REb			
Chr.5	ZNF3	P	Zinc finger protein-3	19451	REa	?relation to ZNF4, ZNF5		

In addition: 1 surface antigen, 1 O'Farrell protein spot, 5 'like' genes, 5 pseudogenes, and 6 fragile sites (HGM10). Critical segment in *cri du chat* syndrome near 5p15.3-p15.2 junction.

5.69

Chromosome No. 6

Location	Symbol	Status	Title	MIM #	Method	Comments	Disorder	Mouse
6pter-p23	ME2	P	Malic enzyme, mitochondrial	15427	F	10cM distal to F13A		7(Mod-2)
6pter-p23	OFC, CL	P	Orofacial cleft (cleft lip with or without cleft palate; isolated cleft palate)	11953	F	linked to F13A	Orofacial cleft (2)	
6pter-p21	TUBB	P	Tubulin, beta, M40	19113	REa			
6pter-q12	PIM1	P	Oncogene PIM1	16496	REa			17(Pim-1)
6p25-p24	F13A1	C	Factor XIII, A1 polypeptide	13457	F, Fd, A	F13A1, F13A2, F13A3 may be clustered loci	Factor XIII, A component deficiency (1)	
6p25-q24	F13A2	P	Factor XIII, A2 polypeptide	13462	F, Fd, A			
6p25-p24	F13A3	P	Factor XIII, A3 polypeptide	13463	F, Fd, A			
6p23-p22.3	FIM1	P	Friend murine leukemia virus integration site 1, homolog of	13675	REa, A			
6p23-q12	HYS, MEA	P	H-Y antigen, structural gene for	14317	REb, REa	male enhanced antigen		
6p23-q12	INSL	P	Insulin-like DNA sequence	14749	REa			
6p23-q12	RNTMI, TRM1	P	Initiator methionine tRNA	18062	REa	2 of 12+ RNTMI genes are on chr. 6		
6p22.2-q21.3	PRL	C	Prolactin	17676	REa, D	?between 6cen and GLO1		
6p22-p21.3	HSPA1, HSP70	C	Heat shock proteins-70	14055	S, REa, A	also 14q22-q24, chr.21, and at least 1 other chromosome		
6p21.3	ASD2	P	Atrial septal defect, secundum type	10880	F	lod = 3.612 at theta 0.0 with HLA	Atrial septal defect, secundum type (2)	
6p21.3	BAT1	P	HLA-B associated transcript-1	14256	RE	5 BATs in 160kb segment including also TNFA, TNFB		
6p21.3	BAT2	P	HLA-B associated transcript-2	14258	RE			
6p21.3	BAT3	P	HLA-B associated transcript-3	14259	RE			
6p21.3	BAT4	P	HLA-B associated transcript-4	14261	RE			
6p21.3	BAT5	P	HLA-B associated transcript-5	14262	RE			
6p21.3	BF	C	Properdin factor B	13847	F, RE	no crossover with C2; less than 1kb from C2, 30kb from C4; C2, BF, C4A, C4B = class III		17(Bf)
6p21.3	C2	C	Complement component-2	21700	F, LD, RE	no crossover with BF; 2% recombination with HLA-B	C2 deficiency (3)	17(C2)
6p21.3	C4A, C4S	C	Complement component 4A, or C4S	12081	F, H, RE, Fd	on HLA-B side of C4B	C4 deficiency (3)	17(C4)
6p21.3	C4B, C4F	C	Complement component 4B, or C4F	12082	F, H, RE, Fd	10kb from C4S	C4 deficiency (3)	17(C4)
6p21.3	COL11A2	C	Collagen, type XI, alpha-2 polypeptide	12029	REa, A, REn	45kb centromeric of HLA-DPB2; 3'-5'-cen		

Location	Symbol	Status	Title	MIM Number	Method	Comments	Disorder	Mouse
6p21.3	CYP21, CA21H, CAH1	C	Congenital adrenal hyperplasia due to 21-hydroxylase deficiency; P450C21	20191	F, RE	linked to C2, C4, BF; 2 loci, A and B; only B active	Adrenal hyperplasia, congenital, due to 21-hydroxylase deficiency (3)	17(P450-21)
6p21.3	GLUR	P	Renal glucosuria	23310	F	closer to HLA-A than HLA-B	[Renal glucosuria] (2)	
6p21.3	HFE	C	Hemochromatosis	23520	LD, F	?between HLA-A and HLA-B or distal to HLA-A	Hemochromatosis (2)	
6p21.3	MHC	C	MAJOR HISTOCOMPATIBILITY COMPLEX		F, S, A, RE, Ch	class I distal to class II		17(Mhc)
6p21.3	HLA60, TCA	P	HLA-6.0	18684	RE			17(Qa)
6p21.3	HLAA	C	HLA-A tissue type	14280	F	HLA-A, -B, -C, -6.0 = class I		17(H-2D)
6p21.3	HLAB	C	HLA-B tissue type	14283	F			17(H-2L)
6p21.3	HLAC	C	HLA-C tissue type	14284	F			
6p21.3	HLADP	C	HLA-DP tissue type	14288	F, RE	2 different alpha, 2 different beta chains		
6p21.3	HLADQ	C	HLA-DQ tissue type	14688	F, RE	1 Dx alpha, 1 Dx beta; 1 DC alpha, 1 DC beta chains		
6p21.3	HLADR	C	HLA-DR tissue type	14286	F, RE	1 alpha, 3 different beta chains		17(H-2I)
6p21.3	HLADZ	C	HLA-DZ tissue type	14293	F, RE	1 alpha, 1 beta chain; DZ, DR, etc. = class II		
6p21.3	HLAE	P	HLA-E tissue type	14301	REn			
6p21.3	HLAF	P	HLA-F tissue type	14311	REn			
6p21.3	IDDM	L	Insulin dependent diabetes mellitus	22210	F, LD	?linkage or association, with HLA	?Diabetes mellitus, insulin dependent (2)	
6p21.3	IGLP1	C	Immune response to synthetic polypeptides-1	14708	F			
6p21.3	IGLP2	C	Immune response to synthetic polypeptides-2	14709	F			
6p21.3	IHG, ITG	P	Blastogenic response to synthetic polypeptides	14695, 14696	F	in A/B segment		
6p21.3	IPHEG, IGAT	P	Blastogenic response to synthetic polypeptides	14681, 14682	F	in B/D segment		
6p21.3	IS, ISCW, ISSCW	P	Immune suppression to streptococcal antigen	14685	H, F	HLA-linked		
6p21.3	LQT	L	Long QT syndrome	19250	F	ca. 5cM from MHC near HLA-A end	?Long QT syndrome (2)	
6p21.3	MLRW	P	Mixed lymphocyte reaction, weak	15786	F	?linkage disequilibrium with HLA-B12	?Kostmann agranulocytosis (2)	
6p21.3	NDF	P	Neutrophil differentiation factor	20270	LD			
6p21.3	NEU, NEU1	I	Neuraminidase-1; sialidosis	16205, 25655	H, F	?linked to HLA; see chr. 10	?Sialidosis (2)	17(Neu-1)
6p21.3	PDB	L	Paget disease of bone	16725	F	?linkage or association, with HLA	?Paget disease of bone (2)	
6p21.3	PLT1	C	Primed lymphocyte test-1	17668	F	near HLA-D		
6p21.3	RDBP	P	RD RNA-binding protein	15404	REn	between C4 and BF		
6p21.3	RWS	L	Ragweed sensitivity	17945	F	?linkage or association, with HLA	?Ragweed sensitivity (2)	17(Rd)

Location		Symbol	Title	MIM #	Method	Comments	Disorder	Mouse
6p21.3-p21.2	L	LAP	Laryngeal adductor paralysis	15027	F	?linkage to HLA and GLO1	?Laryngeal adductor paralysis (2)	
6p21.3-p21.2	P	HMAA	Human monocyte antigen A	14307	F	between HLADQ and GLO		
6p21.3-p21.2	P	HMAB	Human monocyte antigen B	14308	F			
6p21.3-p21.2	C	GLO1	Glyoxalase I	13875	F, S	ca. 3cM proximal to HLA		17(Glo-1)
6p21.3-p21.2	L	CP20	Lymphocyte cytosolic protein, molecular weight 20kD	15338	LD			
6p21.3-p21.2	P	MLN	Motilin	15827	REa, A			
6p21.3-p21.2	C	SCA1	Spinocerebellar ataxia-1	16440	F, Fd	telomeric to GLO1 and centromeric to HLA	Spinocerebellar ataxia-1 (2)	
6p21.3-p21.1	P	B144	B144 protein	10917	RE	10kb 3' from TNFA		
6p21.3-p21.1	C	PGC	Preprogastricsin	16974	REa, F	cen-PGG-GLO1-HLA		17(Upg-1)
6p21.3-p21.1	P	TNFA, TNF1	Tumor necrosis factor, alpha	19116	REa, A, RE	5'-TNFB--TNFA-3' in 7kb segment (pter-cen); 210kb from HLA-B		17(Tnfa)
6p21.3-p21.1	P	TNFB, TNF2	Tumor necrosis factor, beta	15344	REa, A, RE	cen-DR-21OH-C4-BF-C2-TNFA-TNFB-HLA-B		17(Tnfb)
6p21.2-p12	C	MUT, MCM	Methylmalonyl CoA mutase	25100	REa, A, F, D		Methylmalonicaciduria, mutase deficiency type (1)	17(Mut)
6p21.1-p12	C	PGK1P2	Phosphoglycerate kinase-1 pseudogene-2	17227	REa, H, A, REb	proximal to MHC		17(Pgk-2)
6p21	P	MAPT2	Microtubule-associated protein tau-2	see 15713	REa, A	see 17q21		
6p21-qter	P	TPX1	Testis-specific protein TPX1	18743	REa			17(Tpx-1)
6p12	C	GST2	Glutathione S-transferase-2	13836	A, REa	pseudogene		
6p12-p11	C	RASK1, KRAS1P	Oncogene, Kirsten rat sarcoma virus-1	19011	S			
6p	L	CSCI	Corticosterone side-chain isomerase	12255	H			
6p	P	EJM, JME	Epilepsy, juvenile myoclonic	25477	F	?linked to MHC	?Epilepsy, juvenile myoclonic (2)	
6q	I	IFNGR1	Immune interferon, receptor for	10747	S	?linked to BF and HLA both 6 and 18 required		10(Ifgr)
6q	P	TCP10A	T-complex locus TCP10A	18702	Fd, ?A			
6q12	C	ME1	Malic enzyme, cytoplasmic	15425	S			9(Mod-1)
6q12	C	PGM3	Phosphoglucomutase-3	17210	S, F, OT			9(Pgm-3)
6q13	P	COL9A1	Collagen, type IX, alpha-1 polypeptide	12021	A			
6q14-q21	C	NT5, NTE, E5NT	Ecto-5'-nucleotidase	12919	S			
6q21	P	BKMA1	Banded krait minor satellite DNA-1	see 10978	A	related to heterogametic sex		
6q21	C	SOD2	Superoxide dismutase-2, mitochondrial	14746	S, D			17(Sod-2)
6q21	C	SYR	SRC/YES-related oncogene	16499	REa, A			
6q21.1-q23	P	CGA	Chorionic gonadotropin, alpha chain	11885	REa, A	shared with LH, FSH, TSH		4(Tsha)
6q22	C	MYB	Oncogene, avian myeloblastosis virus	18999	REa, A			10(Myb)
6q22	C	ROS1, ROS, MCF3	Oncogene ROS (oncogene MCF3)	16502	REa, A	coamplified with MYB		
6q22-q27	P	CVL	Cytovillin	12390	REa			
6q23	P	ARG1	Arginase, liver	20780	REa		Argininemia (1)	
6q24-q27	C	ESR, ER	Estrogen receptor	13343	REa, A			
6q24-q27	P	MAS1	Oncogene MAS1	16518	A	?same as ESR		

							17(Cp)
6q25-q27	IGF2R, MPRI	P	Insulin-like growth factor-2 receptor (mannose-6-phosphate receptor, cation-independent)	14728	REa, A		
6q25-q27	TCP1	C	T-complex locus TCP-1	18698	REa, H, A, Fd	tightly linked to PLG	
6q25-qter	FUCA2	C	Alpha-L-fucosidase-2	13682	F	linked to PLG	
6q25-qter	VMD2	L	Macular dystrophy, ?vitelline type	15370	Ch		Macular dystrophy, ?vitelline type (2)
6q26-q27	PLG	C	Plasminogen	17335	REa, A, LD, F	20cM from TCP10A	Plasminogen Tochigi disease (1); Dysplasminogenemic thrombophilia (1); Plasminogen deficiency, types I and II (1)
6q26-q27	VIP	C	Vasoactive intestinal peptide	19232	REa, A, REb		
6q27	LPA	C	Apolipoprotein Lp(a)	15220	REa, A, F, Fd		{Coronary artery disease, susceptibility to} (1)
Chr.6	ADCP1	I	Adenosine deaminase complexing protein-1	10271	S		
Chr.6	AMD	P	S-adenosylmethionine decarboxylase	18098	REa	sequences on Xq22-q28	
Chr.6	ASSP2	P	Argininosuccinate synthetase pseudogene-2	10784	REa	others on 8 or more other chromosomes including X and Y	
Chr.6	BEVI	C	Baboon M7 virus replication	10918	S		
Chr.6	DHFRP2	P	Dihydrofolate reductase pseudogene-2	see 12606	S		
Chr.6	FEA	L	F9 embryonic antigen	13701	H		
Chr.6	MRBC	P	Monkey RBC receptor	15805	S		
Chr.6	P	P	P blood group globoside	11140	S		
Chr.6	TS546	P	Temperature sensitivity complementation, cell cycle specific, ts546 cells	18733	S		

In addition: 2 surface antigens, most defined by monoclonals, 6 'like' genes, 6 pseudogenes, and 6 fragile sites (HGM10). Order: 6cen--DR--C2--BF--C4A--CA21HA--C4B--CA21HB--HLA-B--pter (Wilton and Charlton, 1986). Order of class II subregions: 6cen--DP--DZalpha--DObeta--DX--DQ--DRbeta--DRalpha (Hardy et al., 1986). Order in region of class I genes: 6cen--DR--HLA--B--0.01--HLA--C--0.7--HLA--A--pter. HLADP shows relatively high recombination with DQ but the physical distance by molecular studies is about same as DQ-to-DR. Recombinational hotspots probably exist within the DQ subregion and between HLA-A and HLA-C (which molecular data suggest are as close as B and C). Family data show HLA-A to HLA-C = 0.7cM; HLA-C to HLA-B = 0.1cM. Disease/MHC associations of various strengths are probably indicative of pleiotropic effects of specific alleles or haplotypes, not linkage. Two of the strongest are ankylosing spondylitis (10630) with HLA-B27 and narcolepsy (16140) with HLA-DR2.

Chromosome No. 7

Location	Symbol	Status	Title	MIM #	Method	Comments	Disorder	Mouse
7pter-p14	GCTG	P	Gamma-glutamylcyclotransferase	13717	S			
7pter-q22	ACTB	P	Actin, cytoskeletal beta	10263	REa	ca. 20 pseudogenes also		5(Actb)
7pter-q22	NPY	P	Neuropeptide Y	16264	REa			
7p22-q21	PDGFA	C	Platelet-derived growth factor, A chain	17343	REa, A			
7p22-p15	RAL	P	RAS-like protein	17955	REb, A			
7p21.3-p21.2	CRS, CSO	C	Craniosynostosis	12310	Ch		Craniosynostosis (2)	
7p21	IFNB2, IL6, BSF2	C	Interferon, beta-2 (hepatocyte stimulating factor, interleukin-6)	14762	REa, A, Fd	conflicting assignment = 7p31		
7p15.2-p15.1	PSP	C	Phosphoserine phosphatase	17248	S, D			5(Psph)
7p15	TCRG	C	T-cell antigen receptor, gamma subunit	18697	REa	multiple V genes, two J-C duplexes		13(Tcrg)
7p15-p14	HOX1	C	Homeo box-1	14295	A, REa, H			6(Hox-1)
7p15-p13	INHBA, INHB2	P	Inhibin, beta-1	14729	REa			13(Inhba)
7p14-p12	ERBB	C	Oncogene ERBB	19014	REa	?same as EGFR; similar sequences		11(Erbb)
7p14-p12	IBP1	C	Insulin-like growth factor, low molecular weight	14673	REa, A			
7p14-cen	BLVR	C	Biliverdin reductase	10975	S			2(Blvr)
7p13	GCPS	C	Greig craniopolysyndactyly syndrome	17570	Ch, Fd, REn	balanced translocation; same restriction fragment as TCRG	Greig craniopolysyndactyly syndrome (2)	13(Xt)
7p13-p12	PGAMM	C	Phosphoglycerate mutase, muscle form	26167	REa, A		Myopathy due to phosphoglycerate mutase deficiency (1)	
7p13-q22	MDH2	C	Malate dehydrogenase, mitochondrial	15410	S			
7p13-qter	PRKAR1, PKR1	P	Protein kinase, cAMP-dependent, type I regulatory subunit	17689	REa	not in 7q22-q31.3		5(Mor-1)
7p12.3-p12.1	EGFR	C	Epidermal growth factor receptor	13155	S, Fd, D			
7p11.4-q21	ARAF2	P	Oncogene ARAF2	16471	REa, A			
7p11-q11.2	PKS1	P	Oncogene PKS1	16501	REa, A			
7p	GHS	L	Goldenhar syndrome	14140	Ch		?Goldenhar syndrome (2)	
7cen-q11.2	ASL	C	Argininosuccinate lyase	20790	S, REa, A		Argininosuccinicaciduria (1)	5(Asl)
7q11.12-q11.23	ZWS, ZS	P	Zellweger syndrome	21410	Ch		Zellweger syndrome (2)	
7q21	GNAI1	C	Guanine nucleotide binding protein, alpha inhibiting, polypeptide-1	13931	REa, A			
7q21-q22	EPO	C	Erythropoietin	13317	REa, A, REb, Fd	close to COL1A2; no recombination	?Erythremia (1)	5(Epo)
7q21-q22	NKNA	P	Neurokinin A (substance P)	16232	REa, A			
7q21-q31	ASNS, AS	C	Asparagine synthetase	10837	S, REa, A	temperature sensitive G1 mutant		
7q21.1	PGY1, MDR1	C	P-glycoprotein-1 (multidrug resistance)	17105	REa, A, REb	within 500kb of MDR1		5(Mdr-1)
7q21.1	PGY3, MDR3	P	P-glycoprotein-3 (multidrug resistance-3)	17106	RE			
7q21.1	SRL, SCN	P	Sorcin (class 4 gene)	18252	REa, H			
7q21.1-q22	GUSB	C	Beta-glucuronidase	25322	S, D, EM		Mucopolysaccharidosis VII (1)	5(Gus)
7q21.3-q22	CYP3	C	Cytochrome P450C3 (nifedipine oxidase)	12401	REa, D, Fd			6(Cyp-3)

Location	Symbol	Status	Title	MIM	Method	Comments	Disorder	Mouse
7q21.3-q22	PLANH1, PAI	C	Plasminogen activator inhibitor-1	17336	REa, REb, Fd, A		?Thrombophilia due to excessive plasminogen activator inhibitor (1); Hemorrhagic diathesis due to PAI1 deficiency (1)	
7q21.3-q22.1	COL1A2	C	Collagen I, alpha-2 chain	12016	S, REa, D, A	ca. 17cM from CF	Osteogenesis imperfecta, 2 or more clinical forms (3); Ehlers-Danlos syndrome, type VIIA2 (3); Marfan syndrome, atypical (1)	16(Cola-2)
7q22	PON, ESA	C	Paraoxonase	16882	F, Fd	Order: COL1A2-D7S15-PON-CF		
7q22	HCA	I	HISTONE CLUSTER A: H1, H2A, H2B	14271, 14272, 14276	A	7q32-q36 = conflicting localization; others find none on 7		13(Hist-1)
7q22-q31	PRKAR2, PKR2	P	Protein kinase, cAMP-dependent, type II regulatory subunit	17691	REa, Fd, C			
7q22-q32	G7P1	P	Kinase-like protein	14875	REa, Fd			
7q22-q34	BPGM	P	2,3-bisphosphoglycerate mutase	22280	A		Hemolytic anemia due to bisphosphoglycerate mutase deficiency (1)	
7q22-qter	ACTBP5	P	Actin, cytoskeletal beta, pseudogene-5	10264	REa	ca. 20 in all; 1 on X chr.; 2 on chr. 5; 3 on chr. 18; 4 on chr. 5, etc.		
7q22-qter	BCP, CBT	P	Blue cone pigment	19090	REa		Colorblindness, tritan (2)	
7q22-qter	CPA	P	Carboxypeptidase A	11485	REa	both CPA and TRY1 = serine proteases		6(Cpa)
7q22-qter	GP130, NM	C	Neutrophil migration (granulocyte glycoprotein)	16282	D	formerly neutrophil chemotactic response, NCR		
7q22-qter	TRY1, TRP1	P	Trypsin-1	27600	REa	isolated by CMGT with MET	Trypsinogen deficiency (1)	6(Try-1)
7q31	IRP	P	INT1-related protein	14787	C			
7q31	MET	C	Oncogene MET	16486	REa, A, F	ca. 1.2cM from CF		6(Met)
7q31-qter	ODC2	C	Ornithine decarboxylase-2	16565	REa			
7q31.1-q31.3	LAMB1	C	Laminin B1	15024	REa, A	7q22 = conflicting assignment		1(Lamb-1)
7q31.3-q32	CF	C	Cystic fibrosis	21970	F, Fd	distal and 5' to MET	Cystic fibrosis (2)	
7q32-q36	EPH	C	Oncogene EPH	17961	REa, A			
7q32-q36	PIP	P	Prolactin-inducible protein	17672	REa, A			
7q35	TCRB	C	T-cell antigen receptor, beta subunit	18693	REa, A	7q32 by A; cluster of V, D, J, and C genes;many V, two D-J-C triplexes		6(Tcrb)
7q35-q36	MPB3	P	Membrane protein band 3, nonerythroid	10928	REa, A			
7q35-q36	MS3315	P	Minisatellite 33.15	15757	REa, A			
7q36	EN2	P	Engrailed-2	13131	REa, A			
Chr.7	DIA2	L	Diaphorase-2	12587	S			
Chr.7	ERV3	P	Endogenous retrovirus-3	13117	REa			
Chr.7	GCF1	P	Growth rate controlling factor-1	13922	S			
Chr.7	GLI3	P	Oncogene GLI3	16524	REa			
Chr.7	GNB2	P	Guanine nucleotide binding protein, beta-2	13939	REa			

5.76

Chr.7	TTIM1, INM7	P	14783	S	T-cell tumor invasion and metastasis-1 (invasion-metastasis of neoplasms, chromosome 7 determined)	
Chr.7	HADH, ACADL, LCAD	P	14345	S	Hydroxyacyl-CoA dehydrogenase (acyl-CoA dehydrogenase, long chain)	?Acyl-CoA dehydrogenase, long chain deficiency of (1)
Chr.7	DLD, LAD	P	24690	REa	Dihydrolipoamide dehydrogenase	Lipoamide dehydrogenase deficiency (1)
Chr.7	NHCP2	P	11888	S	Nonhistone chromosomal protein-2	
Chr.7	PHKG	P	17247	REa	Phosphorylase kinase, muscle, gamma subunit	presumed pseudogene on 11
Chr.7	UP	C	19173	S	Uridine phosphorylase	

In addition: 1 surface antigen, 2 O'Farrell protein spots, 5 'like' genes, 7 pseudogenes, and 9 fragile sites (HGM10).

Chromosome No. 8

Location	Symbol	Status	Title	MIM #	Method	Comments	Disorder	Mouse
8p23.3-p23.1	F7E, F7R	C	Clotting factor VII expression, or regulator	13445	D			
8p23	DEF, HNP1	P	Defensin-1 (human neutrophil peptide-1)	12522	REa, A			8(Defcr)
8p22	CTSB, CPSB	P	Cathepsin B	11681	REa, A			
8p22	LPL, LIPD	P	Lipoprotein lipase (lipase D)	23860	REa, A	13q14 by rat probe	Hyperlipoproteinemia I (1)	8(Gr-1)
8p21.1	GSR	C	Glutathione reductase	13830	S, D		Hemolytic anemia due to glutathione reductase deficiency (1)	
8p21	NEFL, NFL, NF68	C	Neurofilament, light polypeptide	16228	REa, A	?NFI and NFH on 2 near cen, 7q		
8p21-q11.2	GNRH, LHRH	P	Luteinizing hormone releasing hormone (gonadotropin releasing hormone)	15276	REa, A		?Hypogonadotropic hypogonadism due to GNRH deficiency , 22720 (1)	
8p12	PLAT, TPA	C	Plasminogen activator, tissue type	17337	REa, A, REb		Plasminogen activator deficiency (1)	8(Plat)
8p12-p11.2	FMSL, FLG	P	FMS-like gene	13635	REa, A			
8p12-p11	POLB	C	Polymerase, DNA, beta	17476	REa			
8p11	ANK, SPH2	C	Ankyrin (spherocytosis)	18290	F, Ch, D, REa, A, Fd		Spherocytosis-2 (2)	8(nb)
8q	GPB	C	Beta-glycerol phosphatase	10964	S			
8q11-q12	MOS	C	Oncogene MOS, Moloney murine sarcoma virus	19006	REa, A, REb			4(Mos)
8q12	SGPA, PSA	P	Salivary gland pleomorphic adenoma	18103	Ch	12q13-q15 affected in subset	Salivary gland pleomorphic adenoma (2)	
8q12-q13	IL7	P	Interleukin 7	14666	REa, A			
8q13	CRH	P	Corticotropin releasing hormone	12256	REa, A			
8q13-qter	LYN	P	Oncogene Yamaguchi sarcoma viral related	16512	REa			
8q13.3	BOS	L	Branchiootic syndrome	11365	Ch		Branchiootic syndrome (2)	
8q21	CYP11B1, P450C11	P	11-beta-hydroxylase; corticosteroid methyl oxidase II (CMO II)	20201	REa, A, Ch	?q21.13 multifunctional enzyme	Adrenal hyperplasia, congenital, due to 11-beta-hydroxylase deficiency (1); CMO II deficiency (1)	
8q21	CYP11B2	P	Cytochrome P450 CYP11B2	12408	REa			
8q21	BN51T, TSBN51	P	Temperature sensitive complementation, cell cycle specific, tsBN51	18728	REa, A	block in progression through G1		
8q21.1-q23	MRS	P	Myeloid-related sequence	15956	REa, A		?ANLL-M2 (1)	
8q21.1-qter	GLYB	P	Glycine auxotroph B, complementation of hamster	13848	S	gly(-)B		
8q21.4?	CMD1	L	Campomelic dysplasia with sex reversal	21197	Ch	?or 5q33.1; balanced translocation	?Campomelic dysplasia with sex reversal (2)	
8q22	CAC	C	CARBONIC ANHYDRASE CLUSTER		REa, H, A			
8q22	CA1	C	Carbonic anhydrase I	11480	REa, H, A			3(Car-1)
8q22	CA2	C	Carbonic anhydrase II	11481	REa, H	CA1, CA2 linked in monkey and mouse	Renal tubular acidosis-osteopetrosis syndrome (1)	3(Car-2)
8q22	CA3	C	Carbonic anhydrase III	11475	REa, A			
8q22-q24	HSPG	P	Heparan sulfate proteoglycan, cell surface-associated	14246	REa, A			
8q23-q24	PENK	P	Proenkephalin	13133	REa, A			

5.77

Location	Symbol	Status	Title	MIM number	Method	Comments	Disorder	Mouse
8q23-q24.1?	EXT	L	Multiple exostoses	13370	Ch	closely linked to GPT	?Multiple exostoses (2)	
8q24	EBS1	C	Epidermolysis bullosa, Ogna type	13195	F		Epidermolysis bullosa, Ogna type (2)	
8q24	GPT	C	Glutamate-pyruvate transaminase	13820	S, EM, H, Fd, D			15(Gpt-1)
8q24	PDS	L	Pendred syndrome	27460	Ch		?Pendred syndrome (2)	
8q24	PVT1	P	Oncogene PVT-1 (MYC activator)	16514	RE, Ch			
8q24	VMD1	C	Macular dystrophy, atypical vitelliform	15384	F	5cm from GPT	Macular dystrophy, atypical vitelliform (2)	
8q24.1	MYC	C	Oncogene MYC, avian myelocytomatosis virus	19008	REa, A	cen-5'-3'-ter	Burkitt lymphoma (3)	15(Myc)
8q24.11-q24.13	LGCR, LGS, TRPS2	C	Langer-Giedion syndrome	15023	Ch	?deletion of both EXT and TRP1 in LGS; ?critical segment = 8q24.11-q24.12	Langer-Giedion syndrome (2)	
8q24.12	TRPS1	P	Trichorhinophalangeal syndrome, type I	19035	Ch	?q24.11	Trichorhinophalangeal syndrome, type I (2)	
8q24.2-q24.3	TG	C	Thyroglobulin	18845	A, REa, REb	distal to MYC	Hypothyroidism, hereditary congenital, 1 or more types (1); ?Goiter, adolescent multinodular, 13880 (1)	?15(Tg)
Chr.8	CAB27	P	Calbindin 27kDa	11405	REa			
Chr.8	CYC1	P	Cytochrome c1	12398	REa			
Chr.8	FNZ	L	Fibronectin	13560	S	?concerned with expression on cell surface		
Chr.8	FRV2	P	Full-length endogenous retroviral sequence-2	13687	REa			
Chr.8	GLI4, HKR4	P	GLI-Kruppel family member GLI4 (Oncogene HKR4)	16528	REa			
Chr.8	KRL4	P	Kruppel-like zinc finger protein 4	14944	REa			
Chr.8	SFTP2	P	Pulmonary surfactant apoprotein-2	17862	REa			
Chr.8	ZNF1	P	Zinc finger protein-1	19449	REa			?8(Zfp-2)

In addition: 6 'like' genes, 5 pseudogenes, and 6 fragile sites (HGM10).

5.78

Chromosome No. 9

Location	Symbol	Status	Title	MIM #	Method	Comments	Disorder	Mouse
9pter-p22	ZFY, TDFA	P	ZFY-related autosomal sequence	15423	A			
9pter-q12	RLXH1, RLN1	P	Relaxin, H1	17973	REa			
9pter-q12	RLXH2, RLN2	P	Relaxin, H2	17974	REa			
9pter-q34	LPC2B	P	Lipocortin IIb	15172	REa			
9p24-p13	AK3	C	Adenylate kinase-3, mitochondrial	10303	S, D			
9p22	NKH1	L	Hyperglycinemia, isolated nonketotic, type I	23830	Ch		?Hyperglycinemia, isolated nonketotic, type I (2)	
9p22-p21	LALL	P	Lymphomatous acute lymphoblastic leukemia	24764	Ch			
9p22-p21	MTAP, MSAP	C	Methylthioadenosine phosphorylase	15654	S, D			
9p22-p13	ACO1	C	Aconitase, soluble	10088	S			4(Aco-1)
9p21	IFNB, IFNB1, IFF	C	Fibroblast interferon; beta-interferon	14764	REa, A, Fd, LD	distal to IFL, 9p23-p22 according to Rowley; IFF duplicate in some persons		4(Ifb)
9p21	IFNA, IFL, IFA	C	LEUKOCYTE INTERFERON GENE CLUSTER; ALPHA-INTERFERON	14766	REa, A	very close to IFF by Fd, LD; 15-30 genes	Interferon, alpha, deficiency (1)	4(Ifa)
9p21-p12	RMRPR	P	Mitochondrial RNA-processing endoribonuclease	15766	REa, A			4(Rmrpn)
9p13	GALT	C	Galactose-1-phosphate uridyltransferase	23040	S, D, F		Galactosemia (1)	4(Galt)
9p13	GT1	I	Galactosyltransferase-1	13706	A	?relation to GTB on chr.4		4(Ggt-1)
9cen-q34	FPGS	P	Folylpolyglutamate synthetase	13651	S			2(Fpgs)
9cen-qter	GRP78	P	Glucose-regulated protein	13812	REa			
9q11-q22	LPC1	P	Lipocortin I	15169	REa, A			
9q12	DNCM	P	Cytoplasmic membrane DNA	12633	A	9qh		
9q13-q21.1	FRDA, FAT	C	Friedreich ataxia	22930	Fd		Friedreich ataxia (2)	19(Ahd-1,2)
9q21	ALDH1	P	Aldehyde dehydrogenase-1	10064	REa, A			
9q21-q22	CTSL	P	Cathepsin L	11688	REa, A			
9q22	ALDOB	C	Aldolase B; fructose-1-phosphate aldolase	22960	REb, REa, A, D		Fructose intolerance (1)	
9q31-qter	APPL1	P	Amyloid beta (A4) precursor protein-like-1	10474	A			
9q32-q33	ITIL, ITI, HCP	C	Protein HC (alpha-1-microglobulin); inter-alpha-trypsin inhibitor, light chain	17687	REa, A		?Familial Mediterranean fever, 24910 (1)	
9q32-q34	DYT1	P	Torsion dystonia, autosomal dominant	12810	Fd			
9q33-q34	SPTAN1, NEAS	C	Spectrin, alpha, nonerythrocytic 1 (alpha-fodrin)	18281	REa, A			2(Spma-2)
9q33-34	SURF3	P	Surfeit-3 (L7a ribosomal protein)	18564	REa, A	in cluster with SURF1, SURF2, SURF4		
9q33-q34	TSC1, TSC, TS	C	Tuberous sclerosis-1	19110	F, Fd	linked to ABO, ABL see ch.15	Tuberous sclerosis-1 (2)	
9q33-qter	ITO	I	Hypomelanosis of Ito	14615	X/A		?Hypomelanosis of Ito (2)	
9q34	ABO	C	ABO blood group	11030	F, Fc	linked to AK1		
9q34	ASS	C	Argininosuccinate synthetase	21570	S, D, REa, Fd	14 pseudogenes on 11 chromosomes	Citrullinemia (1)	2(Ass)
9q34	ALAD	C	Delta-aminolevulinate dehydratase	12527	F, S, A, REa	linked to ABO; ORM-13-ALAD-11-AK-13-ABO	Porphyria, acute hepatic (1); [Lead poisoning, susceptibility to) (1)	4(Lv)
9q34	DBH	C	Dopamine-beta-hydroxylase	22336	F, A	tightly linked to ABO		
9q34	GSN	P	Gelsolin	13735	A, REa, RE	40kb proximal to ABL		

Location	Symbol		Name	MIM	Method	Comments	Disorder	Reference
9q34	NPS1	C	Nail-patella syndrome	16120	F	linked to AK1, ABO; no recombination with AK1	Nail-patella syndrome (2)	
9q34.1	ABL	C	Oncogene ABL (Abelson strain, murine leukemia virus)	18998	REa	fusion hybrid gene with BCR1 in CML	Leukemia, chronic myeloid (3)	2(Abl)
9q34.1	AK1	C	Adenylate kinase-1, soluble	10300	F, S, D, Fc	proximal to Ph1 break, 9q34.1; AK1 to ORM =17cM	Hemolytic anemia due to adenylate kinase deficiency (1)	2(Ak-1)
9q34.1	C5	C	Complement component 5	12090	REa, A		C5 deficiency (1)	2(Hc)
9q34.1-q34.3	ORM1, AGP1	C	Orosomucoid-1 (alpha-1-acid glycoprotein-1)	13860	F, S, REa, Fc, A	linked to ABO, AK1, ALAD		4(Agp-1)
9q34.1-q34.3	ORM2	C	Orosomucoid-2	13861	RE, LD			4(Agp-2)
Chr.9	CPRO, CPO	P	Coproporphyrinogen oxidase	12130	S	?on 9p	Coproporphyria (1); Harderoporphyria (1)	
Chr.9	H142T	P	Temperature sensitivity complementation, H142	18729	S			
Chr.9	IGEP2	P	Immunoglobulin epsilon heavy chain pseudogene	14721	A			
Chr.9	VARS	P	Valyl-tRNA synthetase	19215	S			

In addition: 1 antigen, 3 O'Farrell protein spots, 5 'like' genes and 4 pseudogenes, and 6 fragile sites (HGM10).

Chromosome No. 10

Location	Symbol	Status	Title	MIM #	Method	Comments	Disorder	Mouse
10pter-p11.1	PFKF, PFKP	C	Phosphofructokinase, platelet type	17184	S			
10p15	ITIH2	P	Inter-alpha-trypsin inhibitor, heavy chain-2	14664	A			
10p15-p14	IL2R, TAK	C	Interleukin-2 receptor; T-cell growth factor receptor	14773	REa, A			
10p13	VIM	C	Vimentin	19306	REa			
10p12-q23.2	GBM	C	Glioblastoma multiforme	13780	D		Glioblastoma multiforme	
10p11.2	FNRB, VLAB	C	Fibronectin receptor, beta subunit (common unit of very late activator proteins)	13563	REb, REa, F, A, S			
10p11.2	HK1	C	Hexokinase-1	14260	S, D		Hemolytic anemia due to hexokinase deficiency (1)	10(Hk-1)
10p11.2-q11.2	RBP3, IRBP	C	Interstitial retinol-binding protein	18029	REa, A, Fd			
10q11	ERCC6	P	Excision repair cross complementing rodent repair deficiency, complementation group 6	13354	A			
10q11-q12	TST1, PTC, TPC	C	Thyroid papillary carcinoma oncogene	18855	REa, A	?same as MEN2A	Thyroid papillary carcinoma (1)	
10q11-q24	ADK	C	Adenosine kinase	10275	S, D, EM			14(Adk)
10q11-qter	ALDOBP	P	Aldolase B pseudogene	see 22960	REb			
10q11.1-q24	PP	C	Inorganic pyrophosphatase	17903	S, D			10(Pyp)
10q21-q22	LPC2C	P	Lipocortin IIc	15173	REa			
10q21-q22	SAP1, SAP2	C	Sphingolipid activator protein-1; sphingolipid activator protein-2	24990	S, REa, A, D		Metachromatic leukodystrophy due to deficiency of SAP-1 (1)	
10q21-q24	SFTP1, PSAP	C	Pulmonary surfactant-associated protein, 35kD	17863	REa, A			
10q21.1	MEN2A, MEN2	C	Multiple endocrine neoplasia, type II (or IIA)	17140	Fd	19cM from D10S5 at 10q21.1	Multiple endocrine neoplasia II (2)	
10q21.1	MEN3, MEN2B	C	Multiple endocrine neoplasia, type III (or IIB)	16230	Fd	?allelic to MEN2	Multiple endocrine neoplasia III(2)	
10q21.1-q22.1	KROX20, EGR2	P	KROX-20, Drosophila, homolog (early growth response-2)	12901	D, REa, A, F			10(Krox-20; Egr-2)
10q22	COL13A1	P	Collagen, type XIII, alpha-1 polypeptide	12035	REa, A			
10q22.1	PRG	C	Proteoglycan, secretory granule (platelet proteoglycan protein core)	17704	REa, A			
10q23-q24	DNTT, TDT	C	Terminal deoxynucleotidyltransferase	18741	REa, A, Ch			19(Tdt)
10q23-q24	GLUD	P	Glutamate dehydrogenase	13813	REa, A			14(Glud)
10q23-q24	RBP4	C	Retinol binding protein, plasma	18025	REa, A		Retinol binding protein, deficiency of (1)	
10q23-q25	ADRA2R	P	Alpha-2-adrenergic receptor	10421	REa, A			
10q24	TCL3	P	T-cell leukemia-3	18677	Ch		Leukemia, T-cell acute lymphocytic (2)	
10q24-q25	LIPA	P	Lysosomal acid lipase-A	27800	S, H	?close to GOT	Wolman disease (1); Cholesterol ester storage disease (1)	19(Lip-1)
10q24-qter	PLAU, URK	C	Urokinase (plasminogen activator, urinary)	19184	REa, A			9(Plau)
10q24.1-q24.3	CYP2C	C	Cytochrome P450, family II, subfamily C (mephenytoin 4-hydroxylase)	12402	REa, A	multiple genes		19(P450-2c)
10q25-q26	IFNAI1, RNM561	P	Interferon-inducible mRNA 561	14769	REa, A			
10q25.3	GOT1	C	Glutamate oxaloacetate transaminase, soluble	13818	S, D, H	10q24.3 and 26.1 = conflicting localizations		19(Got-1)
10q25.3	PGAMB, PGAM1	P	Phosphoglycerate mutase A, nonmuscle form	17225	D, H			19(Pgam-1)

5.82

Location	Symbol	C	Name	MIM	S, REa, A, Fd	Comments	Disorder	Mouse
10q26	OAT	C	Ornithine aminotransferase	25887	S	pseudogene at Xp11.2	Gyrate atrophy of choroid and retina (1)	7(Oat)
Chr.10	ATPM, OMR	P	Mitochondrial ATPase (oligomycin resistance)	16436	REa			
Chr.10	CDC2	C	Cell cycle controller, CDC2	11694	REa			
Chr.10	CHAT	P	Choline acetyltransferase	11849	REa			
Chr.10	CYP2E	P	Cytochrome P450, family II, subfamily E	12404	REa			15(Cyp2e)
Chr.10	CYP17, P450C17	P	Steroid 17-alpha-hydroxylase / 17,20 lyase	20211	REa	at least 2 genes	Adrenal hyperplasia V (1)	
Chr.10	FUSE	P	Polykaryocyosis inducer	17475	S			
Chr.10	G10P2	P	Interferon, alpha-inducible protein (MW 54kD)	14704	A			
Chr.10	GSAS	P	Glutamate-gamma-semialdehyde synthetase	13825	S	GOT1 and GSAS in same pathway		
Chr.10	HEP10	P	Hepatic protein 10	14239	REa			
Chr.10	M130	P	External membrane protein-130	13371	S			
Chr.10	NCF1	P	Neutrophil cytosolic factor-1	23370	REa	see 6p	Chronic granulomatous disease due to deficiency of NCF-1 (1)	
Chr.10	NEU, NEUG	I	Glycoprotein neuraminidase; sialidosis	25655	S		?Sialidosis (1)	
Chr.10	PROA	P	Proline(-) auxotroph, complementation of	17677	S			

In addition: 2 O'Farrell protein spots 4 'like' genes, and 1 pseudogene (HGM10).

Chromosome No. 11

Location	Symbol	Status	Title	MIM #	Method	Comments	Disorder	Mouse
11pter-p15.5	RMS, RMSCR	P	Rhabdomyosarcoma, embryonal	26821	D	2q37 suggested by translocations	Rhabdomyosarcoma (2)	
11pter-p15.4	BWCR, BWS, WBS	C	Beckwith-Wiedemann syndrome	13065	Ch, Fd	partial trisomy	Beckwith-Wiedemann syndrome (2)	
11pter-p13	CD44, LHR	P	antigen CD44, (homing function), includes MDU2, MDU3, MIC4	15327	S			
11pter-p12	SAA	C	Serum amyloid A	10475	REa, H		?Susceptibility to amyloid in FMF, 24910 (1)	7(Saa)
11pter-p11.2	CDw26, TP250	P	T-cell activation antigen CDw26	18671	S			
11p15.5	ADCR, ADCC	P	Adrenocortical carcinoma region	20230	D		Adrenocortical carcinoma (2)	
11p15.5	NAGC, HBBC	C	NON-ALPHA GLOBIN CLUSTER (HEMOGLOBIN BETA CLUSTER)		HS, REa, A	11p12.08-p12.05 = conflicting localization;		
11p15.5	HBB	C	Hemoglobin beta	14190	LD, AAS, F		Sickle cell anemia (1); Thalassemias, beta- (1); Methemoglobinemias, beta- (1); Erythremias, beta- (1); Heinz body anemias, beta- (1); HPFH, deletion type (1)	7(Hbb)
11p15.5	HBD	C	Hemoglobin delta	14200	AAS			
11p15.5	HBGR	C	Hb gamma regulator	14227	RE		?Hereditary persistence of fetal hemoglobin (3)	
11p15.5	HBG1	C	Hemoglobin gamma 136 alanine	14220	RE		HPFH, nondeletion type A (1)	
11p15.5	HBG2	C	Hemoglobin gamma 136 glycine	14225	RE		HPFH, nondeletion type G (1)	
11p15.5	HBE1	C	Hemoglobin epsilon	14210	RE			
11p15.5	HRAS1, RASH1	C	Oncogene HRAS1, Harvey rat sarcoma-1	19002	S	pseudogene HRAS2 on X		7(Hras1)
11p15.5	IGF2	C	Insulin-like growth factor II, or somatomedin A	14747	REa, A, RE	separate gene for variant, 14741		
11p15.5	INS	C	Insulin	17673	HS, A, REb, Fd, D	5'--INS-12.6kb-IGF2--3'; cen-HBBC-10cM-INS-2cM-HRAS1-3cM-TH	Diabetes mellitus, rare form (1); MODY, one form, 12585 (3); Hyperproinsulinemia, familial (1)	6(Ins-1); 7(Ins-2)
11p15.5	MAFD1, MD1	P	Manic-depressive illness (major affective disorder 1)	12548	Fd	linked to HRAS, INS; in some families, not linked	Manic-depressive illness (2)	
11p15.5	TYH, TH	C	Tyrosine hydroxylase	19129	REa, A, Fd, RE	distal to HRAS1		7(Th)
11p15.5	WTCR2, WT2	P	Wilms tumor region-2	19409	D		Wilms tumor, type 2	
11p15.5-p15.4	HPX	C	Hemopexin	14229	REa, A			
11p15.5-p14.3	LDHC	C	Lactate dehydrogenase C	15015	REa, A, REb	closely linked to LDHB in other species; in man syntenic with LDHA; ?close to LDHA		

Location	Symbol	Status	Title	MIM #	Method	Comments	Disorder	Mouse
11p15.4	CALCA, CALC1	C	Calcitonin/calcitonin gene related peptide, alpha polypeptide	11413	REa, A, REb, D, Fd			7(Calc)
11p15	CTSD, CPSD	P	Cathepsin D	11684	REa, A			
11p15	MUC2	P	Mucin 2, intestinal	15837	REa			
11p15	PTH	C	Parathyroid hormone	16845	REa, REb, A, Fd	ca.9cM distal to CALC1	?Hypoparathyroidism, familial (1)	7(Pth)
11p15	HPFH, FCP, HHPF	L	F-cell production	14247	F	ca.15cM from HBB; ?on Xq28	Heterocellular hereditary persistence of fetal hemoglobin (2)	
11p15	MER2	P	Red blood cell antigen MER2	17962	S			
11p15	RRM1	P	Ribonucleotide reductase, M1 subunit	18041	S, REa, A			
11p15-p14	LDHA	C	Lactate dehydrogenase A	15000	S, D, REb, C, A		Exertional myoglobinuria due to deficiency of LDH-A (1)	7(Ldh-1)
11p15-p13	TPN, TRH1, TRPH	P	Tryptophan hydroxylase (tryptophan-5-monooxygenase)	19106	REa			
11p14.3-p12	ST2	P	Suppressor of transformation/tumorigenicity-2	18544	Ch			
11p14.2-p12	CALCB, CALC2	C	Calcitonin gene related peptide beta	11416	A, REa, REb, D			
11p14-p13	HVBS1, HBVS1	C	Hepatitis B virus integration site-1	11455	REa, A, Ch		Liver cell carcinoma (1)	
11p13	AN2	C	Aniridia-2	10621	Ch, Fd		Aniridia-2 (2)	?2(Sey)
11p13	CAT	C	Catalase	11550	S, D, Fd	cen-CAT-WT-AN-pter distal to AN2	Acatalasemia (1)	2(Cas-1)
11p13	FSHB	C	Follicle stimulating hormone, beta polypeptide	13653	D, REa		?Male infertility, familial (1)	2(Fshb)
11p13	TCL2	P	T-cell leukemia/lymphoma-2	15139	Ch, RE, A, REa	involved in t(11;14)(p15;q11.2); between HRAS1 and INS/IGF2	Leukemia, acute T-cell (2)	
11p12-p11.22	SPI1	P	Oncogene SPI1	16517	REa, A			
11p13	TCR1, WAGR	C	Wilms tumor/aniridia/gonadoblastoma/retardation complex	19407	Ch	actually clump of pter-FSHB-AN2-WT-CAT	Wilms tumor (2); Aniridia of WAGR syndrome (2); Gonadoblastoma (2); Mental retardation of WAGR (2)	
11p12-p11	ACP2	C	Acid phosphatase-2	17165	S, REa		?Lysosomal acid phosphatase deficiency (1)	2(Acp-2)
11p11.2	TYRL	P	Tyrosinase-like	19127	REa, A			
11p11-q12	F2	C	Prothrombin (clotting factor II)	17693	REa, A		Hypoprothrombinemia (1); Dysprothrombinemia (1)	
11p	MDU3, INLU	P	Lutheran inhibitor, dominant (monoclonal antibody A3D8)	11115	S			
11q	DRD2	P	Dopamine receptor, D2	12645	Fd, REa			
11q	NACAE, MDU1, M4F2	C	Sodium-calcium exchanger	15807	S			
11q	PC	P	Pyruvate carboxylase	26615	REa		Pyruvate carboxylase deficiency (1)	
11q11-q13	C1NH, C1I, HANE	C	C1 inhibitor	10610	REa, A		Angioedema, hereditary (1)	
11q11-q23	CLG, EBR1, CLGN	C	Collagenase (recessive epidermolysis bullosa dystrophica)	22660	S, REa		Epidermolysis bullosa dystrophica, recessive (3)	
11q12-q13	CD20	P	Antigen CD20 (differentiation antigen B1)	11221	REa, A			
11q12.1-q13.5	FNL2	P	Fibronectin-like-2	13561	S, A			
11q12-qter	CD15	P	Myeloid-associated surface antigen CD15	15947	S			
11q13	CD5, LEU1	C	Lymphocyte antigen CD5	15334	REa, A			19(Ly-1)

Location	Symbol		Name	MIM	Method	Comments	Disorder	Homology
11q13	GST3	C	Glutathione-S-transferase-3	13837	S, A, REa	formerly called GST1		9(Gsta)
11q13	HST, FGFR, HST1	C	Oncogene HST (fibroblast growth factor related)	16498	A, REa, RE	coamplified with INT2 in melanoma		7(Int-2)
11q13	INT2	C	Oncogene INT2	16495	REa, A, RE	35kb 5' to HST1		
11q13	MEN1	C	Multiple endocrine neoplasia I	13110	Fd, D	linked to PYGM	Multiple endocrine neoplasia I (1)	
11q13	PGA	C	PEPSINOGEN A CLUSTER	16970	REa, A	about 20cM from CAT		
11q13	PGA3	C	Pepsinogen A3	16971	REa, RE, A			
11q13	PGA4	C	Pepsinogen A4	16972	REa, RE, A			
11q13	PGA5	C	Pepsinogen A5	16973	REa, A	pter-5'HRAS--5'INS--cen		
11q13	PYGM, MGP	C	Muscle glycogen phosphorylase	23260	REb, Fd		McArdle disease (1)	19(Pygm)
11q13	SEA	P	Oncogene SEA (S13 avian erythroblastosis)	16511	REa, A			
11q13-q14	BKMA	P	BKM, banded krait minor satellite, DNA	see 10978	A	related to heterogametic sex		
11q13-q22	ESA4	C	Esterase-A4	13322	S			9(Es-17)
11q13-qter	GANAB	P	Neutral alpha-glucosidase AB	10416	S			
11q13-qter	MSK39	P	Antigen defined by monoclonal antibody 5.1H11	10724	S			
11q13.3	BCL1	C	B-cell leukemia-1; chronic lymphocytic leukemia	15140	RE, Ch	t(11;14)--(q13.3;q32.3)	Leukemia/lymphoma, B-cell, 1 (2)	
11q14-q21	CPD3	L	Cerebellar ataxia	21320	F	?linked to ATN		
11q14-q21	TYR, ATN	P	Tyrosinase (albinism, tyrosinase negative)	20310	REa, A, H	?linked to CPD3	Albinism (1)	7(c)
11q22	PGR	C	Progesterone receptor	26408	REa, A, REb	11q13 = earlier regionalization		9(Pgr)
11q22-q23	AT1	P	Ataxia-telangiectasia	20890	Fd	3p22 also deleted	Ataxia-telangiectasia (2)	
11q22-qter	ANC	L	Anal canal carcinoma	10558	Ch	11q23 = conflicting localization	Anal canal carcinoma (2)	
11q22.3	THY1	C	Thy-1 T-cell antigen	18823	REa, H, A, Fd			9(Thy-1)
11q22.3-q23.1	CRYA2	P	Crystallin, alpha-B	12359	REa, A			
11q23	TSC2	P	Tuberous sclerosis-2	19109	Ch, Fd			
11q23	CD3D, CD3, T3D	C	T3 T-cell antigen receptor, delta polypeptide	18679	REa, A, RE	?defect in NCAM		9(T3d)
11q23	CD3G	C	T3 T-cell antigen receptor, gamma polypeptide	18674	RE, REa, A	3 CD3 genes in 60kb		9(T3g)
11q23	NCAM	C	Neural cell adhesion molecule	11693	A, Fd	defective in "staggerer," a form of cerebellar ataxia in mice		9(Ncam)
11q23	TCRE, CD3E, T3D	P	T3 T-cell antigen receptor, epsilon polypeptide	18683	REa, A, RE			9(T3e)
11q23-qter	APOLP1	C	APOLIPOPROTEIN CLUSTER I	10768	RE			
11q23-qter	APOA1	C	Apolipoprotein A-I	10768	REa, RE, Fd, F	11q13 = earlier assignment	ApoA-I and apoC-III deficiency, combined (1); Hypertriglyceridemia, 1 form (1); Hypoalphalipoproteinemia (1); Amyloidosis, Iowa form 10510 (1)	9(Apl-1)
11q23-qter	APOC3	C	Apolipoprotein C-III	10772	REa, RE, F	2.6kb 3' to APOA1		
11q23-qter	APOA4	C	Apolipoprotein A-IV	10769	F, RE	12 kb 3' to APOA1		9(Apoa-4)
11q23.1	EBVM1	P	Epstein-Barr virus modification site-1	13286	V			
11q23.2-qter	PBGD, UPS	C	Porphobilinogen deaminase (uroporphyrinogen I synthase)	17600	S, D		Porphyria, acute intermittent (1)	9(Ups)
11q24	ETS1	C	Oncogene ETS-1	16472	REa, A, Ch	shown by HSR		9(Ets-1)
11q24-q25	SRPR	P	Signal recognition particle receptor	18218	REa, A			
Chr.11	ADX	P	Adrenodoxin	10326	REa	pseudogene on 20		

Chr.11	CD6, TP120	P	T-cell differentiation antigen CD6	18672	S	
Chr.11	FCT3A	P	Alpha-3-fucosyltransferase	10423	S	
Chr.11	FRV1	P	Full-length endogenous retroviral sequence-2	13684	REa	
Chr.11	FTH	P	Ferritin, heavy chain	13477	REa, A	
Chr.11	GLAU1	L	Congenital glaucoma-1	23130	Ch	Glaucoma, congenital (2)
Chr.11	HV1S	P	Herpes virus sensitivity	14245	S	see chr. 3
Chr.11	LEU7, HNK1	P	Leu-7 antigen of natural killer lymphocytes, HNK-1	15129	S	
Chr.11	MYF3	C	Myogenic factor-3	15997	REa	probably on 11p
Chr.11	NANTA3	P	Alpha-3-acetylneuraminyltransferase	10424	S	
Chr.11	STMY	P	Stromelysin	18525	REa	
Chr.11	TRV2	P	Truncated endogenous retroviral sequence-2	19095	REa	

In addition: 19 surface antigens, most defined by monoclonals, 1 O'Farrell protein spot, 5 'like' genes, 6 pseudogenes and 8 fragile sites (HGM10). 11p physical map (HGM9): Cen--CAT--FSHB--LDHA--CALC1--PTH--HBBC--INS--HRAS1--pter. Genetic map (HGM9): Cen--CAT--18%--CALC1--8%--PTH--12%--HBBC--10%--INS--30%--HRAS1--pter; ?linkage heterogeneity (polymorphism) raised by some discrepant results. Map of apolipoprotein cluster I: ?11cen or 11qter 5'--APOA1--3'--(2.6kb)--3'--APOC3-5'--(4.5kb)--5'--APOA4--3'--?11cen or 11qter.

Chromosome No. 12

Location	Symbol	Status	Title	MIM #	Method	Comments	Disorder	Mouse
12pter-p12	F8VWF, VWF	C	von Willebrand factor	19340	A, REa, REb, Fd		von Willebrand disease (1)	
12pter-p12	CD4, T4, LEU3	C	T-cell antigen CD4	18694	REa, A	CD = 'cluster of differentiation' = nomenclature of leukocyte differentiation antigens		
12pter-q12	BCT1	C	Branched chain amino acid transaminase-1	11352	S		?Hyperleucinemia-isoleucinemia or hypervalinemia (1)	
12p13.31-p13.1	GAPD	C	Glyceraldehyde-3-phosphate dehydrogenase	13840	S, D, R			6(Gapd)
12p13.3	GLUT3	P	Glucose transporter-3	13817	A			
12p13.3-p12.3	A2M	P	Alpha-2-macroglobulin	10395	REa, A			
12p13.2	SPC, PRP	P	SALIVARY PROLINE-RICH PROTEIN COMPLEX		F, A, RE	cluster of genes; 6 loci in 2 subfamilies; in 500 kb: PRH1, PRH2, PRB1, PRB2,		6(Prp)
12p13.2	G1, PRB3	C	Parotid salivary glycoprotein	16884	LD, F	= PRB3 in basic subfamily		
12p13.2	PE, PRB1	P	Salivary protein Pe	18097	F			
12p13.2	PM	C	Parotid middle band protein	16878	F	linked to PRH1, PRH2, G1		
12p13.2	PRH1	C	Proline-rich acidic protein, HaeIII type, 1	16873	F, LD, RE	PA, DB, PIF alleles		
12p13.2	PRH2, PR	C	Proline-rich acidic protein, HaeIII type, 2	16879	F, LD, RE			
12p13.2	PS	C	Parotid size variant	16881	F			
12p13.2	PB	C	Parotid basic protein	16875	F			
12p13.2	PPB	C	Post-parotid basic protein	16876	F			
12p13.2	CON1, PRB4	C	Salivary protein CON1	16887	F	close to PS; ?order: PS-PR-PM-G1-DB		
12p13.2	CON2	C	Salivary protein CON2	16888	F	close to PM		
12p13.2	PO	P	Salivary protein Po	18099	F	probably closely linked to CON2		
12p13.2	PCS, PC	P	Parotid proline-rich protein Pc	16871	F	linked to PS		
12p13	MPE, EMP	L	Eosinophils, malignant proliferation of	13144	Ch		Eosinophilic myeloproliferative disorder (2)	
12p13	C1R	C	Complement component C1r	21695	REa, Fd, RE, A		C1r/C1s deficiency, combined (1)	
12p13	C1S	C	Complement component C1s	12058	REa, Fd, RE, A		C1r/C1s deficiency, combined (1)	
12p13	CD9, MIC3	C	Antigen CD9 identified by monoclonal antibodies 602-29, BA-2, et al.	14303	S, REa, A			
12p13	GNB3	P	Guanine nucleotide binding protein, beta-3 polypeptide	13913	REa, A			
12p13	TPI1, TPI	C	Triosephosphate isomerase	19045	S, D, R, REa		Hemolytic anemia due to triosephosphate isomerase deficiency (1)	6(Tpi-1)
12p13-p12.2	PZP	P	Pregnancy zone protein	17642	REa, A			

Location	Symbol		Title	No.	Method	Comments	Disorder	Mouse
12p13-q12	ATPSB, ATPMB	P	Adenosine triphosphate synthase, mitochondrial, beta polypeptide	10291	REa, REb			
12p12.2-p12.1	LDHB	C	Lactate dehydrogenase B	15010	S, D, Fd			6(Ldh-2)
12p12.1	KRAS2, RASK2	C	Oncogene Kras-2, Kirsten rat sarcoma virus	19007	REa, A, Fd		Colorectal adenoma (1); ?Colorectal cancer (1)	6(Kras-2)
12p12.1-p11.2	PTHLH	P	Parathyroid hormone-like hormone	16847	REa, A			6(Pthlh)
12p11.2-q11	KRT4, CYK4	C	Cytokeratin 4	12394	REa, A			
12p11-qter	CS	C	Citrate synthase, mitochondrial	11895	S, D			10(Cs)
12p11-qter	ENO2	C	Enolase-2	13136	S, D, A	by A, pter-p13		
12p	KAR	L	Aromatic alpha-keto acid reductase	10792	S	?same as MDH1 on proximal 12p		
12p	ELA1	C	Elastase-1	13012	REa			15(Ela-1)
12q	VDD1	P	Vitamin D dependency, type I	26470	Fd		Vitamin D dependency, type I (2)	
12cen-q14	IAPP, IAP, DAP	C	Islet amyloid polypeptide (diabetes-associated peptide; amylin)	14794	REa, A	conflicting localization = 12p12.3		
12cen-q14	MIP	P	Major intrinsic protein of lens fiber	15405	REa			
12q11-q13	FNRA	P	Fibronectin receptor, alpha polypeptide	13562	REa, A			
12q11-q21	KRTA	L	Keratin, acid or alpha-	13935	H	close to Hox-3 in mouse		15(Krta)
12q12-q13	INT1	C	Oncogene INT1, murine mammary cancer virus	16482	REa, A			15(Int-1)
12q12-q13	MLA1	P	Melanoma-associated antigen ME491	15574	REa, A			
12q12-q14	SHMT, GLYA	C	Serine hydroxymethyltransferase	13845	S, R	glycine A auxotroph		
12q13	HOX3	C	Homeo box-3	14297	REa, A, H	probably 1 gene; 3 homeoboxes in 1 transcription unit		15(Hox-3)
12q13	LALBA	P	Lactalbumin, alpha	14975	REa, A			
12q13-q14	BABL, LIPO	C	Lipoma (breakpoint in benign lipoma); myxoid liposarcoma	15190	Ch	?recombination 12q13 and 16p11 for myxoid liposarcoma, 15241	Lipoma (2); Myxoid liposarcoma (2); ?Multiple lipomatosis (2)	
12q13-q21	NKNB	P	Neurokinin B	16233	REa, A			
12q13.1-q13.3	COL2A1	C	Collagen II, alpha-1 chain	12014	REa, A	conflicting: 12q14.3	Stickler syndrome (3); Spondyloepiphyseal dysplasia congenita (3); ?Kniest dysplasia (1); Langer-Saldino achondrogenesis-hypochondrogenesis (1); Osteoarthrosis, precocious (3)	
12q13.2-q13.3	GLI1	C	Oncogene GLI1	16522	REa, A			
12q14	GNS, G6S	P	N-acetylglucosamine-6-sulfatase	25294	A, REa		Sanfilippo syndrome D (1)	
12q14	RAP1B	P	RAS-related protein RAP1B	17953	A			
12q21	PEPB	C	Peptidase B	16990	S			10(Pep-2)
12q22-q24.1	IGF1	C	Insulin-like growth factor I, or somatomedin C	14744	REa, A			
12q22-qter	ACADS	P	Acyl-CoA dehydrogenase, short chain	20147	REa		Acyl-CoA dehydrogenase, short chain, deficiency of (1)	5(Bcd-1)
12q24.1	IFNG, IFI, IFG	C	Interferon, gamma or immune type	14757	REa, A	3 introns; IFF, IFL none	Interferon, immune, deficiency (1)	10(Ifg)
12q24.1	PAH, PKU1	C	Phenylalanine hydroxylase	26160	REa, A		Phenylketonuria (3); [Hyperphenylalaninemia, mild] (3)	10(Pah)
12q24.2	ALDH2	C	Aldehyde dehydrogenase, mitochondrial	10065	REa, A		Acute alcohol intolerance (1); ?Fetal alcohol syndrome (1)	4(Aldh-2)
Chr.12	ATP2B	P	ATPase, CA++ dependent, slow twitch/cardiac muscle	10874	REa		?Brody myopathy (1)	
Chr.12	FRV3	P	Full-length endogenous retroviral sequence-3	13689	REa			

Chr.12	GNAI2A, GNAIH	P	Guanine nucleotide binding protein, alpha inhibiting, polypeptide h (or polypeptide 2A)	13918	REa	
Chr.12	GPD1	P	Alpha-glycerophosphate dehydrogenase; glycerol-3-phosphate dehydrogenase	13842	S	15(Gdc-1)
Chr.12	HSTD	P	Histidase	23580	REa	[Histidemia] (1)
Chr.12	LYS, LYZ	P	Lysozyme	15345	REa	
Chr.12	MPRD	P	Mannose-6-phosphate receptor, cation-dependent	15454	REa	
Chr.12	MTRNS, MARS, METRS	P	Methioninyl-tRNA synthetase	15656	S	
Chr.12	MYF5	P	Myogenic factor-5	15999	REa	
Chr.12	NTS	P	Neurotensin	16265	REa, ?A	
Chr.12	OIAS	C	2',5'-oligoisoadenylate synthetase	16435	REa	
Chr.12	PFKX	P	Phosphofructokinase X	17188	REa, REb	15(Pfkx)
Chr.12	PPLA2	P	Phospholipase A2, pancreatic	17241	REa	
Chr.12	TRV3	P	Truncated endogenous retroviral sequence-3	19096	REa	

In addition: 7 surface antigens, most defined by monoclonals, 3 O'Farrell protein spots, 3 'like' genes, 1 pseudogene and 5 fragile sites (HGM10). Probable order: 12pter--TPI1--GAPD--LDHB--ENO2--cen--SHMT--PEPB--12qter

5.89

Chromosome No. 13

Location	Symbol	Status	Title	MIM #	Method	Comments	Disorder	Mouse
13p12	RNR1	C	Ribosomal RNA	18045	A			
13q12	FLT	P	Oncogene FLT (FMS-like tyrosine kinase)	16507	REa, A			
13q14	IGEL	L	Immunoglobulin E level	14705	F	very close to ESD		
13q14	XRS	L	X-ray sensitivity	19437	Ch			
13q14-q21	WND, WD	C	Wilson disease	27790	F, Fd	vs. ESD, max. lod = 5.49, theta = 0.03; distal to RB1	Wilson disease (2)	
13q14-q31	LSD	L	Letterer-Siwe disease	24640	Ch		?Letterer-Siwe disease (2)	
13q14.1	OSRC	P	Osteosarcoma	25950	Ch	probably same locus as retinoblastoma	Osteosarcoma, retinoblastoma-related (2)	
13q14.1-q14.2	RB1	C	Retinoblastoma-1	18020	Ch, F, Fd		Retinoblastoma (2)	14(Rb-1)
13q14.1-q14.3	LCP1	P	Lymphocyte cytosolic protein-1	15343	F, D			
13q14.11	ESD, FGH	C	Esterase D: S-formylglutathione hydrolase	13328	S, F, D	proximal to RB1, WND		14(Es-10)
13q21-q31	ATP1AL2	P	Sodium-potassium-ATPase, alpha-polypeptide-like	18236	REa			
13q22-q34	ERCC5	C	Excision-repair, complementing defective, in Chinese hamster, number 5	13353	S			
13q22.1-q32.1	MGC	L	Megacolon (Hirschsprung disease)	24920	Ch	?linked to factors VII and X	?Megacolon (2)	
13q34	CBT1	L	Carotid body tumor-1	16800	F			
13q34	DJS	L	Dubin-Johnson syndrome	23750	LD	with factor VII deficiency	?Dubin-Johnson syndrome (2)	
13q34	F7	C	Clotting factor VII	22750	D		Factor VII deficiency (1)	
13q34	F10	C	Clotting factor X	22760	D, A, REa		Factor X deficiency (1)	
13q34	COL4A1	C	Collagen IV, alpha-1 chain	12013	REa, A, REb, RE, Fd			
13q34	COL4A2	C	Collagen IV, alpha-2 chain	12009	REa, A, RE, Fd			
13q34	HHHS	L	Hyperornithinemia-hyperammonemia-homocitrullinemia syndrome	23897	?D	associated with deficiency of factors VII and X in 3 unrelated cases	?HHH syndrome (2)	
13q34	RAP2	P	RAP2, member of RAS oncogene family (K-rev)	17954	A			
Chr.13	BRCD, DBC, BCDS1	P	Breast cancer, ductal, suppressor-1	21141	D	evidence of 1p determinants	Breast cancer, ductal (2)	
Chr.13	PCCA	P	Propionyl CoA carboxylase, alpha subunit	23200	REa		Propionicacidemia, type I or pccA type (1)	
Chr.13	TRV5	P	Truncated retroviral sequence-5	19098	REa			
Chr.13	UVDR, ERCM2	P	UV-damage, excision repair of (XP complementation group I)	19206	S	also called UV-135	?Xeroderma pigmentosum, one type (1)	

In addition: 3 'like' genes, 4 pseudogenes and 3 fragile sites (HGM10).

Chromosome No. 14

Location	Symbol	Status	Title	MIM #	Method	Comments	Disorder	Mouse
14p12	RNR2	C	Ribosomal RNA	18045	A			
14q11	TCRD	P	T-cell antigen receptor, delta subunit	18681	RE, Ch			
14q11-q12	TRNP1, TRLPT	C	Transfer RNA cluster-1 (Pro-Leu-Pro-Thr)	18893	REa, A, Fd			
14q11-q13	ANG	P	Angiogenin	10585	A			
14q11.2	CTLA1	P	Cytotoxic-T-lymphocyte-associated serine esterase-1	12391	A			14(Ctla-1)
14q11.2-q13	MYHCA, MYH6	C	Myosin, heavy polypeptide 6, cardiac muscle, alpha	16071	REa, RE, D, A			?11(Myh)
14q11.2-q13	MYHCB, MYH7	P	Myosin, heavy polypeptide-7, cardiac muscle, beta	16076	REa, RE, D, A	5'-B-4.5kb-A-3'		
14q11.2	TCRA	C	T-cell antigen receptor, alpha subunit	18688	H, REa, A	cen--V-C--ter	Leukemia/lymphoma, T-cell (3)	14(Tcra)
14q13.1	NP, NP1	C	Nucleoside phosphorylase	16405	S, D	centromeric to TCRA	Nucleoside phosphorylase deficiency, immunodeficiency due to (2)	14(Np-1,2)
14q21-q31	FOS	C	Oncogene FOS (FBJ murine osteosarcoma virus)	16481	REa, A			12(Fos)
14q21-qter	WARS	C	Tryptophanyl-tRNA synthetase	19105	S			
14q22-q22.2	SPTB, SPH1	C	Beta-spectrin (spherocytosis-1)	18287	REb, F, H, REa, A		Elliptocytosis-3 (2); Spherocytosis-1 (3)	12(Sptb)
14q22-qter	ADEB	P	Phosphoribosylformylglycinamidine synthetase (adenine (-)B auxotroph)	see 17246	S	?separate from ADEE locus		
14q23-q24.2	HOS	L	Holt-Oram syndrome	14290	Ch		?Holt-Oram syndrome (2)	
14q24	CAP	L	Cataract, anterior polar	11565	Fc	see 2p25	?Cataract, anterior polar (2)	
14q24	MTHFD, MTHFC, PGFT	C	5,10-methylenetetrahydrofolate dehydrogenase, 5,10-methylenetetrahydrofolatecyclohydrolase	17246	S, REa, A	trifunctional protein		
14q24	TGFB3	P	Transforming growth factor beta-3	19023	REa, A			12(Tgf-b3)
14q32	CHGA	P	Chromogranin A (parathyroid secretory protein 1)	11891	REa, A			
14q32	CKBB	C	Creatine kinase, brain type	12328	S, REa			
14q32	CKBE	P	Creatine kinase, brain type, ectopic expression of	12327	F	linked to IGH, PI; ?same locus as CKBB	[Creatine kinase, brain type, ectopic expression of] (2)	
14q32	VP, PPOX	P	Porphyria variegata (protoporphyrinogen oxidase)	17620	F		Porphyria variegata (2)	
14q32.1	AACT	C	Alpha-1-antichymotrypsin	10728	REa, A, Fd, REn	gene cluster with PI	?Alpha-1-antichymotrypsin deficiency (1)	
14q32.1	PI, AAT	C	Protease inhibitor (alpha-1-antitrypsin)	10740	F, S, A, D, EM, Fd		Emphysema-cirrhosis (1); Hemorrhagic diathesis due to 'antithrombin' Pittsburgh (1)	12(Aat)
14q32.1	TCL1	C	T-cell lymphoma-1	18696	Ch, RE	proximal to IGH; ?identical to TCL1	Leukemia/lymphoma, T-cell	
14q32.3	AKT1	C	Oncogene AKT1	16473	REa, A			12(Akt)
14q32.3	ELK2	P	Oncogene ELK-2	16535	REa, A			
14q32.33	IGH	C	IMMUNOGLOBULIN HEAVY CHAIN GENE CLUSTER	14412	REa, A		?Combined variable hypogammaglobulinemia (1)	12(Igh)
14q32.33	IGHR	L	Immunoglobulin heavy chain regulator		F		?Hyperimmunoglobulin G1 syndrome (2)	

Location	Symbol	Type	Description	Number	Method	Comments	Mouse
14q32.33	IGHV	C	V (variable) region of heavy chains	14707	REa, RE, A	ca. 250 genes; orientation: cen-PI-D14S1-IGH-IGHV--qter; 3' centromeric, 5' telomeric; IgM telomeric to IgG	
14q32.33	IGD1	C	D (diversity) region of heavy chains	14691	RE, REa, A	many genes	
14q32.33	IGHJ	C	J (joining) region of heavy chains	14701	RE, REa, A	more than 4 genes	
14q32.33	IGHM, MU	C	Constant region of heavy chain of IgM	14702	REa, A		
14q32.33	IGHD	C	Constant region of heavy chain of IgD	14717	REa, A		
14q32.33	IGHG2	C	Constant region of heavy chain of IgG2	14711	REa, A	5'-G2-17kb-G4-3'; closeness of IGG3 and IGG1 known from Lepore-like myeloma protein	
14q32.33	IGHG4	C	Constant region of heavy chain of IgG4	14713	REa, A		
14q32.33	IGHG3	C	Constant region of heavy chain of IgG3	14712	REa, A		
14q32.33	IGHG1	C	Constant region of heavy chain of IgG1	14710	REa, A		
14q32.33	IGHE	C	Constant region of heavy chain of IgE	14718	REa, A		
14q32.33	IGHEP1	C	Constant region of heavy chain of IgEP1	14716	REa, A	IGEP2 on chr. 9; 14721	
14q32.33	IGHA1	C	Constant region of heavy chain of IgA1	14690	REa, A		
14q32.33	IGHA2	C	Constant region of heavy chain of IgA2	14700	REa, A		
Chr.14	CSPB	P	Serine protease B	18213	REa		
Chr.14	ESAT	P	Esterase activator	13325	S		
Chr.14	K12T	C	Temperature sensitivity complementation, K12	18731	S		
Chr.14	LCH	C	Lentil agglutinin binding	15102	S		
Chr.14	M195	P	External membrane protein-195	13374	S		
Chr.14	PYGL, PPYL	P	Liver glycogen phosphorylase	23270	REb	Hers disease, or glycogen storage disease VI (1)	12(Pygl)
Chr.14	RIB1	L	Pancreatic ribonuclease	18044	H	?close to TCRA and NP	14(Rib-1)
Chr.14	TRV1	P	Truncated endogenous retroviral sequence-1	19094	REa		

In addition: 1 antigen defined by monoclonal, 3 'like' genes, 4 pseudogenes and 2 fragile sites (HGM10). A Tunisian deletion indicates order: 5'--G3--G1--psi E1--A1--G2--G4--E--A2--3' (Lefranc et al, Nature 300: 760, 1982). Following information from M. J. Johnson and L. L. Cavalli-Sforza, Stanford Univ., Nov., 1983 and Hofker et al, PNAS **86**: 5567, 1989: 5'(qter)--V--(7cM)--D--J--8kb--mu--5kb--delta--60kb--gamma-3--26kb--gamma-1--19kb--pseudo-epsilon-1 --13kb--alpha--1--80kb--gamma-2--18kb--gamma-4--23kb--epsilon--10kb--alpha-2--3'(centromere). Pseudo-gamma between alpha-1 and gamma-2 (Bech-Hansen et al, PNAS 80:6952, 1983; Migone et al, PNAS 81: 5811, 1984), about 35 kb from alpha-1. Comparable data in mouse for genes for kappa, lambda, and heavy chains of immunoglobulin and for T-cell alpha-, beta-, and gamma-genes are given in Figure 2 of Kronenberg et al., *Ann Rev Immunol* 4:529-591, 1986.

Chromosome No. 15

Location	Symbol	Status	Title	MIM #	Method	Comments	Disorder	Mouse
15p12	RNR3	C	Ribosomal RNA	18045	A			
15p12-q21	SORD	C	Sorbitol dehydrogenase	18250	S, H			2(Sdh-1)
15q	NMB	P	Neuromedin B	16234	REa			
15q11	DLX1	L	Dyslexia-1	12770	Fc, Fd	?near centromere; lod under 3.0 with HGM8 data	Dyslexia-1 (2)	
15q11	PWCR, PWS	C	Prader-Willi syndrome	17627	Ch		Prader-Willi syndrome (2)	
15q11-q12	IGD2	P	Immunoglobulin heavy chain diversity region-2	14699	A	?functional		
15q11-q12	MIC7	P	Attached cell antigen 28.3.7	10899	S			
15q11-q13	ANCR, AGMS	P	Angelman syndrome	23440	Ch		Angelman syndrome (2)	
15q11-q13	ITO	L	Hypomelanosis of Ito	14615	Ch	see chr.9	?Hypomelanosis of Ito (2)	
15q11-q13	MANA	C	Alpha-mannosidase-A, cytoplasmic	15458	S, D			
15q11-qter	ACTC	P	Actin, cardiac alpha	10254	REa			2(Actc-1)
15q11-qter	CVS, HCVS	P	Coronavirus 229E sensitivity	12246	S			
15q13-q15	B2MR	C	Beta-2-microglobulin regulator	10971	D			
15q13-qter	PEPN, APN	P	Aminopeptidase N	10463	REa			
15q14-q15	IVD	P	Isovaleryl CoA dehydrogenase	24350	REa		Isovalericacidemia (1)	
15q15-q21	CHR39B	P	Cholesterol repressible protein 39B	11848	REa			
15q21-q22	B2M	C	Beta-2-microglobulin	10970	S, D, H	on 15q+ in APL	Hemodialysis-related amyloidosis (1)	2(B2m)
15q21-q22	LPC2D	P	Lipocortin IId	15174	REa			
15q21-q23	LIPC, LIPH, HL, HTGL	C	Hepatic triglyceride lipase	15167	REa, A		?Hepatic lipase deficiency (1)	
15q21-qter	IDH2	C	Isocitrate dehydrogenase, mitochondrial	14765	S			7(Idh-2)
15q21.1	CYP19, ARO	C	Cytochrome P450 aromatization of androgen (aromatase)	10791	REa, A		?Gynecomastia, familial, due to increased aromatase activity (1)	
15q22	PK3, PKM2	C	Pyruvate kinase-3	17905	S, D, A	on 15q+ in APL		9(Pk-3)
15q22-q25.1	HEXA, TSD	C	Beta-hexosaminidase A, alpha chain	27280	S, D		Tay-Sachs disease (1); GM2-gangliosidosis, juvenile, adult (1); [Hex A pseudodeficiency] (1)	
15q22-q26	ACP5	P	Phosphatase, acid, type 5, tartrate-resistant	17164	A			
15q22-qter	MPI	C	Mannosephosphate isomerase	15455	S			9(Mpi-1)
15q22-qter	CYP1A1, CYP1, P450C1	P	Dioxin-inducible P1-450 (TCDD-inducible P1-450)	10833	S, REa, H	CYP2 = earlier symbol		9(P450-1)
15q22-qter	CYP1A2	P	Dioxin-inducible P3-450	12406	REa, H	both CYP1 genes close to MPI in rodents		9(P450-1)
15q23-q25	ETFA, GA2	P	Electron transfer flavoprotein, alpha subunit	23168	REa, A		Glutaricaciduria, type II (1)	
15q23-q25	FAH	P	Fumarylacetoacetate hydrolase	27670	A		Tyrosinase I(1)	
15q24-q25	CTSH	P	Cathepsin H	11682	REa, A			
15q25-q26	FES	C	Oncogene FES, feline sarcoma virus	19003	S, A	?15q26; far from breakpoint in acute promyelocyte leukemia:t(15;17)(q22;q21)	?Batten disease, 1 form, 20420 (1)	7(Fes)
15q25-q26	FUR	C	Furin membrane associated receptor protein	13695	RE	less than 1.1kb 5' to FES		
15q25-q26	IGF1R	P	Insulin-like growth factor-1 receptor	14737	REa, A	?relation to FES		

15q25-q26	CD13	P	Antigen CD13	15153	REa, A		
Chr.15	CHRNA, ACHRA	L	Cholinergic receptor, nicotinic, alpha polypeptide	10069	H	linked to Actc in mouse	17(?Achr-?)
Chr.15	CKMT	P	Creatine kinase, mitochondrial	12329	REa		
Chr.15	COL1AR, COLR	P	Collagen, type 1, alpha, receptor	12034	S		
Chr.15	CSPG1	P	Chondroitin sulfate proteoglycan core protein	15576	S		
Chr.15	CYP11A, P450SCC, P450C11A1	P	P450 side chain cleavage enzyme (20,22 desmolase)	20171	REa		Lipoid adrenal hyperplasia, congenital (1)
Chr.15	GANC	P	Neutral alpha-glucosidase C	10418	S		
Chr.15	XPF	P	Xeroderma pigmentosum, group F	27876	M		Xeroderma pigmentosum, group F (2)

In addition: 5 surface antigens, most defined by monoclonals, 2 O'Farrell protein spots, 2 pseudogenes and 1 fragile site (HGM10). Hemodialysis-related amyloidosis is presumably not genetic in a specific sense.

Chromosome No. 16

Location	Symbol	Status	Title	MIM #	Method	Comments	Disorder	Mouse
16pter-p13.3	HBAC, ABC	C	ALPHA GLOBIN GENE CLUSTER		HS, RE, A, D	order: cen-APKD-HBZ1--HBA1-3'HVR-pter; distal to PGP		
16pter-p13.3	HBA1	C	Hemoglobin alpha-1	14180	HS	1, 2, or 3 loci; 5'-zeta-pseudozeta-pseudoalpha-alpha-2-alpha-1-3'	Thalassemias, alpha- (1); Methemoglobinemias, alpha- (1); Erythremias, alpha- (1); Heinz body anemias, alpha- (1)	11(Hba)
16pter-p13.3	HBA2	C	Hemoglobin alpha-2	14185	HS			11(Hba)
16pter-p13.3	HBQ1	P	Hemoglobin theta-1	14224	RE			
16pter-p13.3	HBZ1	C	Hemoglobin zeta pseudogene (formerly zeta-1)	see 14231	RE			
16pter-p13.3	HBZ2	C	Hemoglobin zeta (formerly zeta-2)	14231	RE			
16pter-p13.3	HBHR	L	Hb H mental retardation syndrome	14175	F		Hb H mental retardation syndrome (2)	
16p13.31-p13.12	PGP	C	Phosphoglycolate phosphatase	17228	S	no recombination with PKD1		
16p13.31-p13.12	PKD1, APKD	C	Adult polycystic kidney disease	17390	F, Fd	tightly linked to PGP; rare form unlinked	Polycystic kidney disease (2)	
16p13.3-p13.11	ERCC4	P	Excision-repair, complementing defective, in Chinese hamster, number 4	13352	S			
16p13.1-p11	LAAC	C	LEUKOCYTE ADHESION, ALPHA, CLUSTER		REa, A, RE			
16q13.1-p11	CD11A, LFA1A	C	Lymphocyte function associated antigen-1, alpha subunit	15337	S, REa, A			
16q13.1-p11	CD11C	P	Leukocyte surface antigen p150,95, alpha subunit	15151	REa, A			
16q13.1-p11	CR3A, CD11B, MAC1A, MO1A	P	Complement component receptor-3 (Macrophage antigen-1, Mac-1, alpha subunit)	12098	RE, A	?in same restriction fragment as LFA1A		
16p13	HAGH, GLO2	C	Glyoxalase II; hydroxyacyl glutathione hydrolase	13876	S		[Glyoxalase II deficiency] (1)	
16p11.2	PRKCB, PKCB	P	Protein kinase C, beta polypeptide	17697	REa, A			
16p11.2	SPN, LSN, CD43	C	Sialophorin (leukosialin)	18216	REa, A			
16p11-q23	CHE2	I	Cholinesterase, serum, 2	17750	A, F	see chr. 2		
16p11-q24	UVO	P	Uvomorulin	19209	REa			
16cen-q22	GOT2	C	Glutamate oxaloacetic transaminase, mitochondrial	13815	S, F, H	on 16q by homology to mouse		8(Um) 8(Got-2)
16q12-q13.1	PHKB	P	Phosphorylase kinase, beta polypeptide	17249	REa, A		?Phosphorylase kinase deficiency of liver and muscle, 26175 (2)	
16q12-q22	DIA4	C	Diaphorase-4	12586	S, REa			
16q21	CETP	C	Cholesterol ester transfer protein, plasma	11847	REa, A		CETP deficiency (1)	
16q21	CLG4	P	Collagenase, type IV	12036	REa, A			
16q21	PRM1	P	Sperm protamine P1	18288	REa, A			16(Prm-1)
16q21	PRM2	P	Sperm protamine P2	18289	H			16(Prm-2)
16q22	MT1	C	METALLOTHIONEIN I CLUSTER	15635	REa, A			8(Mt-1)
16q22	MT2	C	METALLOTHIONEIN II CLUSTER	15636	REa, A			8(Mt-2)
16q22	NCL, BD	C	Neuronal ceroid-lipofuscinosis (Batten disease)	20420	F	linked to HP	Batten disease (2)	

5.96

Location	Symbol	Status	Name	MIM	Method	Comments	Disorder	Mouse
16q22-q24	ALDOA, ALDA	C	Aldolase A	10385	REa, REb, A		?Aldolase A deficiency (1)	
16q22.1	CA4	P	Carbonic anhydrase IV	11476	REa, A	not proved to be expressed		
16q22.1	CPM, CAM	C	Cataract, Marner type	11680	F		Cataract, Marner type (2)	
16q22.1	HP	C	Haptoglobin	14010	Fc	just distal to fra16q22.1		8(Hp)
16q22.1	HPR	C	Haptoglobin-related locus	14021	REa	2.2kb 3' to HP; multiple tandem genes in blacks		
16q22.1	LCAT	C	Lecithin-cholesterol acyltransferase	24590	F, LD, A, REa	very close to HP	Norum disease (3)	8(Lcat)
16q22.1-q22.3	TAT	C	Tyrosine aminotransferase, cytosolic	27660	REa, A, H, D		Tyrosinemia, type II (1)	8(Tat-1)
16q22.3	CTRB	C	Chymotrypsinogen B	11889	REa, H, D, Fd	HP-7cM-TAT-9cM-CTRB		8(Ctrb)
16q24	APRT	C	Adenine phosphoribosyltransferase	10260	S, D	distal to GOT2, DIA4; earlier mapped to 16q22.2-q22.3	Urolithiasis, 2,8-dihydroxyadenine (1)	8(Aprt)
Chr.16	ATP2A	P	ATPase, Ca++ transporting, fast-twitch, muscle	10873	REa			
Chr.16	CTH	P	Cystathionase	21950	S		[Cystathioninuria] (1)	
Chr.16	DIPl, VDI	P	Vesicular stomatitis virus defective interfering particle repressor	12526	S			
Chr.16	ESB3	P	Esterase-B3	13329	S			
Chr.16	GCF2	P	Growth rate controlling factor-2	13923	S			
Chr.16	GRLL	P	Glucocorticoid receptor, lymphocyte, like	13806	REb			
Chr.16	LIPB	P	Lysosomal acid lipase-B	24798	S			
Chr.16	NHCP1	P	Nonhistone chromosomal protein-1	11887	S			
Chr.16	TK2	P	Thymidine kinase, mitochondrial	18825	S			

In addition: 3 'like' genes, 3 pseudogenes and 5 fragile sites (HGM10). Order: pter--PGP--0.25--16qh--0.17--GOT2--0.08--HP--qter (Jeremiah et al., Ann. Hum. Genet. 46: 145, 1982).

Chromosome No. 17

Location	Symbol	Status	Title	MIM #	Method	Comments	Disorder	Mouse
17pter-p21	TRNP2	P	Transfer RNA cluster-2 (Leu-Gln-Lys)	18992	REa, A			11(Zfp-1)
17pter-p12	ZFP3	P	Zinc finger protein-3	19448	REa			11(Mds)
17p13.3	MDCR, MDLS, MDS	C	Miller-Dieker lissencephaly syndrome	24720	Ch, D		Miller-Dieker lissencephaly syndrome (2)	
17p13.105-p12	POLR2, RPOL2, RNP2	C	RNA polymerase II, large subunit	18066	REa, A, C			11(Rpol-2)
17p13.105-p12	TP53	C	Tumor protein p53	19117	REa, A		Colorectal carcinoma, 11450 (1)	11(Trp53)
17p13	GLUT4	P	Glucose transporter, insulin-responsive	13819	REa, A			
17p13	MYH1	C	MYOSIN, HEAVY CHAIN CLUSTER		REa, C	a myosin gene cluster also on 7		11(Myh)
17p13	MYHSA1	C	Myosin heavy chain, skeletal, adult-1	16073	REa, C			
17p13	MYHSA2	C	Myosin heavy chain, skeletal, adult-2	16074	REa, C	17p13.105-p12		
17p13	MYH3, MYHSE1	C	Myosin heavy chain, embryonic-1	16072	REa		Colorectal cancer, 11450 (2)	
17p12	CRCR2, CRC17	P	Colorectal caarcinoma 2 (colorectal cancer-related sequence-17)	12046	D	?same as TP53		
17p12-p11	CHRNB, ACHRB	C	Cholinergic receptor, nicotinic, beta polypeptide	10071	H, REa, A	linked to Myh on mouse 11		11(?Achr-?)
17p11.2	SMCR	C	Smith-Magenis syndrome chromosome region	18229	Ch			
17p11.1-qter	PNP, PPY	P	Pancreatic polypeptide	16778	REa, Fd			
17p11-qter	ACTG	P	Cytoskeletal gamma-actin	10256	REa			
17cen-q12	ALDOC, ALDC	C	Aldolase C	10387	REb, REa, A			
17cen-q25	ADXR	P	Adrenodoxin reductase	10327	REa			
17cen-qter	GAS	C	Gastrin	13725	REa			
17cen-qter	HTLVR	P	Receptor for HTLV-1 and HTLV-2	14309	REa			
17q11-q12	EDHB17	P	Estradiol 17-beta-dehydrogenase	13344	A			
17q11-q21	WSS	P	Watson syndrome	19352	Fd	?allelic to NF1	Watson syndrome (2)	
17q11-q23	KRTB, CYK15	C	Keratin, basic or beta- (cytokeratin 15)	14803	H, REa, A	tightly linked to Hox-2 in mouse; probable cluster of CYK genes		11(Krb)
17q11-q21	TCP228	P	T-cell specific protein p288	18071	A	?related gene at 5q31-q34		
17q11.1-q12	CRYB1	C	Crystallin, beta-B1	12361	REa, A			
17q11.2	ERBA1, THRA	C	Oncogene ERB-A1 (avian erythroblastic leukemia virus)	19012	REa, Ch	centromeric to NF1		11(Erba)
17q11.2	NF1, VRNF	C	Neurofibromatosis, von Recklinghausen type	16220	Fd, EM, Ch, F	?linked to GALK	Neurofibromatosis, von Recklinghausen (2)	
17q11.2-q12	CSF3, GCSF	C	Granulocyte colony-stimulating factor-3	13897	A, REa, REb, RE			
17q12-q21.32	ERBB2	C	Oncogene ERB-B2	19015	REb, A			
17q21	ACC	P	Acetyl-CoA carboxylase	20035	A	proximal to q21.33	Acetyl-CoA carboxylase deficiency (1)	
17q21	MTBT1, MAPT1	P	Microtubule, beta, associated protein tau	15714	REb, A	see 6p21		
17q21-q22	A12M4	C	Adenovirus-12 chromosome modification site-17	10297	V	in RNU2		
17q21-q22	HOX2, HU2	C	Homeo box-2	14296	REa, A, H, Fd, RE	5 or 6 genes, e.g., HOX2.1, HOX2.2		11(Hox-2)

5.98

Location	Symbol	P/C	Title	MIM #	Method	Comments	Disorder	Mouse
17q21-q22	INT4	P	Oncogene INT-4 (murine mammary tumor virus integration site, v-int-4, oncogene homologue)	16533	REa, S, A, H			11(Int-4)
17q21-q22	NGFR	C	Nerve growth factor receptor	16201	REa, A, S, Fd, C	distal to APL breakpoint, q21; < 0.5mb from HOX2		
17q21-q22	NGL, NEU	C	Oncogene NGL (oncogene NEU, neuro- or glioblastoma derived; HER2; TKR1)	16487	REa, A			
17q21-q22	RNU2	C	U2 snRNA GENE CLUSTER	18069	REa, A, C			
17q21-q22	GALK	C	Galactokinase	23020	S, Ch, R, C	by CMGT, order = cen-GALK-TK1-COL1A1	Galactokinase deficiency (1)	11(Glk)
17q21-q22	EPB3, EMPB3	C	Erythroid membrane protein band 3	10927	REa, RE		[Acanthocytosis, 1 form] (1)	
17q21-q22	TOP2	P	DNA topoisomerase II	12643	REa, A			
17q21.1	RARA	P	Retinoic acid receptor, alpha type	18024	A			
17q21.1-q21.3	GP3A	C	Platelet glycoprotein IIIa	17347	REa, REb, A, RE, F, LD	in same 260kb fragment as GP2B; PL(A) antigen		
17q21.3-q22	GIP	P	Gastric inhibitory polypeptide	13724	REa, A			
17q21.3-q23	MPO	C	Myeloperoxidase	25460	REa, A, F, Ch, C	translocated in t(15;17)(q22;q11.2)	Myeloperoxidase deficiency (1)	
17q21.31-q22.05	COL1A1	C	Collagen I, alpha-1 chain	12015	C, M, A, REa	?proximal to GH1	Osteogenesis imperfecta, 2 or more clinical forms (3); Ehlers-Danlos syndrome, type VIIA1 (3); ?Marfan syndrome, atypical, 15470(1)	11(Cola-1)
17q21.32	GP2B	C	Platelet glycoprotein IIb	27380	A, REb, REa, RE, F, LD	3' to GP3A; BAK antigen	Glanzmann thrombasthenia (1)	
17q22-q24	GHC	C	GROWTH HORMONE/PLACENTAL LACTOGEN GENE CLUSTER		REa, A, C			
17q22-q24	GH1, GHN	C	Growth hormone, normal	13925	REa, A	5'-GH1-CSHP1-CSH1-GH2-CSH2-3'	Isolated growth hormone deficiency, Illig type with absent GH and Kowarski type with bioinactive GH (3)	11(Gh)
17q22-q24	GH2, GHV	C	Growth hormone, variant	13924	REa, A			
17q22-q24	CSHP1, CSL	C	Chorionic somatomammotropin pseudogene	see 15020	REa, A			
17q22-q24	CSA, PL, CSH1	C	Chorionic somatomammotropin A	15020	REa, A			
17q22-q24	CSB, CSH2	C	Chorionic somatomammotropin B	11882	REa, A		[Placental lactogen deficiency] (1)	
17q22-q24	PKCA	P	Protein kinase C, alpha form	17696	REa, A	at 3' end		11(Pkca)
17q23	DCP	P	Dipeptidylcarboxypeptidase-1 (angiotensin I converting enzyme)	10618	A			
17q23	GAA	C	Acid alpha-glucosidase	23230	S, A, D, C		Pompe disease (1); Acid-maltase deficiency, adult (1)	
17q23-q24	UMPH2	C	Uridine 5'-monophosphate phosphohydrolase-2; uridine monophosphatase-2	19172	S, C	between GAA and GHC		11(Umph-2)
17q23-qter	PEPE	C	Peptidase E	17020	S			
17q23.2-q25.3	TK1	C	Thymidine kinase-1	18830	S, Ch, R, C, Fd, A	to chromosome 15 in APL		11(Tk-1)
17q25	P4HB, PROHB	P	Prolyl-4-hydroxylase, beta subunit; disulfide isomerase; cellular thyroid hormone-binding protein p55	17679	S, REa, A			

Chr.17	ALDH3	P	Aldehyde dehydrogenase-3	10066	S		
Chr.17	ALPPL	P	Placental alkaline phosphatase-like gene	see 17180	REa		
Chr.17	APOH	P	Apolipoprotein H (beta-2-glycoprotein I)	13870	Fd	[Apolipoprotein H deficiency] (1)	
Chr.17	CD7	C	T-cell antigen CD7	18682	S		
Chr.17	G6PDL	P	Glucose-6 phosphate dehydrogenase-like	13811	REa, REb		
Chr.17	GALC	P	Galactocerebrosidase	24520	S	Krabbe disease (1)	?11(tw)
Chr.17	MYL4	P	Myosin light polypeptide-4, alkali; atrial, embryonic	16077	REa		11(Myl-4)
Chr.17	PENT, PNMT	P	Phenylethanolamine N-methyltransferase	17119	REa		
Chr.17	PFN1	P	Profilin-1	17661	REa		
Chr.17	SMPD1, NPD	P	Sphingomyelinase (Niemann-Pick disease)	25720	S	Niemann-Pick disease (1)	
Chr.17	TSE1	P	Tissue-specific extinguisher-1	18883	S, M		11(Tse-1)

In addition: 3 surface antigens, most defined by monoclonals, 5 'like' genes, 3 pseudogenes and 1 fragile site (HGM10). Order by CMGT (Xu et al., 1988): pter--(TP53--POLR2--D17S1)--(MYHSA2--MYHSA1)--D17Z1--CRYB1--acute promyelocytic leukemia breakpoint--RNU2--HOX2--(NGFR--COL1A1--MPO)--GAA--UMPH2--GHC--TK1--GALK--qter.

Chromosome No. 18

Location	Symbol	Status	Title	MIM #	Method	Comments	Disorder	Mouse
18p11.32	MCL	L	Multiple hereditary cutaneous leiomyomata	15080	Ch		?Leiomyomata, multiple hereditary cutaneous (2)	
18p11.31	LAMA	P	Laminin A	15032	A			
18p11.1-q11.2	PLI	P	Alpha-2-plasmin inhibitor	26285	A		Plasmin inhibitor deficiency (1)	
18q11-q12	JK	P	Kidd blood group	11100	Fd, EM	previous suggestion of chr.7 or chr.2		
18q11-q12	LCFS2	L	Lynch cancer family syndrome II	11440	F	?linked to JK	?Lynch cancer family syndrome II (2)	
18q11.2-q12.1	PALB, TTR, TBPA	C	Thyroxine-binding prealbumin (transthyretin)	17630	REa, A		Amyloid neuropathy, familial, several allelic types (3); [Dystransthyretinemic hyperthyroxinemia](1)	
18q21	GRP	C	Gastrin releasing peptide	13726	REa, A	mammalian equivalent of bombesin		
18q21	SSAV1	P	Simian sarcoma-associated virus-1/Gibbon ape leukemia virus	18209	REa, A			
18q21.2-q22	PLANH2, PAI2	C	Plasminogen activator inhibitor, type II	17339	REa, A			
18q21.3	BCL2, BCL3	C	B-cell CLL/lymphoma-2	15143	Ch, RE	most frequent hematologic malignancy t(14;18)(q22;q21)	Leukemia/lymphoma, B-cell, 2 (2)	1(bcl-2)
18q21.3	YES1	C	Oncogene YES-1	16488	REa			
18q21.31-qter	TS, TMS	C	Thymidylate synthase	18835	S	?on 18p or proximal 18q		
18q22-q23	ERV1	C	Oncogene ERV1; endogenous retrovirus-1	13115	REa, A			
18q22-qter	MBP	C	Myelin basic protein	15943	REa, A	defective in "shiverer," neurologic mutant in mouse		18(Mbp)
18q22.1	GTS	L	Gilles de la Tourette syndrome	13758	Ch	t(7;18)(q22;q22.1)	?Tourette syndrome (2)	
18q23	PEPA	C	Peptidase A	16980	S, D			18(Pep-1)
18q23.3	CRC18	P	Colorectal cancer-related sequence-18	12047	D		Colorectal cancer (1)	
Chr.18	DD	L	Diastrophic dysplasia	22260	Ch		?Diastrophic dysplasia (2)	
Chr.18	DHFRP1	P	Dihydrofolate reductase pseudogene-1	see 12606	REa			
Chr.18	IFNGR2	P	Interferon, gamma, receptor for	10747	S	shows +/- polymorphism		
Chr.18	NARS, ASNRS	P	Asparaginyl-tRNA synthetase	10841	REa	both 6 and 18 required		

In addition: 1 'like' gene, 4 pseudogenes and 2 fragile sites (HGM10). Subband critical to trisomy 18 phenotype = 18q12.2.

Chromosome No. 19

Location	Symbol	Status	Title	MIM #	Method	Comments	Disorder	Mouse
19pter-p13.2	OK	P	Blood group OK	11138	S			
19pter-q13	CXB3S	P	Coxsackie virus B3 sensitivity	12005	S			
19pter-q12	EF2	C	Elongation factor-2	13061	S			
19p13.3-p13.2	AMH, MIF	P	Anti-Mullerian hormone	26155	REa, A		Persistent Mullerian duct syndrome (1)	
19p13.3-p13.2	INSR	C	Insulin receptor	14767	REa, A, REb	1 gene for alpha and beta subunits	?Leprechaunism (1); Acanthosis nigricans, insulin-resistant (1); ?Rabson-Mendenhall syndrome (2)	8(Insr)
19p13.3-p13.2	C3	C	Complement component-3	12070	F, S, A, REa	LE ca. 7cM in males vs. C3 RFLP	C3 deficiency (1)	17(C3)
19p13.2	RAB3A	P	RAS-associated protein RAB3A	17949	REb, A			
19p13.2-p13.1	LDLR, FHC	C	Familial hypercholesterolemia (LDL receptor)	14389	F, REa, A	ca. 20cM distal to C3	Hypercholesterolemia, familial (3)	9(Ldlr)
19p13.2-q12	MANB	C	Lysosomal alpha-D-mannosidase-B	24850	S		Mannosidosis (1)	
19p13.2-q13.4	DNL	P	Lysosomal DNA-ase	12635	S			
19p13.1-cen	LW	C	LW (Landsteiner-Weiner) blood group	11125	F, Fd	close to C3, LU		
19p13.1-q13.11	LE, LES	C	Lewis blood group	11110	F	linked to C3; order:FHC-C3-LE-DM-SE-LU		
19p13.1-q13.11	FUT1, H, HH	P	Fucosyltransferase-1 (Bombay phenotype)	21110	F	SE tightly linked		
19p13.1-q13.11	HCL1, BRHC	P	Brown hair color	11375	F			
19p13.1-q13.11	GEY	P	Green/blue eye color	22724	F	different locus for brown/blue		
19p13	LYL1	P	Leukemia, lymphoid, 1	15144	Ch			
19p13	RFX	P	HLA class II regulatory factor RF-X	20992	A		Severe combined immunodeficiency (SCID), HLA class II-negative type (1)	
19cen-q12	GPI	C	Glucosephosphate isomerase; neuroleukin	17240	S, D		Hemolytic anemia due to glucosephosphate isomerase deficiency (1); Hydrops fetalis (1)	7(Gpi-1)
19cen-q13.11	LU	C	Lutheran blood group	11120	F	linked to SE		
19cen-q13.11	PEPD	C	Peptidase D (prolidase)	17010	S, F, H, Fd	closely linked to APOC2	?Prolidase deficiency, 26413 (1)	7(Pep-4)
19cen-q13.11	FUT2, SE	C	Fucosyltransferase-2 (secretor)	18210	F	H, SE = alpha-L-fucosyltransferases; from common ancestral genes		
19cen-q13.2	MEL	C	Oncogene MEL	16504	REa, Fd			
19cen-q13.3	LIPE, LHS	P	Lipase, hormone-sensitive	15175	REa			
19q12-q13.2	ATP1A3	C	Sodium-potassium-ATPase, alpha-3 polypeptide	18235	REa, H, Fd			7(Atpa-2)
19q12-q13.2	PVS	C	Polio virus sensitivity	17385	S		[Polio, susceptibility to] (2)	
19q13	APS	C	Prostate-specific antigen	17682	REa, RE, A	probably with cluster KLK1, KLK2		7(Aps)
19q13	BCL3	P	B-cell CLL/lymphoma-3	10956	Ch, S			
19q13	CKMM, CKM	C	Creatine kinase, muscle type	12331	REa, A	distal to APOLP2		

Location	C/P	Disease/Gene	Symbol	MIM No.	Method	Comments	Disorder	Mouse
19q13	C	Myotonic dystrophy	DM	16090	F, Fd	distal to APOLP2; ?proximal to CKM 12kb from APS	Myotonic dystrophy (2)	
19q13	P	Kallikrein, glandular	KLK2	14796	REa, RE			
19q13	P	Normal cross-reacting antigen	NCA	16398	REa, A			
19q13.1	C	APOLIPOPROTEIN CLUSTER II	APOLP2		F, REa, LD	closely related to CEA 5'--APOE-4.3kb-APOC1-6kb-APOC1 pseudogene-22kb-APOC2--3'		
19q13.1	C	Apolipoprotein E	APOE	20776	F, REa, LD, A, Fd		Hyperlipoproteinemia, type III (1)	7(Apoe)
19q13.1	C	Apolipoprotein C-I	APOC1	10771	REa, RE, A			
19q13.1	C	Apolipoprotein C-II	APOC2	20775	REa, F, LD, A, Fd		Hyperlipoproteinemia, type Ib (1)	7(Pkcc)
19q13.1	C	Pregnancy-specific beta-1-glycoprotein-1	PSBG1, B1G1, SP1	17639	REa			
19q13.1-q13.3	C	Carcinoembryonic antigen	CEA	11489	REa, A	inconsistent regionalization = q31-q32		
19q13.1-q13.3	C	Cytochrome P-450, family II, subfamily A, phenobarbitol inducible	CYP2A, P450C2A	12396	REa, A, Fd, REn	CYP1 = earlier symbol		
19q13.1-q13.3	C	Cytochrome P-450, family II, subfamily B, phenobarbital inducible	CYP2B	12393	REa, REn, Fd	same NoI fragment as CYP2A		7(Coh)
19q13.1-q13.3	P	Cytochrome P-450, family II, subfamily F	CYP2F	12407	F, REa, A			
19q13.1-q13.3	P	Transforming (or tumor) growth factor, beta form	TGFB1	19018	REa, A			7(Tgfb)
19q13.1-qter	C	Echo 11 sensitivity	E11S	12915	S			
19q13.1-qter	C	RD114 virus receptor (Baboon M7 virus receptor)	RDRC, M7V1, M7VS1	10919	Ch, S			
19q13.2-q13.3	C	Complementation of CHO DNA-repair defect UV20	ERCC1, UV20	12638	S		?Xeroderma pigmentosum, 1 form (1)	
19q13.2-q13.3	P	Complementation of CHO DNA-repair defect EM9	ERCC2, EM9	12634	S			
19q13.2-q13.3	P	X-ray-repair, complementing defective, in Chinese hamster	XRCC1	19436	S			
19q13.2-q13.4	P	Kallikrein, renal/pancreas/salivary	KLK1, KLKR	14791	REa, A			7(Klk-1)
19q13.2-q13.4	C	Protein kinase C, gamma form	PKCC, PKCG	17698	REa, A, Fd			
19q13.3	P	Myeloid differentiation antigen	CD33	15959	REa, A			
19q13.3-q13.4	C	Ferritin, light chain	FTL	13479	S, A, REa, REb			
19q13.32	C	CHORIONIC GONADOTROPIN, BETA CHAIN	CGB	11886	REa, H, A	at least 5 genes		
19q13.32	C	Luteinizing hormone, beta chain	LHB	15278	RE	beta chains of FSH, TSH on 11p, 1p, respectively	?Male pseudohermaphroditism due to defective LH (1)	7(Lhb)
Chr.19	P	Branched chain amino acid transaminase-2	BCT2	11353	S		?Hypervalinemia or hyperleucinemia-isoleucinemia (1)	
Chr.19	C	Branched chain ketoacid dehydrogenase, E1-alpha subunit	BCKDE1A, MSUD1	24860	REa, REb		Maple syrup urine disease (1)	
Chr.19	P	ERBA-related gene-2	ERBAL2, EAR2	13288	REb			
Chr.19	P	Beta-glucuronidase, mouse, modifier of	GUSM	23161	S			
Chr.19	P	Oncogene HKR1	HKR1	16525	REa			

Chr.19	HKR2	P	Oncogene HKR2	16526	REa	
Chr.19	KRL1	P	Kruppel-like zinc finger protein-1	14941	REa	
Chr.19	KRL2	P	Kruppel-like zinc finger protein-2	14942	REa	
Chr.19	MAG, GMA	P	Myelin-associated glycoprotein	15946	REa	7(Gma)
Chr.19	OTF2, OCT2	P	Lymphoid-specific octamer-binding transcription factor-2	15326	REa	
Chr.19	PGK2	C	Phosphoglycerate kinase-2 (testicular PGK)	17227	REa, A, REb	
Chr.19	G19P1, PKCSH	P	Protein kinase C substrate, H form	17706	REb	
Chr.19	U1RNP, RNPU1Z, RPU1	P	U1 small nuclear ribonucleoprotein 70kD	18074	REa	
Chr.19	RRAS	P	Oncogene RRAS	16509	REa	7(Rras)
Chr.19	TRSP	P	Opal suppressor phosphoserine tRNA	16506	REa	pseudogene on 22

In addition: 3 surface antigens, most defined by monoclonals, 1 O'Farrell protein spot, 2 'like' genes, 2 pseudogenes and 2 fragile sites (HGM10). Map (HGM9): pter--LDLR--(C3--LE)--LW--(PEPD--DM)--(SE--APOC2)--APOE--LU--qter. LDLR, distal 19p; C3, mid 19p; APOE, 19q. Order (Eiberg et al., 1983): LE--C3--DM--(SE--PEPD)--LU. Order (Breslow, 1984):FHC--C3--APOE/APOC2. APOC2. APOC1 6kb 3'to APOE. Location 19p13.1-q13.11 for LE, HH, LW, PEPD, SE, GEY, LU is estimated from collation of physical and meiotic data.

Chromosome No. 20

Location	Symbol	Status	Title	MIM #	Method	Comments	Disorder	Mouse
20pter-p12	SCG1, CHGB	P	Chromogranin B (secretogranin B)	11892	REa, A			
20pter-p12	PDYN	P	Prodynorphin	13134	REa, A			
20pter-p12	PRIP, PRNP	C	Prion protein	17664	REa, REb, A		Creutzfeld-Jacob disease, 12340 (2); Gerstmann-Straussler disease, 13744 (2)	2(Pm-p)
20p12-cen	THBD, THRM	C	Thrombomodulin	18804	REb, A			
20p11.23-qter	GHRF	C	Growth hormone releasing factor, somatocrinin	13919	REa, REb, Ch		?Isolated growth hormone deficiency due to defect in GHRF (1)	
20p11.2	AGS, AHD	L	Arteriohepatic dysplasia (Alagille syndrome)	11845	D		?Arteriohepatic dysplasia (2)	2(Itp)
20p	ITPA	C	Inosine triphosphatase-A	14752	S		[Inosine triphosphatase deficiency] (1)	
20cen-q13.1	SAHH, AHCY	C	S-adenosylhomocysteine hydrolase	18096	S, F	˜13cM from ADA		
20q11-q12	HCK	P	Hemopoietic cell kinase	14237	REb, A			
20q12-q13	SRC, ASV, SRC1	C	Protooncogene SRC, Rous sarcoma	19009	REa, A, REb			2(Src)
20q12-q13.2	TOP1	P	DNA topoisomerase I	12642	REa, A			
20q13.11	ADA	C	Adenosine deaminase	10270	S, D, REa, F		Severe combined immunodeficiency due to ADA deficiency (1); Hemolytic anemia due to ADA excess (1)	2(Ada)
Chr.20	ARVP, VP	P	Arginine vasopressin-neurophysin II	19234	REa, RE		?Diabetes insipidus, neurohypophyseal, 12570 (1)	
Chr.20	EBN, BNS	P	Seizures, benign neonatal	12120	Fd		Seizures, benign neonatal (2)	
Chr.20	DCE	P	Desmosterol-to-cholesterol enzyme	12565	S			
Chr.20	GNAS1, GNAS, GPSA	P	G-protein, stimulatory, alpha subunit (Gs-alpha)	13932	REa, H		Albright hereditary osteodystrophy (1)	2(Gs-a)
Chr.20	GSL, NGBE	P	Neuraminidase/beta-galactosidase expression (galactosialidosis)	25654	S		Galactosialidosis (1)	
Chr.20	LEUT, HTL, HLT	P	Leucine transport, high	15131	S			
Chr.20	OT	P	Oxytocin-neurophysin I	16705	RE	separated from VP by 12kb		
Chr.20	PYGB	P	Glycogen phosphorylase, brain	13855	REa, REb			2(Pygb)
Chr.20	RPN2	P	Ribophorin II	18049	REa			2(Rpn-2)

In addition: 1 antigen, 2 'like' genes and 2 fragile sites (HGM10).

Chromosome No. 21

Location	Symbol	Status	Title	MIM #	Method	Comments	Disorder	Mouse
21p12	RNR4	C	Ribosomal RNA	18045	A			
21q11.2-q21	AD1	C	Alzheimer disease-1	10430	Fd	near centromere	Alzheimer disease (2)	16(Ifrc)
21q21-qter	IFNAR	C	Antiviral protein; alpha-interferon receptor	10745	S, D			
21q21-qter	IFNBR	C	Antiviral protein; beta-interferon receptor	10746	S, D			16(Cvap)
21q21.3-q22.05	APP, AAA, CVAP	C	Amyloid beta A4 precursor protein	10476	REa, A, Fd, RE	proximal to SOD; very distal q21 or boundary with q22	?Amyloidosis, cerebroarterial, Dutch type, 10516 (1)	
21q22	S100B	P	S100 protein, beta polypeptide	17699	REa			10(S100b)
21q22.1	PGFT, PAIS, GART	C	Phosphoribosylaminoimidazole synthetase; phosphoribosylglycineamide synthetase; phosphoribosylglycineamide formyltransferase	13844	S, H	multifunctional protein: Ade(-)C, Ade(-)G, GART		16(Prgs)
21q22.1	SOD1	C	Superoxide dismutase-1, soluble	14745	S, D, Fd	mid q22.1		16(Sod-1)
21q22.1-q22.3	ETS2	C	Oncogene ETS2	16474	REa, A, Fd	proximal q22.3		16(Ets-2)
21q22.1-qter	CD18, LCAMB, LAD	C	Cell adhesion molecule, leukocyte, beta subunit	11692	S, A	common subunit for CR3, LFA1, and P150,95	Leukocyte adhesion deficiency (2)	7(Ly-15)
21q22.3	BCEI	C	Breast cancer estrogen-inducible sequence	11371	REa, A, Fd			
21q22.3	CBS	C	Cystathionine beta-synthase	23620	S, D, A	subtelomeric	Homocystinuria, B6-responsive and nonresponsive types (1)	17(Cbs)
21q22.3	COL6A1	P	Collagen VI, alpha-1 chain	12022	REa, A, REn, Fd			10(Col6a1)
21q22.3	COL6A2	P	Collagen VI, alpha-2 chain	12024	REa, A, REn, Fd			10(Col6a2)
21q22.3	CRYA1	C	Crystallin, alpha A	12358	REa, A	alpha B on another chr., ?chr. 16		17(Crya-1)
21q22.3	ERG	C	Oncogene ERG	16508	REa, Fd, A	related to ETS2; proximal to ETS2		
21q22.3	PFKL	C	Phosphofructokinase, liver type	17186	S, D		Hemolytic anemia due to phosphofructokinase deficiency (1)	17(Pfkl)
Chr.21	AABT	L	Beta-amino acids, renal transport of	10966	D			
Chr.21	BAS	L	Beta-adrenergic stimulation, response to	10967	D			
Chr.21	HMG14	P	Nonhistone chromosomal protein HMG-14	16392	REa			
Chr.21	HTOR	L	5-hydroxytryptamine oxygenase regulator	14346	D			
Chr.21	MX1, MX, IFI78	P	Myxovirus (influenza) resistance-1 (interferon induced protein p78)	14715	REa, D			16(Mx-1)
Chr.21	MX2	L	Myxovirus (influenza) resistance-2	14789	REa			16(Mx-2)

In addition: 1 surface antigen, 5 O'Farrell protein spots, 1 'like' gene and 1 pseudogene (HGM10). Band critical to Down syndrome phenotype = 21q22, probably 21q22.3. Hyperuricemia, leukemia, Alzheimer disease, and cataract of Down syndrome may be explained by the presence of specific genes on chromosome 21. Concentration of expressed genes in q22. DSCR, Down syndrome chromosome region = 21q22.3.

Chromosome No. 22

Location	Symbol	Status	Title	MIM #	Method	Comments	Disorder	Mouse
22pter-q11.2	PVALB	P	Parvalbumin	16889	REa, D	?role in DiGeorge syndrome		
22p12	RNR5	C	Ribosomal RNA	18045	A			
22q11	IDUA, IDA	P	Alpha-L-iduronidase	25280	S, D	on Ph1 chr.	Hurler syndrome (1); Mucopolysaccharidosis I (1); Hurler-Scheie syndrome (1); Scheie syndrome (1)	
22q11	CECR, CES	C	Cat eye syndrome	11547	Ch, A, D	partial tetrasomy of 22q11	Cat eye syndrome (2)	
22q11	DGCR, DGS	C	DiGeorge syndrome	18840	Ch, D	proximal to Ph1 break	DiGeorge syndrome (2)	
22q11	NAGA	C	N-acetyl-alpha-D-galactosaminidase (alpha-galactosidase B)	10417	S, Ch		Alpha-NAGA deficiency (1)	
22q11-q13	TSHR	C	Thyroid stimulating hormone receptor	27520	REa, F	tight linkage to CYP2D	Hypothyroidism, nongoitrous, due to TSH resistance (1)	
22q11.1-q11.2	GGT, GTG	C	Gamma-glutamyl transpeptidase	23195	A, S, F, RE	minor peak, q13.1	Glutathioninuria (1)	
22q11.12	IGL	C	IMMUNOGLOBULIN LAMBDA LIGHT CHAIN GENE CLUSTER		REa, A	on Ph1 chr.; order 5' to 3':cen-V-C-ter		16(Igl)
22q11.12	IGLV	C	Variable region of lambda light chains	14724	REa, A	many genes		
22q11.12	IGLJ	C	J region of lambda light chains	14723	REa, A	nine J-C duplexes		
22q11.12	IGLC	C	Constant region of lambda light chains	14722	REa, A	several genes		
22q11.2-q12.2	CRYB2	P	Crystallin, beta-B2	12363	REa, A			
22q11.2-q12.2	CRYB3	P	Crystallin, beta-B3	12363	RE			
22q11.2-q12.2	CYP2D, P450C2D	C	Cytochrome P450, family II, subfamily D	12403	F	debrisoquine 4-hydroxylase	(?Parkinsonism, susceptibility to) (1)	10(Cyp2d)
22q11.2-q13	MB	C	Myoglobin	16000	REa, Fd			
22q11.2-qter	P1	P	P1 blood group	11141	F, Fd	?linked to DIA1 and SIS		
22q11.2-qter	SGLT1, NAGT	P	Sodium-glucose transporter-1	18238	REa		Glucose/galactose malabsorption (1)	
22q11.2-qter	TCN2, TC2	C	Transcobalamin II	27535	F, S, D	linked to P1	Transcobalamin II deficiency (1)	11(Tcn-2)
22q11.21	BCR, CML, PHL	C	Chronic myeloid leukemia; breakpoint cluster region-1	15141	Ch, RE	distal to IGL; Ph1=t(9;22)(q34.1:q11.21); fusion gene with ABL in CML; cluster of 4 loci: cen-BCR2, BCR4, IGL-BCR1-BCR3-SIS	Leukemia, chronic myeloid (3)	
22q11.21	VPREB	P	Pre-B lymphocyte-specific protein	14677	REa	between BCR2 and BCR4		
22q11.21-q13.1	NF2, ACN	C	Acoustic neuroma	10100	RE, F	deletion of chr.22 markers	Acoustic neuroma (2)	
22q11.21-q13.31	ACO2	C	Aconitase, mitochondrial	10085	S, Ch	distal to Ph1 break		
22q12	ES	P	Ewing sarcoma	13345	Ch	(t11;22)(q24;q12)	Ewing sarcoma (2)	
22q12.1-q12.2	LIF	P	Leukemia inhibitory factor	15954	REa, A			
22q12.1-q13.1	NEFH	P	Neurofilament, heavy polypeptide	16223	A			
22q12.3-q13.1	SIS, PDGFB	C	Oncogene SIS (platelet derived growth factor, B chain)	19004	REa, Fd			15(Sis)
22q12.3-qter	MGCR, MGM	C	Meningioma	15610	Ch, RE, D		Meningioma (2)	

Location	Symbol		Title	MIM #	Method		Disorder	Mouse
22q13-qter	ACR, ACRS	P	Acrosin (proacrosin)	10248	REa			
22q13-qter	GLB2, PPGB	C	Beta-galactosidase-2 (GLB protective protein)	10968	S			
22q13.31-qter	ARSA	C	Arylsulfatase A	25010	S		Metachromatic leukodystrophy (1)	15(As-2)
22q13.31-qter	DIA1	C	NADH-diaphorase-1 (cytochrome b5 reductase)	25080	S, REa		Methemoglobinemia, enzymopathic (1)	15(Dia-1)
Chr.22	ADSL, ADS	P	Adenylosuccinase (adenylosuccinate lyase)	10305	S, REa	ade(-)I	Adenylosuccinase deficiency (1)	
Chr.22	ASLP	P	Argininosuccinate lyase pseudogene	see 20790	REa			
Chr.22	COMT	P	Catechol-O-methyltransferase	11679	S			
Chr.22	CST3	P	Cystatin C	10515	REa		Cerebral amyloid angiopathy (1)	
Chr.22	GNAZ	P	Guanine nucleotide-binding protein, alpha Z polypeptide	13916	REa			
Chr.22	HCF2, HC2	P	Heparin cofactor II	14236	REb		Thrombophilia due to heparin cofactor II deficiency (1)	
Chr.22	MSK41	P	Antigen MSK41 identified by monoclonal antibody E3	10726	S			

In addition: 3 surface antigens, 3 'like' genes, 4 pseudogenes and 2 fragile sites (HGM10).

5.107

Gene Map of the X Chromosome

About 160 separate expressed genetic loci have been assigned to the X chromosome; for about an equal number of loci, X-chromosomal location has been suggested but not proved. Most of these loci have been placed on the X chromosome because of pedigree patterns and other characteristics of X-linked traits in families. Some have been assigned to the X chromosome by the same methods used in autosomal mapping: interspecies somatic cell hybridization (S, REa), in situ hybridization (A), or small, microscopically visible deletions (Ch). Some methods unique to the X chromosome have corroborated X-linkage or in some instances have given the first information on X-linkage or regional mapping: lyonization (L), Ohno's law of the evolutionary conservatism of the X chromosome in mammals (H), and X-autosome translocations in females affected by X-linked recessive disorders (X/A). The 'status' information in this case refers to certainty of regional assignment.

In addition: 3 surface antigens, 7 O'Farrell protein spots, 10 'like' genes, 9 pseudogenes and 2 fragile sites (HGM10).

Location	Symbol	Status	Title	MIM #	Method	Comments	Disorder	Mouse
Xpter-p22.32	ARSC	P	Arylsulfatase C	30178	D			
Xpter-p22.32	KAL, KMS	C	Kallmann syndrome	30870	F, Fd, D	with ichthyosis in probable microdeletion syndrome	Kallmann syndrome (2)	
Xpter-p22.32	MIC2, MIC2X	C	MIC2 (monoclonal antibody 12E7)	31347	S, A, D	distal to STS		
Xpter-p22.32	STS, SSDD	C	Steroid sulfatase	30810	F, S, D	in nonlyonizing segment	Ichthyosis, X-linked (3); Placental steroid sulfatase deficiency (3)	X,Y(Sts)
Xpter-p22.32	XG	C	Xg blood group	31470	F, D	nonlyonizing		
Xpter-p22.32	CPXR	P	Chondrodysplasia punctata, X-linked recessive	30295	D		Chondrodysplasia punctata, X-linked recessive (2)	
Xpter-q21	PRPS2	P	Phosphoribosylpyrophosphate synthetase II	31186	REa			
Xp22.31	DHOF, FODH	P	Focal dermal hypoplasia	30560	Ch		Focal dermal hypoplasia (2)	
Xp22	HYP, HPDR1	C	Hereditary hypophosphatemia	30780	Fd	linked to DXS41	Hypophosphatemia, hereditary (2)	X(Hyp)
Xp22	AIC	C	Aicardi syndrome	30405	X/A, Ch		Aicardi syndrome (2)	
Xp22	GY	L	Hereditary hypophosphatemia II (gyro equivalent)	30781	H	close to hyp in mouse	?Hypophosphatemia with deafness (2)	
Xp22	HOMG, HSH, HMGX	C	Hypomagnesemia, X-linked primary	30760	X/A		Hypomagnesemia, X-linked primary (2)	
Xp22	MRX1	P	Mental retardation, X-linked nonspecific, I	30953	F, Fd	?11cM from XG	Mental retardation, X-linked nonspecific, I (2)	
Xp22	OA1	C	Ocular albinism, Nettleship-Falls type	30050	F, Fd	linked to XG	Ocular albinism, Nettleship-Falls type (2)	
Xp22	RS	C	Retinoschisis	31270	F, Fd	25cM from XG	Retinoschisis (2)	
Xp22	SEDL, SEDT	C	Spondyloepiphyseal dysplasia tarda	31340	Fd		Spondyloepiphyseal dysplasia tarda (2)	
Xp22	XLA2, IMD6	P	X-linked agammaglobulinemia, type 2	30031	Fd		Agammaglobulinemia, type 2, X-linked (2)	X(xid)
Xp22-p21	GDXY, TDFX	P	Gonadal dysgenesis, XY female type	30610	F, Ch	?same as ZFX	Gonadal dysgenesis, XY female type (2)	
Xp22.3	HYR	P	H-Y regulator, or repressor	30697	Ch	structural HY locus on chr.6, 14317		
Xp22.3-p22.1	AMGS, AMG, ALGN, AIH1	C	Amelogenin (amelogenesis imperfecta, hypoplastic type I)	30120	REa, A, Fd	?also Y	Amelogenesis imperfecta (1)	X(Amel)
Xp22.3-p21.1	POLA	C	Polymerase, DNA, alpha	31204	S			
Xp22.2-p21.2	ZFX	P	Zinc finger protein, X-linked	31498	Ch, REa			X(Xfx)
Xp22.2-p22.1	CLS	P	Coffin-Lowry syndrome	30560	Fd	distal to DMD	46,XY female (2)	
Xp22.2-p22.1	CND	L	Corneal dermoids	30473	Fd	linked to DXS43	Coffin-Lowry syndrome (2)	

Location	Symbol	Status	Title	MIM No.	Method	Comments	Disorder	Mouse
Xp22.2-p22.1	PDHA1, PHE1A	P	Pyruvate dehydrogenase, E1-alpha polypeptide	31217	REa, A		Pyruvate dehydrogenase deficiency (1)	X(Phe-1a)
Xp22.1-p21.2	GLR	P	Glycine receptor	30599	REa, Fd	distal to GK		
Xp21.3-p21.2	AHC, AHX	C	Primary adrenal hypoplasia	30020	D, Fd	2Mb distal to DMD	Adrenal hypoplasia, primary (2)	
Xp21.3-p21.2	GK	C	Glycerol kinase	30703	D, Fd	?linked to XG	Glycerol kinase deficiency (2)	
Xp21.3-p21.1	OA2	C	Ocular albinism, Forsius-Eriksson type	30060	F, D		Ocular albinism, Forsius-Eriksson type (2)	
Xp21.2	DMD, BMD	C	Duchenne muscular dystrophy; Becker muscular dystrophy	31020	X/A, Fd, D	dystrophin gene; cen-5'-3'-pter; 2Mb; ?Xp21.13	Duchenne muscular dystrophy (3); Becker muscular dystrophy (3)	X(mdx)
Xp21.2-q21.1	LUS, XS	L	Lutheran suppressor, X-linked	30905	Fd			
Xp21.2-p21.1	XK	C	Xk	31485	F, D	~500kb distal to CGD	[McLeod phenotype] (2)	
Xp21.1	CYBB, CGD	C	Chronic granulomatous disease	30640	F, D	proximal to DMD	Granulomatous disease, chronic, X-linked (3)	X(Cybb)
Xp21.1	OTC	C	Ornithine transcarbamylase	31125	L, REa, A, D	proximal to DMD, CGD	Ornithine transcarbamylase deficiency (3)	X(spf; Otc)
Xp21.1-p11.3	COD1, PCDX	P	Progressive cone dystrophy, X-linked	30402	Fd		Progressive cone dystrophy (2)	
Xp21.1-p11.1	A1S9T, A1S9	C	Temperature-sensitive mutation, mouse, complementation of (ts A1S9)	31366	S, Ch	escapes inactivation		
Xp21	GTD	L	Gonadotropin deficiency	30619	D	distal to AHC	?Gonadotropin deficiency; ?Cryptorchidism	
Xp21	RP3	C	Retinitis pigmentosa 3 (RP with metallic sheen in heterozygotes)	31261	Fd, D	probably between OTC and CGD	Retinitis pigmentosa 3 (2)	
Xp21-p11	GAPDP1	C	Glyceraldehyde-3-phosphate dehydrogenase pseudogene-1	30598	REa, A			
Xp13-p11	ARAF1, RAFA1	P	Oncogene ARAF1	31101	REa, A			X(Araf)
Xp11.4	NDP, ND	C	Norrie disease	31060	Fd, D	close to DXS7	Norrie disease (2)	
Xp11.4	PKS2	P	Oncogene PKS2	31102	REa, A			
Xp11.4	PPP	C	Properdin P deficiency	31206	Fd, REa, A		Properdin deficiency, X-linked (3)	X(Prop)
Xp11.4-p11.23	EPA, TIMP	C	Erythroid-potentiating activity (tissue inhibitor of metalloproteinases)	30537	REa, A		?Menkes disease (1)	X(Timp)
Xp11.3	RP2	C	Retinitis pigmentosa 2	31260	Fd		Retinitis pigmentosa 2 (2)	
Xp11.3-p11	IMD2, WAS	C	Wiskott-Aldrich syndrome	30100	Fd, X/A		Wiskott-Aldrich syndrome (2)	
Xp11.23	MAOA	C	Monoamine oxidase A	30985	Fd, REa, D, A, REn	t(18;X)(q11.2;p11.2) NDP, MAOA, MAOB closely linked		
Xp11.23	MAOB	C	Monoamine oxidase B	30986	REa, D, A, REn			
Xp11.2	SSRC	P	Sarcoma, synovial	31282	Ch, RE		Sarcoma, synovial (2)	
Xp11.2	IP1, IP	C	Incontinentia pigmenti-1, sporadic	30830	X/A	Xq21 = conflicting localization; both excluded by Fd	Incontinentia pigmenti (2)	
Xp11	SYN1	P	Synapsin I	31344	A			X(Syn-1)
Xp	CCT	L	Cataracts, congenital total	30220	Fd		?Cataracts, congenital total (2)	
Xp	NHS	L	Nance-Horan syndrome	30235	Fd		?Nance-Horan syndrome (2)	
Xq	A11	P	A-11 gene	30001	REc			
Xcen-q13	AR, DHTR, TFM	C	Testicular feminization (androgen receptor)	31370	S, Fd, REa, ?A		Testicular feminization (1); Reifenstein syndrome (1); Infertile male syndrome (1)	X(Tfm)

5.109

Location	Symbol	Status	Title	MIM No.	Method	Comments	Disorder	Mouse
Xq11-q12	MRX2	L	Mental retardation, X-linked nonspecific, II	30954	Fd		Mental retardation, X-linked nonspecific, II (2)	
Xq11-q13	PGK1P1	C	Phosphoglycerate kinase-1 pseudogene-1	31181	REa, A			
Xq12-q13	MNK, MK	C	Menkes disease	30940	Fc, X/A, H		Menkes disease (2)	X(Mo)
Xq12-q13	PHKA	P	Phosphorylase kinase, alpha subunit	30600	H, REa, A	probably Xq13.1 ?proximal and close to PGKA	Glycogen storage disease VIII (1)	X(Phka)
Xq12-q26	MGCN	P	Megalocornea, X-linked	30930	Fd		Megalocornea, X-linked (2)	
Xq12.2-13.1	EDA, HED	C	Anhidrotic ectodermal dysplasia	30510	X/A, H, Fd	~10cM distal to DXS1, proximal to DXYS1	Anhidrotic ectodermal dysplasia (2)	X(Ta)
Xq13	ALAS2, ASB, ANH1	P	Aminolevulinate, delta, synthase (sideroblastic anemia)	30130	Ch, REa	somatic cell chromosome rearrangement	Anemia, sideroblastic/hypochromic (3)	
Xq13	CMTX, CMT2	C	Charcot-Marie-Tooth disease, X-linked	30280	Fd	5cM to DXYS1; very close to PGK	Charcot-Marie-Tooth disease, X-linked (2)	
Xq13	FGDY, AAS	P	Aarskog-Scott syndrome (faciogenital dysplasia)	30540	X/A		Aarskog-Scott syndrome (2)	
Xq13	PGKA, PGK1	C	Phosphoglycerate kinase-1	31180	S, R, REb, Fd		Hemolytic anemia due to PGK deficiency (1)	X(Pgk-1)
Xq13	XCE, XIC	P	X chromosome controlling element (X-inactivation center)	31467	Ch, S	q13-q21: metaphase bend, or fold, at q13.3-q21.1		X(Xce)
Xq13-q13.3	SCAR	P	Single copy abundant mRNA	31288	REa			
Xq13-q21	CHR39C	P	Cholesterol repressible protein 39C	30292	REa			
Xq13-q21	SDYS, DGSX, GDS	L	Simpson dysmorphia syndrome (dysplasia-giantism syndrome, X-linked)	31287	Fd	?linked to DXYS1	Dysplasia-gigantism syndrome, X-linked (2)	
Xq13-q21	WWS	P	Wieacker-Wolff syndrome	31458	Fd	linked to DXYS1	Wieacker-Wolff syndrome (2)	
Xq13-q21.1	DFN3	C	Conductive deafness with stapes fixation	30440	Fd, D	linked to DXS51	Conductive deafness with stapes fixation (2)	
Xq13-q21.1	ZNF6	P	Zinc finger protein-6	31499	REa			
Xq13-q27	CCG1, BA2R, C1HR	C	Temperature sensitivity, mouse and hamster, complement	31365	S, H	?near HPRT		X(Ccg1)
Xq13.1-q21.1	IMD4, SCIDX	C	Severe combined immunodeficiency, X-linked	30040	Fd	linked to DXS159	SCID, X-linked (2)	
Xq21	CPX	C	Cleft palate, X-linked	30340	Fd, D		Cleft palate, X-linked (2)	
Xq21-q22	SPG2, SPPX2	P	Spastic paraplegia, X-linked, uncomplicated	31292	Fd		Spastic paraplegia, X-linked, uncomplicated (2)	
Xq21-q22	TBG	P	Thyroxine-binding globulin	31420	REa, A		[Euthyroidal hyper- and hypothyroxinemia] (1)	
Xq21.2	TCD	C	Choroideremia	30310	Fd, LD, D, A, Ch	0.0 recombination with DXYS1, DXYS12	Choroideremia (2)	
Xq21.3-q22	SBMA, KD, SMAX1	C	Kennedy spinal muscular atrophy	31320	Fd	with DXYS1, lod=3.63, theta .05	Kennedy disease (2)	
Xq21.3-q22	AGMX1, IMD1, XLA	C	X-linked agammaglobulinemia	30030	H, Fd	?Ig V-D-J recombinase	Agammaglobulinemia, X-linked (2)	
Xq22	GLA	C	Alpha-galactosidase A	30150	S, R, A, Fd		Fabry disease (3)	X(Ags)
Xq22	PLP, PMD	C	Myelin proteolipid protein	31208	REa, A, Ch, R, Fd		Pelizaeus-Merzbacher disease (3)	X(Plp(jp))
Xq22-q24	ATS, ASLN	C	Alport syndrome	30105	Fd	distal to DXS3	Alport syndrome (2)	
Xq22-q26	PRPS1	C	Phosphoribosylpyrophosphate synthetase	31185	S, R, REa		Phosphoribosylpyrophosphate synthetase-related gout (1)	
Xq22-q28	MYCL2	P	MYCL-related processed gene	31031	REa			
Xq24-q26	GLUDP1	C	Glutamate dehydrogenase pseudogene-1	30591	REa, A			

Location	Symbol	Status	Title	MIM	Method	Comments	Disorder	Mouse
Xq24-q27	IMD3, XHM	P	X-linked immunodeficiency with hyper-IgM	30823	Fd		Immunodeficiency, X-linked, with hyper-IgM (2)	
Xq25	OCRL, LOCR	C	Lowe oculocerebrorenal syndrome	30900	X/A, Fd		Lowe syndrome (2)	
Xq26	IMD5, XLP, XLPD	C	Lymphoproliferative syndrome, X-linked	30824	Fd	1cM from DXS42; no recombination with DXS37	Lymphoproliferative syndrome, X-linked (2)	
Xq26-q27	HPTX, HYPX	P	Hypoparathyroidism, X-linked	30770	Fd	?distal to F9	Hypoparathyroidism, X-linked (2)	
Xq26-q27	POF	L	Premature ovarian failure	31136	Ch		Ovarian failure, premature (2)	
Xq26-q27.2	CDR	C	Cerebellar degeneration-related autoantigen	30265	REa, A, Fd	between HPRT and F9		
Xq26-q27.2	HPRT	C	Hypoxanthine-guanine phosphoribosyltransferase	30800	S, M, C, R, REa, Fd		Lesch-Nyhan syndrome (3); HPRT-related gout (1)	X(Hprt)
Xq26.3-q27.1	ALDS, ADFN	P	Albinism-deafness syndrome	30070	Fd	5cM proximal to F9	Albinism-deafness syndrome (2)	
Xq27	DBL	P	Oncogene DBL	16503	REa, A	?same as MCF2		
Xq27	MCF2	P	Oncogene MCF2	31103	REa, A, RE	<270kb from F9 and telomeric		
Xq27-q28	IP2	P	Incontinentia pigmenti-2 (familial, male-lethal type)	30831	Fd		Incontinentia pigmenti, familial (2)	X?streaked)
Xq27.1-q27.2	HEMB, F9	C	Hemophilia B; clotting factor IX	30690	REa, A, Fd, D, X/A, RE	distal to HPRT; proximal part of Xq27	Hemophilia B (3)	X(Cf-9)
Xq27.3	FRAXA	C	Fragile site Xq27.3	30955	Ch, F, Fd		Martin-Bell syndrome (2)	
Xq27.3	IDS, MPS2, SIDS	C	Hunter syndrome (sulfoiduronate sulfatase deficiency)	30990	X/A, Fd, F		Mucopolysaccharidosis II (2)	
Xq28	ALD	C	Adrenoleukodystrophy	30010	F, Fd, D	cone pigment gene deleted in some ALD males	Adrenoleukodystrophy (2)	
Xq28	CAML1, L1CAM	P	L1 cell adhesion molecule	30884	A			
Xq28	CBBM, BCM	C	Blue-monochromatic colorblindness (blue cone monochromacy)	30370	F, Fd, RE		Colorblindness, blue-monochromatic (3)	
Xq28	CPXD, CDPX, CPX	L	Chondrodysplasia punctata, X-linked dominant	30296	H	in mouse Bpa, bare-patches, close to G6pd and mdx	Chondrodysplasia punctata, X-linked dominant (2)	X(?Bpa)
Xq28	DIR, DI1	C	Nephrogenic diabetes insipidus	30480	Fd		Diabetes insipidus, nephrogenic (2)	
Xq28	DKC	P	Dyskeratosis congenita	30500	Fd		Dyskeratosis congenita (2)	
Xq28	EMD	C	Emery-Dreifuss muscular dystrophy	31030	F, Fd, H	combined with ALD in some cases; distal to DXS15	Emery-Dreifuss muscular dystrophy (2)	
Xq28	FCPX	L	F-cell production	30543	F, Fd		?Heterocellular hereditary persistence of fetal hemoglobin (2)	
Xq28	G6PD	C	Glucose-6-phosphate dehydrogenase	30590	F, S, REb, ?A, RE	telomeric to GDX; proximal to F8, in same .29mb PFGE fragment	G6PD deficiency (3); Favism (1); Hemolytic anemia due to G6PD deficiency (1)	X(G6pd)
Xq28	GABRA3	L	Gamma-aminobutyric acid receptor, alpha-3 polypeptide	30566	H			
Xq28	GCP, CBD	C	Deutan colorblindness (green cone pigment)	30380	F, RE, A	linked to G6PD; multiple genes	Colorblindness, deutan (2)	
Xq28	GDX	P	Protein GDX	31207	RE	40kb 3' to G6PD		

				MIM				
Xq28	C	HEMA, F8C	Hemophilia A (clotting factor VIII)	30670	F, Fd, REa, A, RE	linked to G6PD, CB; proximal q28; DX13 and St14 distal	Hemophilia A (3)	X(Cf-8)
Xq28	P	MAFD2, MDX	Manic-depressive illness, X-linked	30920	F	linkage to G6PD,CB in non-Ashkenazi Jews	Manic-depressive illness, X-linked (2)	
Xq28	P	MASA	MASA syndrome	30925	Fd		MASA syndrome (2)	
Xq28	P	MRSD, CHRS	Mental retardation-skeletal dysplasia	30962	Fd		Mental retardation-skeletal dysplasia (2)	
Xq28	P	MTM1, MTMX	Myotubular myopathy, X-linked	31040	Fd		Myotubular myopathy, X-linked (2)	
Xq28	P	NBM1	Nyctalopia (night blindness)	31050	Fd	some families not linked to Xq28 markers		
Xq28	P	P3	Protein P3	31209	RE	order: G6PD-3'-(7kb)-5'-P3-3'-(0.5kb)-5'-GDX		
Xq28	C	RCP, CBP	Protan colorblindness (red cone pigment)	30390	F, RE, A	linked to G6PD; 5' to CBD	Colorblindness, protan (2)	X(Rsvp)
Xq28	C	TKCR	Goeminne TKCR syndrome	31430	X/A	distal to G6PD	Goeminne TKCR syndrome (2)	
Xq28	P	XM	Xm	31490	F	linked to DCB, PCB		

5.112

Gene Map of the Y Chromosome

1. According to the classical model (using the term of Goodfellow et al., J. Med. Genet. 22: 329-344, 1985), the Y chromosome has been thought to have several subregions: (1) an X-Y homologous, meiotic-pairing region occupying most of Yp and perhaps including a pseudoautosomal region of X-Y exchange; (2) a pericentric region containing the sex determining gene(s); and (3) a long arm heterochromatic, genetically inert region. Some recent findings support the classical model, whereas others refute it. Molecular studies indicate that Yp contains many sequences not homologous to Xp but with homology to Yq, Xq, or an autosome.

2. From the study of normal males and females, of persons with abnormal numbers of sex chromosomes, and of those carrying variant Y chromosomes, a factor (or factors) that determines the differentiation of the indifferent gonads into testes is known to be located on the Y chromosome, probably on the short arm; this may be called testis determining factor (TDF). (See *Mendelian Inheritance in Man*, 4th edition, figure 1, p.lix, 1975.) Translocation of this locus to Xp as the cause of XX males was suggested by Ferguson-Smith (1966) and found confirmation in several observations. Location of TDF near the centromere was suggested by Davis (J. Med. Genet. 18: 161-195, 1981) and others; translocation to an X in XX males may indicate a somewhat more distal location, probably the junction of pseudoautosomal and Y-specific segments of Yp. Deletion of TDF in XY females was suggested by the observation of Disteche et al. (PNAS 83: 7841, 1986). Page et al., (Cell 51: 1091-1104, 1987) cloned part or all of the TDF gene, found that some sequences were highly conserved in mammals and even birds, and showed that the nucleotide sequence of the conserved DNA on the human Y chromosome corresponds to a protein with multiple 'finger' domains. The product probably binds to nucleic acids in a sequence-specific manner and may regulate transcription. TDF is probably located in band Yp11.2. ZFY (zinc finger protein, Y-linked) is the designation approved by HGM workshop committee with ZFX being the X-linked counterpart.

3. A pseudoautosomal segment (PAS) of distal Yp and distal Xp (between which crossing-over occurs) has been suspected from microscopic observations and has been confirmed by studies using polymorphic DNA markers. (See Fig. 1 of Polani, Hum. Genet. 60: 207, 1982, and of Burgoyne, Hum. Genet. 61: 95, 1982, for a suggested homologous segment of X and Y.) DNA polymorphisms in the homologous segment of X and Y show 'pseudoautosomal' inheritance (Cooke et al., Nature 317: 687, 1985; Simmler et al., Nature 317: 692, 1985). It appears that TDF is just proximal to PAS and that in PAS there is

one, but only one, obligatory crossingover. The following order has been observed with regard to recombination with sex: Ypter--DXYS14--(50% recombination with sex)--DXYS15--(35.5%)--DXYS17--(11.5%)--MIC2--(2.7%)--TDF. These values are strictly additive indicating no double crossovers. Thus, the sequence can be written: Ypter--DXYS14--14.5%--DXYS15--24%--DXYS17--8.8%--MIC2--2.7%--nonexchanging segment of Yp containing TDF.

4. Until identification of the TDF gene product, the only specific structural gene confidently identified on the Y chromosome was that homologous to the X-linked gene for surface antigen MIC2 (Goodfellow et al., Nature 298: 346, 1983). The so-called MYC2Y (see 31347) gene is the first Y chromosome gene for which a gene product was identified, the first gene proved to have pseudoautosomal inheritance and the first gene proved to escape lyonization. The locus is in the Ypter-Yq11.2 segment. MIC2X (see 31347) is a homologous locus at Xp22.32 (Buckle et al., Nature 317: 739, 1985).

5. Histocompatibility antigens determined by the Y chromosome were first found in the mouse (Eichwald, E. J. and Silmser, C. R., Transplant Bull. 2: 148-149, 1955; see review by Gasser, D. L. and Silver, W. K., Adv. Immun. 15: 215-217, 1972) and later in the rat, guinea pig, and many other species. Their existence in man was first shown by the fact that mouse antisera react with human male lymphocytes but not with female lymphocytes (Wachtel et al., PNAS 71: 1215-1218, 1974). The possibility that the locus that determines heterogametic sex determination and that for the H-Y antigen are one and the same was suggested by Wachtel et al. (New Eng. J. Med. 293: 1070-1072, 1975). Subsequent evidence ruled out this possibility. (At the 1986 Cold Spring Harbor Symposium, 6 speakers discussed the Y chromosome at length, but H-Y was not once mentioned.) From the study of XX males and XY females, it can be concluded that the H-Y determinant on the Y (whatever its nature) and TDF are separate entities and not closely situated (Simpson, E., Cell 44: 813, 1986; Simpson et al., Nature 326: 876-878, 1987). H-Y (structural gene or regulator) is coded near the centromere, possibly on Yq.

6. The existence of factors controlling spermatogenesis on the nonfluorescent part of the long arm of Y (distal part of Yq11) was suggested by study of 6 men with deletion of this segment and azoospermia (Tiepolo, L. and Zuffardi, O., Hum. Genet. 34: 110-124, 1976). This has been called azoospermia third factor (symbolized Sp-3), or more recently (HGM9), AZF (for azoospermia factor). This might be identical to H-Y because it maps to the same region. In mice, H-Y or a closely linked gene has been implicated in spermatogenesis (Burgoyne et al., Nature 320: 170, 1986).

7. That one or more genes concerned with stature are on the Y chromosome is suggested by the comparative heights of the XX, XY and XYY genotypes; that the effect of the Y chromosome on stature is mediated through a mechanism other than androgen is suggested by the tall stature of persons with XY gonadal dysgenesis (30610). See also the argument, from XO and XXY cases, that genes determining slower maturation must be on the Y (Tanner et al., Lancet II: 141-144, 1959). The postulated locus is symbolized STA (for 'stature'). Yamada et al. (Hum. Genet. 58:268-270, 1981) found a correlation between the length of heterochromatic band Yq12 and height. The STA locus may be identical to the TSY (or GCY) locus (see later).

8. Alvesalo and de la Chapelle (Ann. Hum. Genet. 43: 97-102, 1979; HGM5, Edinburgh, 1979) suggested, on the basis of tooth size in males of various Y chromosome constitutions, that a Y-chromosomal gene controlling tooth size is independent of the testis-determining gene and is carried by Yq11 (symbolized TS for 'tooth size,' or, more recently, in HGM8, GCY for 'growth control Y'. See Alvesalo and Portin, Am. J. Hum. Genet. 32: 955-959, 1980; Alvesalo and de la Chapelle, Ann. Hum. Genet. 54: 49-54, 1981. The dental growth factors are thought to coincide with determinants of stature.

9. An argininosuccinate synthetase pseudogene is on the Y (Daiger et al., Nature 298: 682, 1982), as is also an actin pseudogene (Heilig et al., EMBO J. 3:1803, 1984).

10. The Howard Hughes Medical Institute's Human Gene Mapping Library (HGML) has cataloged 103 seemingly low copy number anonymous DNA segments mapped exclusively to specific regions of the Y chromosome, as well as at least 60 DNA segments that map to both the Y and the X. Some segments map to both the Y and an autosome, and some repetitive DNA segments map to the Y exclusively or to both the Y and the X.

11. Repetitive sequences located exclusively or predominantly to the Y chromosome (e.g., Kunkel, Smith and Boyer, Biochemistry 18: 3343-3353, 1979) map to the heterochromatic portion of Yq and are presumably genetically inert because persons lacking these are phenotypically normal and normally fertile.

Unassigned Linkage Groups (ULGs)

Linked autosomal loci for which assignment to a specific chromosome has not yet been achieved. The tightness of the linkage is stated in general terms defined as follows:

v = very close; recombination less than 2% (NR = no recombinant observed)
c = close; recombination 2-6%
m = medium; recombination 6-22%
l = loose; recombination more than 22%

? = lods 2.0-3.0

1. Phenylthiocarbamide taste, PTC (17120) (m, F)
   Kell blood group, K, KEL (11090)
   YT blood group (11210) (l,F)
   Hyperreflexia, HRX (14529)
       (Spence: Hum. Genet. 67: 183-186, 1984. All published data on PTC vs. KEL: lod = 8.94 at theta 0.14, but evidence of heterogeneity.)

2. Ii blood group, II (11080)
   Congenital cataract, CCAT (21250) (?c, F)

3. Epidermolysis bullosa progressiva, EBR3 (22650)
   Hypoacusis (HOAC; a recessive partial deafness) (22070) (?c, F)
   ?Red hair, RHA (26630) - linked to MNS (11130) on chr.4
   ?Ataxia-deafness-retardation syndrome (20885)

4. Marinesco-Sjogren syndrome, MSS (24880)
   Hypergonadotropic hypogonadism, HRGHG (23832) (m, F)
       (Lod score more than +30; however, HRGHG probably pleiotropic effect of MSS.)

5. Xeroderma pigmentosum, group D, XPD (27873)
   Trichothiodystrophy (24217) (?c,F)

6. Blepharophimosis, epicanthus inversus, ptosis (11010)
   Premature ovarian failure (17644)
   ?contiguous gene syndrome

For a collection and collation of published linkage data, see the following: Keats, B. J. B., Morton, N. E., Rao, C. C. and Williams, W. R.: *A Source Book for Linkage in Man*. Baltimore: Johns Hopkins University Press, 1979; Keats, B. J. B.: *Linkage and Chromosome Mapping in Man*. Honolulu: University Press of Hawaii, 1981.

The Morbid Anatomy of the Human Genome (see Figure 2, A-D)

The following is an alphabetized list of disorders for which the mutation has been mapped to a specific site. In some instances, the disease phenotype was mapped by finding of linkage to a marker (e.g., Huntington disease) or by finding a specific chromosome change (e.g., Prader-Willi syndrome). In other instances, the disorder was located by virtue of mapping the 'wildtype' gene, combined with the presumption or proof that the given disorder represents a mutation in that structural gene (e.g., Tay-Sachs disease and hexosaminidase A). Cancers, which represent a form of somatic cell genetic disease, are, in selected instances, included here when a specific chromosomal change and/or relation to an oncogene indicates a specific localization of determinant(s). Malformations which are related to specific localized chromosomal change and again represent a form of somatic cell genetic disease are also included (e.g., Langer-Giedion syndrome). Certain 'nondiseases,' mainly genetic variations that lead to apparently abnormal laboratory test values (e.g., dysalbuminemic euthyroidal hyperthyroxinemia) are included in **brackets**. For some loci, multiple quite dissimilar disorders are caused by different mutations in the same gene; examples of allelic disorders are placed in **boxes** in the figure. **Braces** indicate examples of mutations which lead to universal susceptibility to a specific infection (diphtheria, polio), to frequent resistance to a specific infection (vivax malaria), as well as some other susceptibilities. **?** before the disease name is the equivalent of L (in limbo). The number in parentheses after the name of each disorder indicates whether the mutation was positioned by mapping the 'wildtype' gene (1), by mapping the disease phenotype itself (2), or by both approaches (3).

Disorder	Symbol	MIM #	Location
46,XY female (2)	ZFX	31498	Xp22.2-p21.2
Aarskog-Scott syndrome (2)	FGDY, AAS	30540	Xq13
?Abetalipoproteinemia (1)	APOB	10773	2p24
[Acanthocytosis, 1 form] (1)	EPB3, EMPB3	10927	17q21-q22
Acanthosis nigricans, insulin-resistant (1)	INSR	14767	19p13.3-p13.2
Acatalasemia (1)	CAT	11550	11p13
Acetyl-CoA carboxylase deficiency (1)	ACC	20035	17q21
Acid-maltase deficiency, adult (1)	GAA	23230	17q23
Acoustic neuroma (2)	NF2, ACN	10100	22q11.21-q13.1
?Acrokeratoelastoidosis (2)	AKE	10185	2p
ACTH deficiency (1)	POMC	17683	2p25
Acute alcohol intolerance (1)	ALDH2	10065	12q24.2
?Acyl-CoA dehydrogenase, long chain deficiency of (1)	HADH, ACADL, LCAD	14345	Chr.7
Acyl-CoA dehydrogenase, medium chain, deficiency of (1)	ACADM, MCAD	20145	1p31
Acyl-CoA dehydrogenase, short chain, deficiency of (1)	ACADS	20147	12q22-qter
Adenylosuccinase deficiency (1)	ADSL, ADS	10305	Chr.22
Adrenal hyperplasia, congenital, due to 11-beta-hydroxylase deficiency (1)	CYP11B1, P450C11	20201	8q21
Adrenal hyperplasia, congenital, due to 21-hydroxylase deficiency (3)	CYP21, CA21H, CAH1	20191	6p21.3
Adrenal hyperplasia II (1)	HSDB	20181	1p13
Adrenal hyperplasia V (1)	CYP17, P450C17	20211	Chr.10
Adrenal hypoplasia, primary (2)	AHC, AHX	30020	Xp21.3-p21.2
Adrenocortical carcinoma (2)	ADCR, ADCC	20230	11p15.5
Adrenoleukodystrophy (2)	ALD	30010	Xq28
Agammaglobulinemia, type 2, X-linked (2)	XLA2, IMD6	30031	Xp22
Agammaglobulinemia, X-linked (2)	AGMX1, IMD1, XLA	30030	Xq21.3-q22
Aicardi syndrome (2)	AIC	30405	Xp22
Albinism (1)	TYR, ATN	20310	11q14-q21
Albinism-deafness syndrome (2)	ALDS, ADFN	30070	Xq26.3-q27.1
Albright hereditary osteodystrophy (1)	GNAS1, GNAS, GPSA	13932	Chr.20
?Aldolase A deficiency (1)	ALDOA,ALDA	10385	16q22-q24
?Alpha-1-antichymotrypsin deficiency (1)	AACT	10728	14q32.1
Alpha-NAGA deficiency (1)	NAGA	10417	22q11
Alport syndrome (2)	ATS, ASLN	30105	Xq22-q24
Alzheimer disease (2)	AD1	10430	21q11.2-q21

Amelogenesis imperfecta (1)	AMGS, AMG, ALGN, AIH1	30120	Xp22.3-p22.1
Amyloid neuropathy, familial, several allelic types (3)	PALB, TTR, TBPA	17630	18q11.2-q12.1
?Amyloidosis, cerebroarterial, Dutch type, 10516 (1)	APP, AAA, CVAP	10476	21q21.3-q22.05
Amyloidosis, Iowa form 10510 (1)	APOA1	10768	11q23-qter
{?Amyloidosis, secondary, susceptibility to} (2)	APCS, SAP	10477	1q12-q23
Anal canal carcinoma (2)	ANC	10558	11q22-qter
Analbuminemia (1)	ALB	10360	4q11-q13
Anemia, sideroblastic/hypochromic (3)	ALAS2, ASB, ANH1	30130	Xq13
Angelman syndrome (2)	ANCR, AGMS	23440	15q11-q13
Angioedema, hereditary (1)	C1NH, CII, HANE	10610	11q11-q13
Anhidrotic ectodermal dysplasia (2)	EDA, HED	30510	Xq12.2-13.1
Aniridia of WAGR syndrome (2)	TCR1, WAGR	19407	11p13
Aniridia-1 (2)	AN1	10620	2p25
Aniridia-2 (2)	AN2	10621	11p13
?ANLL-M2 (1)	MRS	15956	8q21.1-q23
Anterior segment mesenchymal dysgenesis (2)	ASMD	10725	4q28-q31
Antithrombin III deficiency (3)	AT3	10730	1q23.1-q23.9
ApoA-I and apoC-III deficiency, combined (1)	APOA1	10768	11q23-qter
Apolipoprotein B-100, defective (1)	APOB	10773	2p24
[Apolipoprotein H deficiency] (1)	APOH	13870	Chr.17
Argininemia (1)	ARG1	20780	6q23
Argininosuccinicaciduria (1)	ASL	20790	7cen-q11.2
?Arteriohepatic dysplasia (2)	AGS, AHD	11845	20p11.2
Aspartylglucosaminuria (1)	AGA	20840	4q21-qter
Ataxia-telangiectasia (2)	AT1	20890	11q22-q23
Atransferrinemia (1)	TF	19000	3q21
Atrial septal defect, secundum type (2)	ASD2	10880	6p21.3
?Batten disease, 1 form, 20420 (1)	CTSH	11682	15q24-q25
Batten disease (2)	NCL, BD	20420	16q22
Becker muscular dystrophy (3)	DMD, BMD	31020	Xp21.2
Beckwith-Wiedemann syndrome (2)	BWCR, BWS, WBS	13065	11pter-p15.4
Branchiootic syndrome (2)	BOS	11365	8q13.3
Breast cancer, ductal (2)	BCDS2	21142	1p36
Breast cancer, ductal (2)	BRCD, DBC, BCDS1	21141	Chr.13
?Brody myopathy (1)	ATP2B	10874	Chr.12
Burkitt lymphoma (3)	MYC	19008	8q24.1
?C1q deficiency (1)	C1QA	12055	1p
?C1q deficiency (1)	C1QB	12057	1p
C1r/C1s deficiency, combined (1)	C1R	21695	12p13
C1r/C1s deficiency, combined (1)	C1S	12058	12p13
C2 deficiency (3)	C2	21700	6p21.3
C3 deficiency (1)	C3	12070	19p13.3-p13.2
C3b inactivator deficiency (1)	IF	21703	4q25
C4 deficiency (3)	C4A, C4S	12081	6p21.3
C4 deficiency (3)	C4B, C4F	12082	6p21.3
C5 deficiency (1)	C5	12090	9q34.1
C6 deficiency (1)	C6	21705	Chr.5
C7 deficiency (1)	C7	21707	Chr.5
C8 deficiency, type I (2)	C8A	12095	1p22

Disorder	MIM	Gene/Locus	Location
C8 deficiency, type II (2)	12096	C8B	1p22
C9 deficiency (1)	12094	C9	5p13
?Campomelic dysplasia with sex reversal (2)	21197	CMD1	5q33.1?
?Campomelic dysplasia with sex reversal (2)	21197	CMD1	8q21.4?
Carbamoylphosphate synthetase I deficiency (1)	23730	CPS1	2p
Cat eye syndrome (2)	11547	CECR, CES	22q11
?Cataract, anterior polar (2)	11565	CAP	14q24
?Cataract, anterior polar (2)	11565	CAP	2p25
Cataract, Coppock-like (3)	12366	CRYG1	2q33-q35
Cataract, Mamer type (2)	11680	CPM, CAM	16q22.1
Cataract, zonular pulverulent (2)	11620	CAE	1q2
Cerebral amyloid angiopathy (1)	10515	CST3	Chr.22
CETP deficiency (1)	11847	CETP	16q21
Charcot-Marie-Tooth disease, slow nerve conduction type (2)	11820	CMT1	1p22-q23
Charcot-Marie-Tooth disease, X-linked (2)	30280	CMTX, CMT2	Xq13
Cholesterol ester storage disease (1)	27800	LIPA	10q24-q25
Chondrodysplasia punctata, X-linked dominant (2)	30296	CPXD, CDPX, CPX	Xq28
Chondrodysplasia punctata, X-linked recessive (2)	30295	CPXR	Xpter-p22.32
Choroideremia (2)	30310	TCD	Xq21.2
Chronic granulomatous disease due to deficiency of NCF-1 (1)	23370	NCF1	Chr.10
Chronic granulomatous disease due to deficiency of NCF-2 (1)	23371	NCF2	1cen-q32
Citrullinemia (1)	21570	ASS	9q34
Cleft palate, X-linked (2)	30340	CPX	Xq21
CMO II deficiency (1)	20201	CYP11B1, P450C11	8q21
Coffin-Lowry syndrome (2)	30360	CLS	Xp22.2-p22.1
?Coloboma of iris (2)	12020	COI	2pter-p25.1
Colorblindness, blue-monochromatic (3)	30370	CBBM, BCM	Xq28
Colorblindness, deutan (2)	30380	GCP, CBD	Xq28
Colorblindness, protan (2)	30390	RCP, CBP	Xq28
Colorblindness, tritan (2)	19090	BCP, CBT	7q22-qter
Colorectal adenoma (1)	19007	KRAS2, RASK2	12p12.1
?Colorectal cancer (1)	19007	KRAS2, RASK2	12p12.1
Colorectal cancer, 11450 (2)	12047	CRC18	18q23.3
Colorectal carcinoma, 11450 (1)	12046	CRCR2, CRC17	17p12
		TP53	17p13.105-p12
Combined C6/C7 deficiency (1)	19117	C6	Chr.5
?Combined variable hypogammaglobulinemia (1)	21705	IGH	14q32.33
Conductive deafness with stapes fixation (2)	30440	DFN3	Xq13-q21.1
Coproporphyria (1)	12130	CPRO,CPO	Chr.9
[Coronary artery disease, susceptibility to] (1)	15220	LPA	6q27
Cortisol resistance (1)	13804	GRL	5q31
CR1 deficiency (1)	12062	CR1, C3BR	1q32
Craniosynostosis (2)	12310	CRS, CSO	7p21.3-p21.2
[Creatine kinase, brain type, ectopic expression of] (2)	12327	CKBE	14q32
Creutzfeld-Jacob disease, 12340 (2)	17664	PRIP, PRNP	20pter-p12
?Cryptorchidism	30619	GTD	Xp21
[Cystathioninuria] (1)	21950	CTH	Chr.16
Cystic fibrosis (2)	21970	CF	7q31.3-q32
Dentinogenesis imperfecta-1 (2)	12549	DGI1	4q13-q21
Diabetes insipidus, nephrogenic (2)	30480	DIR, DI1	Xq28

Disorder	Gene	MIM	Location
?Diabetes insipidus, neurohypophyseal, 12570 (1)	ARVP, VP	19234	Chr.20
?Diabetes mellitus, insulin dependent (2)	IDDM	22210	6p21.3
Diabetes mellitus, rare form (1)	INS	17673	11p15.5
?Diaphragmatic hernia (2)	DH	22240	1q32.3-q42.3
?Diastrophic dysplasia (2)	DD	22260	Chr.18
DiGeorge syndrome (2)	DGCR, DGS	18840	22q11
{Diphtheria, susceptibility to} (1)	DTS	12615	5q23
?Dubin-Johnson syndrome (2)	DJS	23750	13q34
Duchenne muscular dystrophy (3)	DMD, BMD	31020	Xp21.2
[Dysalbuminemic hyperthyroxinemia] (1)	ALB	10360	4q11-q13
[Dysalbuminemic hyperzincemia] (1)	ALB	10360	4q11-q13
Dysfibrinogenemia, alpha types (1)	FGA	13482	4q26-q28
Dysfibrinogenemia, beta types (1)	FGB	13483	4q26-q28
Dysfibrinogenemia, gamma types (1)	FGG	13485	4q26-q28
Dyskeratosis congenita (2)	DKC	30500	Xq28
Dyslexia-1 (2)	DLX1	12770	15q11
Dysplasia-gigantism syndrome, X-linked (2)	SDYS, DGSX, GDS	31287	Xq13-q21
Dysplasminogenemic thrombophilia (1)	PLG	17335	6q26-q27
Dysprothrombinemia (1)	F2	17693	11p11-q12
[Dystransthyretinemic hyperthyroxinemia](1)	PALB, TTR, TBPA	17630	18q11.2-q12.1
Ehlers-Danlos syndrome, type IV (3)	COL3A1	12018	2q31
Ehlers-Danlos syndrome, type VIIA1 (3)	COL1A1	12015	17q21.31-q22.05
Ehlers-Danlos syndrome, type VIIA2 (3)	COL1A2	12016	7q21.3-q22.1
?Ehlers-Danlos syndrome, type X (3)	FN1	13560	2q34-q36
Elliptocytosis-1 (3)	EL1	13050	1p36.2-p34
?Elliptocytosis-2 (2)	EL2	13060	1q2
Elliptocytosis-2 (2)	SPTA1	18286	1q21
Elliptocytosis-3 (2)	SPTB, SPH1	18287	14q22-q23.2
Emery-Dreifuss muscular dystrophy (2)	EMD	31030	Xq28
Emphysema-cirrhosis (1)	PI, AAT	10740	14q32.1
Enolase deficiency (1)	ENO1, PPH	17243	1pter-p36.13
Eosinophilic myeloproliferative disorder (2)	MPE, EMP	13144	12p13
Epidermolysis bullosa dystrophica, recessive (3)	CLG, EBR1, CLGN	22660	11q11-q23
Epidermolysis bullosa, Ogna type (2)	EBS1	13195	8q24
Epidermolysis bullosa progressiva (2)	EBR3	22650	ULG
?Epilepsy, juvenile myoclonic (2)	EJM, JME	25477	6p
?Erythremia (1)	EPO	13317	7q21-q22
Erythremias, alpha- (1)	HBA1	14180	16pter-p13.3
Erythremias, beta- (1)	HBB	14190	11p15.5
Erythroblastosis fetalis (1)	RH		1p36.2-p34
Erythrokeratodermia variabilis (2)	EKV	13320	1p36.2-p34
[Euthyroidal hyper- and hypothyroxinemia] (1)	TBG	31420	Xq21-q22
Ewing sarcoma (2)	ES	13345	22q12
Exertional myoglobinuria due to deficiency of LDH-A (1)	LDHA	15000	11p15-p14
Fabry disease (3)	GLA	30150	Xq22
Factor V deficiency (1)	F5	22740	1q23
Factor VII deficiency (1)	F7	22750	13q34
Factor X deficiency (1)	F10	22760	13q34
Factor XI deficiency (1)	F11	26490	Chr.4
Factor XII deficiency (1)	F12, HAF	23400	5q33-qter

Disease	Gene	MIM	Location
Factor XIII, A component deficiency (1)	F13A1	13457	6p25-p24
?Familial colorectal cancer (2)	APC, GS, FPC	17510	5q22-q23
?Familial Mediterranean fever, 24910 (1)	ITIL, ITI, HCP	17687	9q32-q33
?Fanconi anemia (1)	PPOL, PARP	17387	1q41-q42
Favism (1)	G6PD	30590	Xq28
?Fetal alcohol syndrome (1)	ALDH2	10065	12q24.2
?Fetal hydantoin syndrome (1)	EPHX, EPOX	13281	1p22-qter
Focal dermal hypoplasia (2)	DHOF, FODH	30560	Xp22.31
Friedreich ataxia (2)	FRDA, FAT	22930	9q13-q21.1
Fructose intolerance (1)	ALDOB	22960	9q22
Fucosidosis (1)	FUCA1,FUCA	23000	1p34
?Fumarase deficiency (1)	FH	13685	1q42.1
G6PD deficiency (3)	G6PD	30590	Xq28
Galactokinase deficiency (1)	GALK	23020	17q21-q22
Galactose epimerase deficiency (1)	GALE	23035	1pter-p32
Galactosemia (1)	GALT	23040	9p13
Galactosialidosis (1)	GSL, NGBE	25654	Chr.20
Gardner syndrome (2)	APC, GS, FPC	17510	5q22-q23
Gaucher disease (1)	GBA	23080	1q21
Gerstmann-Straussler disease, 13744 (2)	PRIP, PRNP	17664	20pter-p12
Glanzmann thrombasthenia (1)	GP2B	27380	17q21.32
Glaucoma, congenital (2)	GLAU1	23130	Chr.11
Glioblastoma multiforme	GBM	13780	10p12-q23.2
Glucose/galactose malabsorption (1)	SGLT1, NAGT	18238	22q11.2-qter
Glutaricaciduria, type II (1)	ETFA, GA2	23168	15q23-q25
Glutathioninuria (1)	GGT, GTG	23195	22q11.1-q11.2
Glycerol kinase deficiency (2)	GK	30703	Xp21.3-p21.2
Glycogen storage disease, type VII (1)	PFKM	23280	1cen-q32
Glycogen storage disease VIII (1)	PHKA	30600	Xq12-q13
[Glyoxalase II deficiency] (1)	HAGH,GLO2	13876	16p13
GM1-gangliosidosis (1)	GLB1	23050	3p21-p14.2
GM2-gangliosidosis, AB variant (1)	GM2A	27275	Chr.5
GM2-gangliosidosis, juvenile, adult (1)	HEXA, TSD	27280	15q22-q25.1
Goeminne TKCR syndrome (2)	TKCR	31430	Xq28
?Goiter, adolescent multinodular, 13880 (1)	TG	18845	8q24.2-q24.3
?Goldenhar syndrome (2)	GHS	14140	7p
Gonadal dysgenesis, XY female type (2)	GDXY, TDFX	30610	Xp22-p21
Gonadoblastoma (2)	TCR1, WAGR	19407	11p13
?Gonadotropin deficiency	GTD	30619	Xp21
Granulomatous disease, chronic, X-linked (3)	CYBB, CGD	30640	Xp21.1
Greig craniopolysyndactyly syndrome (2)	GCPS	17570	7p13
?Gynecomastia, familial, due to increased aromatase activity (1)	CYP19, ARO	10791	15q21.1
Gyrate atrophy of choroid and retina (1)	OAT	25887	10q26
Harderoporphyria (1)	CPRO,CPO	12130	Chr.9
Hb H mental retardation syndrome (2)	HBHR	14175	16pter-p13.3
Heinz body anemias, alpha- (1)	HBA1	14180	16pter-p13.3
Heinz body anemias, beta- (1)	HBB	14190	11p15.5
Hemochromatosis (2)	HFE	23520	6p21.3
Hemodialysis-related amyloidosis (1)	B2M	10970	15q21-q22
Hemolytic anemia due to ADA excess (1)	ADA	10270	20q13.11

Disease	Gene	MIM	Location
Hemolytic anemia due to adenylate kinase deficiency (1)	AK1	10300	9q34.1
Hemolytic anemia due to bisphosphoglycerate mutase deficiency (1)	BPGM	22280	7q22-q34
Hemolytic anemia due to G6PD deficiency (1)	G6PD	30590	Xq28
Hemolytic anemia due to glucosephosphate isomerase deficiency (1)	GPI	17240	19cen-q12
Hemolytic anemia due to glutathione peroxidase deficiency (1)	GPX1	13832	3p13-q12
Hemolytic anemia due to glutathione reductase deficiency (1)	GSR	13830	8p21.1
Hemolytic anemia due to hexokinase deficiency (1)	HK1	14260	10p11.2
Hemolytic anemia due to PGK deficiency (1)	PGKA,PGK1	31180	Xq13
Hemolytic anemia due to phosphofructokinase deficiency (1)	PFKL	17186	21q22.3
Hemolytic anemia due to triosephosphate isomerase deficiency (1)	TPI1,TPI	19045	12p13
Hemophilia A (3)	HEMA, F8C	30670	Xq28
Hemophilia B (3)	HEMB, F9	30690	Xq27.1-q27.2
Hemorrhagic diathesis due to 'antithrombin' Pittsburgh (1)	PI, AAT	10740	14q32.1
Hemorrhagic diathesis due to PAI1 deficiency (1)	PLANH1, PAI1	17336	7q21.3-q22
?Hepatic lipase deficiency (1)	LIPC, LIPH, HL, HTGL	15167	15q21-q23
Hepatocellular carcinoma (2)	HVBS6, HCC2	14238	4q32.1
[Hereditary persistence of alpha-fetoprotein] (3)	HPAFP	10414	4q11-q13
?Hereditary persistence of fetal hemoglobin (3)	HBGR	14227	11p15.5
Hers disease, or glycogen storage disease VI (1)	PYGL, PPYL	23270	Chr.14
Heterocellular hereditary persistence of fetal hemoglobin (2)	HPFH, FCP, HHPF	14247	11p15
?Heterocellular hereditary persistence of fetal hemoglobin (2)	FCPX	30543	Xq28
[Hex A pseudodeficiency] (1)	HEXA, TSD	27280	15q22-q25.1
?HHH syndrome (2)	HHHS	23897	13q34
[Histidemia] (1)	HSTD	23580	Chr.12
?Holt-Oram syndrome (2)	HOS	14290	14q23-q24.2
Homocystinuria, B6-responsive and nonresponsive types (1)	CBS	23620	21q22.3
HPFH, deletion type (1)	HBB	14190	11p15.5
HPFH, nondeletion type A (1)	HBG1	14220	11p15.5
HPFH, nondeletion type G (1)	HBG2	14225	11p15.5
HPRT-related gout (1)	HPRT	30800	Xq26-q27.2
Huntington disease (2)	HD	14310	4p16.3
Hurler syndrome (1)	IDUA, IDA	25280	22q11
Hurler-Scheie syndrome (1)	IDUA, IDA	25280	22q11
Hydrops fetalis (1)	GPI	17240	19cen-q12
Hyperbetalipoproteinemia (1)	APOB	10773	2p24
Hypercholesterolemia, familial (3)	LDLR,FHC	14389	19p13.2-p13.1
?Hyperglycinemia, isolated nonketotic, type I (2)	NKH1	23830	9p22
?Hyperimmunoglobulin G1 syndrome (2)	IGHR	14412	14q32.33
?Hyperleucinemia-isoleucinemia or hypervalinemia (1)	BCT1	11352	12pter-q12
Hyperlipoproteinemia I (1)	LPL, LIPD	23860	8p22
Hyperlipoproteinemia, type Ib (1)	APOC2	20775	19q13.1
Hyperlipoproteinemia, type III (1)	APOE	20776	19q13.1
[Hyperphenylalaninemia, mild] (3)	PAH, PKU1	26160	12q24.1
[?Hyperproglucagonemia] (1)	GCG	13803	2q36-q37
Hyperproinsulinemia, familial (1)	INS	17673	11p15.5
Hypertriglyceridemia, 1 form (1)	APOA1	10768	11q23-qter
?Hypervalinemia or hyperleucinemia-isoleucinemia (1)	BCT2	11353	Chr.19
Hypoalphalipoproteinemia (1)	APOA1	10768	11q23-qter
Hypobetalipoproteinemia (1)	APOB	10773	2p24
[Hypoceruloplasminemia, hereditary] (1)	CP	11770	3q21-q24

Hypofibrinogenemia, gamma types (1)	FGG	13485	4q26-q28
?Hypogonadotropic hypogonadism due to GNRH deficiency , 22720 (1)	GNRH,LHRH	15276	8p21-q11.2
Hypomagnesemia, X-linked primary (2)	HOMG, HSH, HMGX	30760	Xp22
?Hypomelanosis of Ito (2)	ITO	14615	15q11-q13
?Hypomelanosis of Ito (2)	ITO	14615	9q33-qter
?Hypoparathyroidism, familial (1)	PTH	16845	11p15
Hypoparathyroidism, X-linked (2)	HPTX, HYPX	30770	Xq26-q27
?Hypophosphatasia, adult 14630 (1)	ALPL, HOPS	17176	1p36.2-p34
Hypophosphatasia, infantile 24150 (3)	ALPL, HOPS	17176	1p36.2-p34
Hypophosphatemia, hereditary (2)	HYP, HPDR1	30780	Xp22
?Hypophosphatemia with deafness (2)	GY	30781	Xp22
Hypoprothrombinemia (1)	F2	17693	11p11-q12
Hypothyroidism, hereditary congenital, 1 or more types (1)	TG	18845	8q24.2-q24.3
Hypothyroidism, nongoitrous (1)	TSHB	18854	1p22
Hypothyroidism, nongoitrous, due to TSH resistance (1)	TSHR	27520	22q11-q13
?Ichthyosis vulgaris, 14670 (1)	FLGR	13594	1q21
Ichthyosis, X-linked (3)	STS,SSDD	30810	Xpter-p22.32
Immunodeficiency, X-linked, with hyper-IgM (2)	IMD3, XHM	30823	Xq24-q27
Incontinentia pigmenti (2)	IP1, IP	30830	Xp11.2
Incontinentia pigmenti, familial (2)	IP2	30831	Xq27-q28
Infertile male syndrome (1)	AR, DHTR, TFM	31370	Xcen-q13
[Inosine triphosphatase deficiency] (1)	ITPA	14752	20p
Interferon, alpha, deficiency (1)	IFNA, IFL,IFA	14766	9p21
Interferon, immune, deficiency (1)	IFNG, IFL, IFG	14757	12q24.1
?Isolated growth hormone deficiency due to defect in GHRF (1)	GHRF	13919	20p11.23-qter
Isolated growth hormone deficiency, Illig type with absent GH and Kowarski type with bioinactive GH (3)	GH1, GHN	13925	17q22-q24
Isovalericacidemia (1)	IVD	24350	15q14-q15
Kallmann syndrome (2)	KAL, KMS	30870	Xpter-p22.32
Kennedy disease (2)	SBMA, KD, SMAX1	31320	Xq21.3-q22
?Kniest dysplasia (1)	COL2A1	12014	12q13.1-q13.3
?Kostmann agranulocytosis (2)	NDF	20270	6p21.3
Krabbe disease (1)	GALC	24520	Chr.17
?Lactase deficiency, adult, 22310 (1)	LTS, LAC, LPH	22300	Chr.2
Lactase deficiency, congenital (1)	LTS, LAC, LPH	22300	Chr.2
Langer-Giedion syndrome (2)	LGCR, LGS, TRPS2	15023	8q24.11-q24.13
Langer-Saldino achondrogenesis- hypochondrogenesis (1)	COL2A1	12014	12q13.1-q13.3
Laron dwarfism, 26250 (1)	GHR	13917	5p13-p12
?Laryngeal adductor paralysis (2)	LAP	15027	6p21.3-p21.2
{Lead poisoning, susceptibility to} (1)	ALAD	12527	9q34
?Leiomyomata, multiple hereditary cutaneous (2)	MCL	15080	18p11.32
?Leprechaunism (1)	INSR	14767	19p13.3-p13.2
Lesch-Nyhan syndrome (3)	HPRT	30800	Xq26-q27.2
?Letterer-Siwe disease (2)	LSD	24640	13q14-q31
?Leukemia, acute lymphocytic, with 4/11 translocation (3)	IGCJ, JCH	14779	4q21
Leukemia, acute T-cell (2)	TCL2	15139	11p13
Leukemia, chronic myeloid (3)	BCR, CML, PHL	15141	22q11.21
Leukemia, chronic myeloid (3)	ABL	18998	9q34.1
Leukemia, T-cell acute lymphocytic (2)	TCL3	18677	10q24
Leukemia/lymphoma, B-cell, 1 (2)	BCL1	15140	11q13.3

5.121

Disorder	Symbol	MIM	Location
Leukemia/lymphoma, B-cell, 2 (2)	BCL2, BCL3	15143	18q21.3
Leukemia/lymphoma, T-cell	TCL1	18696	14q32.1
Leukemia/lymphoma, T-cell (3)	TCRA	18688	14q11.2
Leukocyte adhesion deficiency (2)	CD18, LCAMB, LAD	11692	21q22.1-qter
Lipoamide dehydrogenase deficiency (1)	DLD, LAD	24690	Chr.7
Lipoid adrenal hyperplasia, congenital (1)	CYP11A, P450SCC, P450C11A1	20171	Chr.15
Lipoma (2)	BABL, LIPO	15190	12q13-q14
Liver cell carcinoma (1)	HVBS1, HBVS1	11455	11p14-p13
?Long QT syndrome (2)	LQT	19250	6p21.3
Lowe syndrome (2)	OCRL, LOCR	30900	Xq25
Lymphoproliferative syndrome, X-linked (2)	IMD5, XLP, XLPD	30824	Xq26
?Lynch cancer family syndrome II (2)	LCFS2	11440	18q11-q12
?Lysosomal acid phosphatase deficiency (1)	ACP2	17165	11p12-p11
Macrocytic anemia of 5q- syndrome, refractory (2)	MAR	15355	5q12-q32
Macular dystrophy, atypical vitelliform (2)	VMD1	15384	8q24
Macular dystrophy, ?vitelline type (2)	VMD2	15370	6q25-qter
?Male infertility, familial (1)	FSHB	13653	11p13
?Male pseudohermaphroditism due to defective LH (1)	LHB	15278	19q13.32
Malignant melanoma, cutaneous (2)	CMM, HCMM	15560	1p36
Manic-depressive illness (2)	MAFD1, MD1	12548	11p15.5
Manic-depressive illness, X-linked (2)	MAFD2, MDX	30920	Xq28
Mannosidosis (1)	MANB	24850	19p13.2-q12
Maple syrup urine disease (1)	BCKDE1A, MSUD1	24860	Chr.19
Marfan syndrome, atypical (1)	COL1A2	12016	7q21.3-q22.1
?Marfan syndrome, atypical, 15470(1)	COL1A1	12015	17q21.31-q22.05
Maroteaux-Lamy syndrome (1)	ARSB	25320	5q11-q13
Martin-Bell syndrome (2)	FRAXA	30955	Xq27.3
MASA syndrome (2)	MASA	30925	Xq28
McArdle disease (1)	PYGM, MGP	23260	11q13
[McLeod phenotype] (2)	XK	31485	Xp21.2-p21.1
?Megacolon (2)	MGC	24920	13q22.1-q32.1
?Megaloblastic anemia (1)	DHFR	12606	5q11.1-q13.2
Megalocornea, X-linked (2)	MGCN	30930	Xq12-q26
Meningioma (2)	MGCR, MGM	15610	22q12.3-qter
?Menkes disease (1)	EPA, TIMP	30537	Xp11.4-p11.23
Menkes disease (2)	MNK, MK	30940	Xq12-q13
Mental retardation of WAGR (2)	TCR1, WAGR	19407	11p13
Mental retardation, X-linked nonspecific, I (2)	MRX1	30953	Xp22
Mental retardation, X-linked nonspecific, II (2)	MRX2	30954	Xq11-q12
Mental retardation-skeletal dysplasia (2)	MRSD, CHRS	30962	Xq28
Metachromatic leukodystrophy (1)	ARSA	25010	22q13.31-qter
Metachromatic leukodystrophy due to deficiency of SAP-1 (1)	SAP1, SAP2	24990	10q21-q22
Methemoglobinemia, enzymopathic (1)	DIA1	25080	22q13.31-qter
Methemoglobinemias, alpha- (1)	HBA1	14180	16pter-p13.3
Methemoglobinemias, beta- (1)	HBB	14190	11p15.5
Methylmalonicaciduria, mutase deficiency type (1)	MUT, MCM	25100	6p21.2-p12
?Microcephaly, true (2)	MCT	25120	1q31-q32.1
Miller-Dieker lissencephaly syndrome (2)	MDCR, MDLS, MDS	24720	17p13.3
MODY, one form, 12585 (3)	INS	17673	11p15.5

Disease	Symbol	Number	Location
Mucolipidosis II (1)	GNPTA	25250	4q21-q23
Mucolipidosis III (1)	GNPTA	25250	4q21-q23
Mucopolysaccharidosis I (1)	IDUA, IDA	25280	22q11
Mucopolysaccharidosis II (2)	IDS, MPS2, SIDS	30990	Xq27.3
Mucopolysaccharidosis IVB (1)	GLB1	23050	3p21-p14.2
Mucopolysaccharidosis VII (1)	GUSB	25322	7q21.1-q22
Multiple endocrine neoplasia I (1)	MEN1	13110	11q13
Multiple endocrine neoplasia II (2)	MEN2A, MEN2	17140	10q21.1
Multiple endocrine neoplasia III(2)	MEN3, MEN2B	16230	10q21.1
?Multiple exostoses (2)	EXT	13370	8q23-q24.1?
?Multiple lipomatosis (2)	BABL, LIPO	15190	12q13-q14
Myeloperoxidase deficiency (1)	MPO	25460	17q21.3-q23
Myopathy due to phosphoglycerate mutase deficiency (1)	PGAMM	26167	7p13-p12
Myotonic dystrophy (2)	DM	16090	19q13
Myotubular myopathy, X-linked (2)	MTM1, MTMX	31040	Xq28
Myxoid liposarcoma (2)	BABL, LIPO	15190	12q13-q14
Nail-patella syndrome (2)	NPS1	16120	9q34
?Nance-Horan syndrome (2)	NHS	30235	Xp
Neuroblastoma (2)	NB, NBS	25670	1p36.3-p36.1
Neurofibromatosis, von Recklinghausen (2)	NF1, VRNF	16220	17q11.2
?Nevoid basal cell carcinoma syndrome (2)	NBCCS, BCNS, NBCS	10940	1p
Niemann-Pick disease (1)	SMPD1, NPD	25720	Chr.17
Norrie disease (2)	NDP, ND	31060	Xp11.4
Norum disease (3)	LCAT	24590	16q22.1
Nucleoside phosphorylase deficiency, immunodeficiency due to (2)	NP, NP1	16405	14q13.1
Ocular albinism, Forsius-Eriksson type (2)	OA2	30060	Xp21.3-p21.1
Ocular albinism, Nettleship-Falls type (2)	OA1	30050	Xp22
?Optic atrophy, Kjer type (2)	OAK	16550	2p
Ornithine transcarbamylase deficiency (3)	OTC	31125	Xp21.1
Orofacial cleft (2)	OFC, CL	11953	6pter-p23
Oroticaciduria (1)	UMPS, OPRT	25890	3q13
Osteoarthrosis, precocious (3)	COL2A1	12014	12q13.1-q13.3
Osteogenesis imperfecta, 2 or more clinical forms (3)	COL1A1	12015	17q21.31-q22.05
Osteogenesis imperfecta, 2 or more clinical forms (3)	COL1A2	12016	7q21.3-q22.1
Osteosarcoma, retinoblastoma-related (2)	OSRC	25950	13q14.1
Ovarian failure, premature (2)	POF	31136	Xq26-q27
?Paget disease of bone (2)	PDB	16725	6p21.3
{?Parkinsonism, susceptibility to} (1)	CYP2D, P450C2D	12403	22q11.2-q12.2
Pelizaeus-Merzbacher disease (3)	PLP, PMD	31208	Xq22
?Pendred syndrome (2)	PDS	27460	8q24
Periodontitis, juvenile (2)	JPD, JP	17065	4q11-q13
Persistent Mullerian duct syndrome (1)	AMH, MIF	26155	19p13.3-p13.2
Phenylketonuria (3)	PAH, PKU1	26160	12q24.1
Phenylketonuria due to dihydropteridine reductase deficiency (1)	QDPR, DHPR	26163	4p15.3
Phosphoribosylpyrophosphate synthetase-related gout (1)	PRPS1	31185	Xq22-q26
?Phosphorylase kinase deficiency of liver and muscle, 26175 (2)	PHKB	17249	16q12-q13.1
?Piebaldism (2)	PBT	17280	4q12
PK deficiency hemolytic anemia (1)	PKLR, PK1, PKR	26620	1q21-q22
[Placental lactogen deficiency] (1)	CSA, PL, CSH1	15020	17q22-q24
Placental steroid sulfatase deficiency (3)	STS,SSDD	30810	Xpter-p22.32

Disease	Gene	MIM	Location
Plasmin inhibitor deficiency (1)	PLI	26285	18p11.1-q11.2
Plasminogen activator deficiency (1)	PLAT, TPA	17337	8p12
Plasminogen deficiency, types I and II (1)	PLG	17335	6q26-q27
Plasminogen Tochigi disease (1)	PLG	17335	6q26-q27
[Polio, susceptibility to] (2)	PVS	17385	19q12-q13.2
Polycystic kidney disease (2)	PKD1, APKD	17390	16p13.31-p13.12
Polyposis coli, familial (2)	APC, GS, FPC	17510	5q22-q23
Pompe disease (1)	GAA	23230	17q23
Porphyria, acute hepatic (1)	ALAD	12527	9q34
Porphyria, acute intermittent (1)	PBGD, UPS	17600	11q23.2-qter
Porphyria cutanea tarda (1)	UROD	17610	1p34
Porphyria, hepatoerythropoietic (1)	UROD	17610	1p34
Porphyria variegata (2)	VP, PPOX	17620	14q32
Postanesthetic apnea (1)	CHE1	17740	3q25.2
Prader-Willi syndrome (2)	PWCR, PWS	17627	15q11
Progressive cone dystrophy (2)	COD1, PCDX	30402	Xp21.1-p11.3
?Prolidase deficiency, 26413 (1)	PEPD	17010	19cen-q13.11
Properdin deficiency, X-linked (3)	PPP	31206	Xp11.4
Propionicacidemia, type I or pccA type (1)	PCCA	23200	Chr.13
Propionicacidemia, type II or pccB type (1)	PCCB	23205	3q13.3-q22
Protein C deficiency (1)	PROC	17686	2q13-q14
Protein S deficiency (1)	PSA, PROS	17688	3p11.1-q11.2
Pseudohypoaldosteronism (1)	MLR, MCR, MR	26435	4q31.1
Pseudo-Zellweger syndrome (1)	ACAA	26151	3p23-p22
?Pyridoxine dependency with seizures (1)	GAD	26610	Chr.2
Pyropoikilocytosis (1)	SPTA1	18286	1q21
Pyruvate carboxylase deficiency (1)	PC	26615	11q
Pyruvate dehydrogenase deficiency (1)	PDHA1, PHE1A	31217	Xp22.2-p22.1
?Rabson-Mendenhall syndrome (2)	INSR	14767	19p13.3-p13.2
?Ragweed sensitivity (2)	RWS	17945	6p21.3
Reifenstein syndrome (1)	AR, DHTR, TFM	31370	Xcen-q13
Renal cell carcinoma (2)	RCC1, RCC	14470	3p14.2
[Renal glucosuria] (2)	GLUR	23310	6p21.3
Renal tubular acidosis-osteopetrosis syndrome (1)	CA2	11481	8q22
Retinitis pigmentosa 2 (2)	RP2	31260	Xp11.3
Retinitis pigmentosa 3 (2)	RP3	31261	Xp21
Retinitis pigmentosa-1 (2)	RP1	18010	3q21-q23
Retinoblastoma (2)	RB1	18020	13q14.1-q14.2
Retinol binding protein, deficiency of (1)	RBP4	18025	10q23-q24
Retinoschisis (2)	RS	31270	Xp22
Rhabdomyosarcoma (2)	RMS, RMSCR	26821	11pter-p15.5
Rhabdomyosarcoma, aveolar (2)	RMSA	26822	2q37
Rh-null disease (1)	MER6, RHN	26815	3cen-q22
?Rh-null hemolytic anemia (1)	RH		1p36.2-p34
Rieger syndrome (2)	RGS	18050	4q23-q27
Salivary gland pleomorphic adenoma (2)	SGPA, PSA	18103	8q12
Sandhoff disease (1)	HEXB	26880	5q13
Sanfilippo syndrome D (1)	GNS, G6S	25294	12q14
Sarcoma, synovial (2)	SSRC	31282	Xp11.2
Scheie syndrome (1)	IDUA, IDA	25280	22q11

Disease	Gene	MIM	Location
Schizophrenia (2)	SCZD1	18151	5q11.2-q13.3
SCID, X-linked (2)	IMD4, SCIDX	30040	Xq13.1-q21.1
Sclerotylosis (2)	TYS	18160	4q28-q31
Seizures, benign neonatal (2)	EBN, BNS	12120	Chr.20
Severe combined immunodeficiency due to ADA deficiency (1)	ADA	10270	20q13.11
Severe combined immunodeficiency (SCID), HLA class II-negative type (1)	RFX	20992	19p13
?Sialidosis (1)	NEU, NEUG	25655	Chr.10
?Sialidosis (2)	NEU, NEU1	16205, 25655	6p21.3
Sickle cell anemia (1)	HBB	14190	11p15.5
?SLE (1)	CR1, C3BR	12062	1q32
Small-cell cancer of lung (2)	SCLC1, SCCL	18228	3p23-p21
Spastic paraplegia, X-linked, uncomplicated (2)	SPG2, SPPX2	31292	Xq21-q22
Spherocytosis, recessive (1)	SPTA1	18286	1q21
Spherocytosis-1 (3)	SPTB, SPH1	18287	14q22-q23.2
Spherocytosis-2 (2)	ANK, SPH2	18290	8p11
Spinocerebellar ataxia-1 (2)	SCA1	16440	6p21.3-p21.2
Spondyloepiphyseal dysplasia congenita (3)	COL2A1	12014	12q13.1-q13.3
Spondyloepiphyseal dysplasia tarda (2)	SEDL, SEDT	31340	Xp22
Stickler syndrome (3)	COL2A1	12014	12q13.1-q13.3
Sucrose intolerance (1)	SI	22290	3q25-q26
?Susceptibility to amyloid in FMF, 24910 (1)	SAA	10475	11pter-p12
Tay-Sachs disease (1)	HEXA, TSD	27280	15q22-q25.1
Testicular feminization (1)	AR, DHTR, TFM	31370	Xcen-q13
Thalassemias, alpha- (1)	HBA1	14180	16pter-p13.3
Thalassemias, beta- (1)	HBB	14190	11p15.5
Thrombophilia due to elevated HRG	HRG	14264	Chr.3
?Thrombophilia due to excessive plasminogen activator inhibitor (1)	PLANH1, PAI1	17336	7q21.3-q22
Thrombophilia due to heparin cofactor II deficiency (1)	HCF2, HC2	14236	Chr.22
Thyroid hormone resistance, 27430, 18857 (3)	THR, THRB, THR1, ERBA2	19016	3p24.1-p22
Thyroid iodine peroxidase deficiency (1)	TPO, TPX	27450	2pter-p12
Thyroid papillary carcinoma (1)	TST1, PTC, TPC	18855	10q11-q12
?Tourette syndrome (2)	GTS	13758	18q22.1
Transcobalamin II deficiency (1)	TCN2, TC2	27535	22q11.2-qter
?Treacher Collins mandibulofacial dysostosis (2)	MFD1	15450	5q11
Trichorhinophalangeal syndrome, type I (2)	TRPS1	19035	8q24.12
Trypsinogen deficiency (1)	TRY1, TRP1	27600	7q22-qter
Tuberous sclerosis-1 (2)	TSC1, TSC, TS	19110	9q33-q34
Tyrosinemia, type I (1)	FAH	27670	15q23-q25
Tyrosinemia, type II (1)	TAT	27660	16q22.1-q22.3
Urolithiasis, 2,8-dihydroxyadenine (1)	APRT	10260	16q24
van der Woude syndrome (2)	LPS, PIT, VDWS	11930	1q32
Vitamin D dependency, type I (2)	VDD1	26470	12q
{Vivax malaria, susceptibility to} (1)	FY	11070	1q21-q22
von Hippel-Lindau syndrome (2)	VHL	19330	3p25-p24
von Willebrand disease (1)	F8VWF, VWF	19340	12pter-p12
?Waardenburg syndrome, type I (2)	WS1	19350	2q35
Watson syndrome (2)	WSS	19352	17q11-q21
Wieacker-Wolff syndrome (2)	WWS	31458	Xq13-q21
Wilms tumor (2)	TCR1, WAGR	19407	11p13

Wilms tumor, type 2	WTCR2, WT2	19409	11p15.5
Wilson disease (2)	WND, WD	27790	13q14-q21
Wiskott-Aldrich syndrome (2)	IMD2, WAS	30100	Xp11.3-p11
Wolman disease (1)	LIPA	27800	10q24-q25
?Xeroderma pigmentosum (1)	PPOL, PARP	17387	1q41-q42
?Xeroderma pigmentosum, 1 form (1)	ERCC1, UV20	12638	19q13.2-q13.3
?Xeroderma pigmentosum A (1)	XPAC, XPA	27870	1q42-qter
?Xeroderma pigmentosum, group F (2)	XPF	27876	Chr.15
?Xeroderma pigmentosum, one type (1)	UVDR,ERCM2	19206	Chr.13
Zellweger syndrome (2)	ZWS, ZS	21410	7q11.12-q11.23

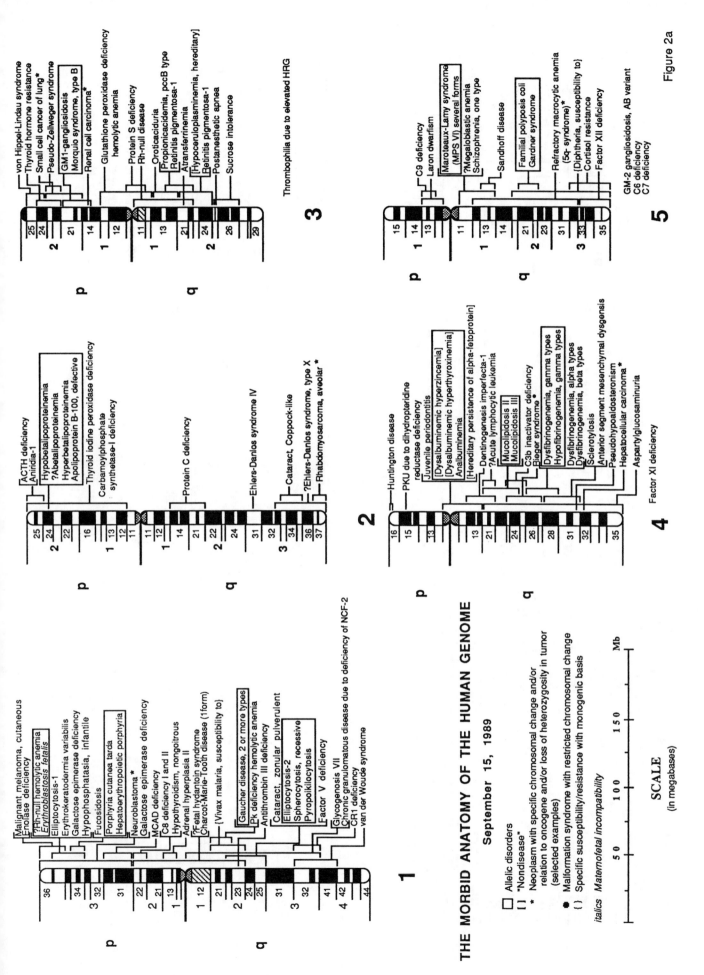

THE MORBID ANATOMY OF THE HUMAN GENOME

September 15, 1989

Figure 2a

□ Allelic disorders
| | "Nondisease"
* Neoplasm with specific chromosomal change and/or
relation to oncogene and/or loss of heterozygosity in tumor
(selected examples)
● Malformation syndrome with restricted chromosomal change
{} Specific susceptibility/resistance with monogenic basis

italics Maternofetal incompatibility

SCALE
(in megabases)

5.127

Figure 2b

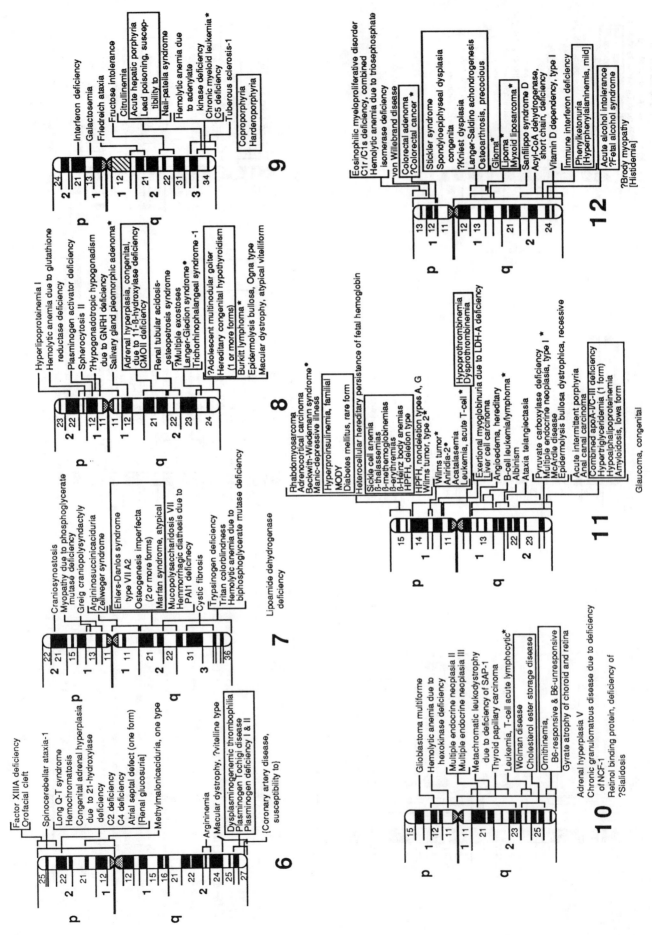

5.128

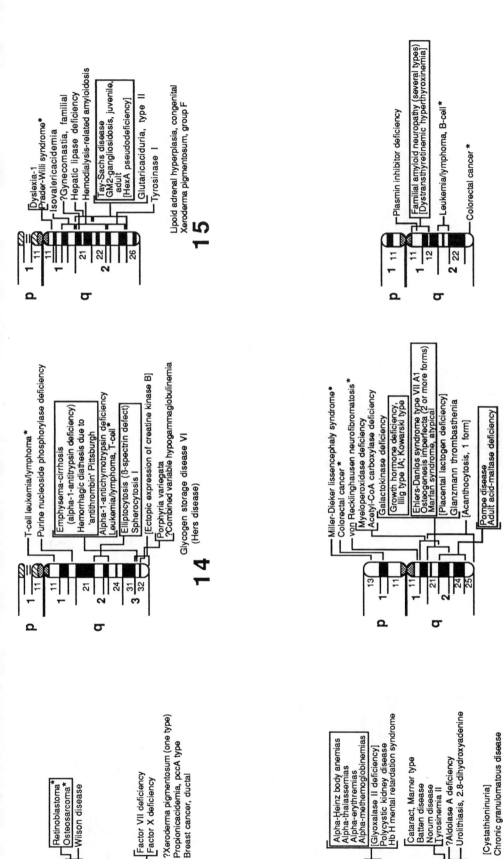

Figure 2c

The Oxford Grid

Courtesy of J.H. Edwards, June 1989. (Earlier versions: Searle et al., Ann. Hum. Genet. **53**: 89-140, 1989; Searle et al., Genomics **1**: 3-18, 1987; Buckle et al., Clin. Genet. **26**: 1-11, 1984.)

In this ingenious diagram, the length of the horizontal side of each rectangle represents the proportionate length of the given chromosome of mouse, whereas the length of the vertical side represents the proportionate length of the given chromosome of man. The triangles or circles indicate the chromosomal location of a given homologous locus in mouse and man. Triangles point down or up depending on whether the locus is on the short arm or the long arm, respectively, of

the human chromosome. Circles indicate that the arm location is unknown. Large triangles refer to 5 or more loci. Open symbols refer to assignments since the last published version. Note that for every human chromosome at least 2 loci are syntenic also in the mouse. The approximately 330 homologous loci plotted here demonstrate 51 conserved segments. Note that longer human chromosomes show a tendency to homology of synteny with larger mouse chromosomes. This diagram is useful in predicting the location of a human gene whose murine homolog has been mapped. Furthermore, by looking at genes mapped to homologous chromosome segments in the mouse, candidate genes for human disorders of unknown mechanism may come to attention, or mouse diseases for use as models may be identified.

Figure 3

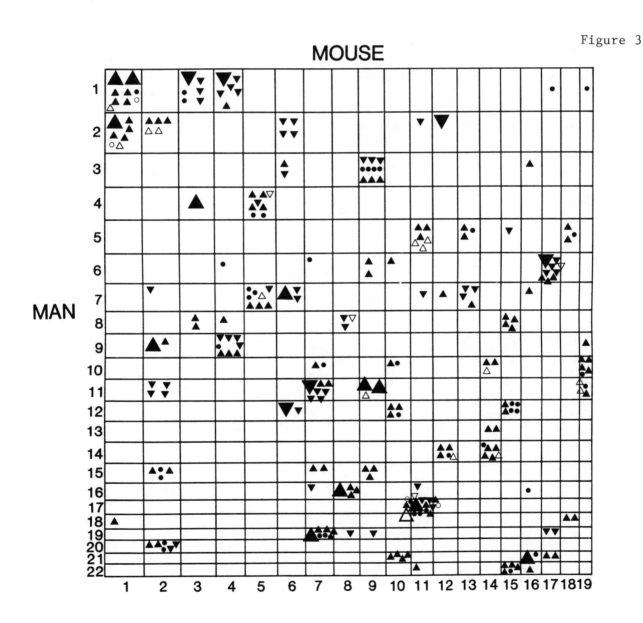

The Mitochondrial Chromosome
(Chromosome M or Chromosome 25)

Each mitochondrion contains several circular chromosomes. The most important function of the mitochondria is synthesis of ATP by the process of oxidative phosphorylation (OXPHOS). OXPHOS involves 5 multi-polypeptide enzyme complexes in the mitochondrial inner membrane. The biogenesis of 4 of these 5 complexes is under the combined control of nuclear DNA and mitochondrial DNA (mtDNA). At least 69 separate polypeptides are known to be required for OXPHOS; only 13 of them are coded by mtDNA.

The 16,569 basepairs of the mitochondrial chromosome are the equivalent of 5,523 codons. Most of the mtDNA serves a coding function. The genes contain no intervening sequences, and little in the way of flanking sequences is present. The ribosomal and transfer RNAs are those involved in the synthesis of protein in the mitochondrion. Of the 22 tRNAs, 14 are coded on the L (light) strand. For all 13 reading frames, the function of the specific protein coded is known. The mtDNA code differs from that of the nuclear DNA and the genetic code of any presentday prokaryote. UGA codes for tryptophan (not termination), AUA codes for methionine (not isoleucine), and AGA and AGG code for termination (not arginine). The mitochondrial genome is transcribed as a single mRNA transcript which is subsequently cleaved into its several component genes.

Mitochondrial inheritance ('cytoplasmic inheritance' in the old terminology) is exclusively matrilineal. Restriction fragment length polymorphisms (RFLPs) are known in mitochondrial DNA as in nuclear DNA. Mitochondrial restriction patterns have been used to construct a biological history of the human species.

Whereas each nuclear chromosome is present in only 2 copies per cell at the most, the mitochondrial chromosome is present in thousands of copies. The behavior of a mitochondrial mutation in inheritance might be expected, therefore, to be galtonian rather than mendelian.

In cultured cells, chloramphenicol resistance is demonstrably the result of mutation in the mtDNA gene for 16S rRNA (see 21465). Indeed, the specific nucleotide changes have been identified. A point mutation (substitution of nucleotide 1178) has been identified as a cause of Leber optic atrophy (Wallace et al., Science **242**: 1427-1430, 1988). Deletions (which always avoid the origins of replication) have been observed in many patients with progressive external ophthalmoplegia alone or as part of Kearns-Sayre syndrome (Moraes et al., New Eng. J. Med. **320**: 1293-1299, 1989). Deletion in the mitochondrial chromosome has also been found in mitochondrial myopathy (e.g., Holt et al., Nature **221**: 717-719, 1988), aplastic anemia, and oncocytomas, and even a relation to aging has been suggested.

Figure 4

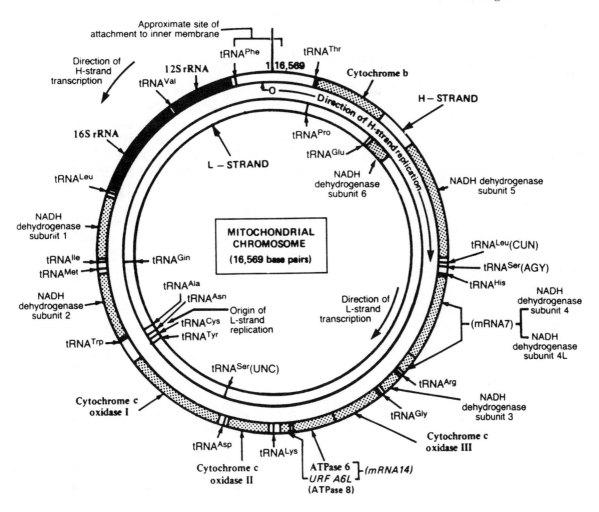

GENE	LOCATION (nucleotide pair)	GENE	LOCATION (nucleotide pair)
(L-strand promoter	about 392-435)	tRNA asparagine	7518-7585
(major H-strand promoter	about 545-567)	cytochrome c oxidase subunit II	7586-8262
(minor L-strand start site	561)	tRNA lysine	8295-8364
tRNA phenylalanine	577-647	ATPase subunit 8 (URF A–L)	8366-8572
(minor H-strand start site	about 645)	ATPase subunit 6	8527-9207
12S rRNA	648-1601	cytochrome c oxidase subunit III	9207-9990
tRNA valine	1602-1670	tRNA glycine	9991-10058
16S rRNA	1671-3229	NADH dehydrogenase subunit 3	10059-10404
tRNA leucine(UUR)	3230-3304	tRNA arginine	10405-10469
NADH dehydrogenase subunit 1	3307-4262	NADH dehydrogenase subunit 4L	10470-10766
tRNA isoleucine	4263-4331	NADH dehydrogenase subunit 4	10760-12137
tRNA glutamine	4329-4400	tRNA histidine	12138-12206
tRNA methionine (including fMET)	4402-4469	tRNA serine (AGY)	12207-12265
NADH dehydrogenase subunit 2	4470-5511	tRNA leucine (CUN)	12266-12336
tRNA tryptophan	5512-5576	NADH dehydrogenase subunit 5	12337-14148
tRNA alanine	5587-5655	NADH dehydrogenase subunit 6 (on L)	14149-14673
tRNA asparagine	5657-5729	tRNA glutamic acid (L-strand)	14674-14742
(origin of L-strand replication	5729-5805)	cytochrome b	14747-15887
tRNA cysteine	5761-5826	tRNA threonine	15888-15953
tRNA tyrosine	5826-5891	tRNA proline (L-strand)	15955-16023
cytochrome c oxidase subunit I	5904-7444	(membrane attachment site	about 15925-499)
tRNA serine	7445-7516		

Linkage Maps of Man (<u>Homo sapiens</u>)   2N=46

September, 1989

Ray White, Jean-M. Lalouel, Mark Lathrop, Mark Leppert,
Yusuke Nakamura, Peter O'Connell
Howard Hughes Medical Institute and Department of Human Genetics
University of Utah
Salt Lake City, Utah 84132

Linkage maps of individual chromosomes were derived from RFLP analysis of human pedigrees over the past several years. Specific data on each chromosome map is found in indicated citations.

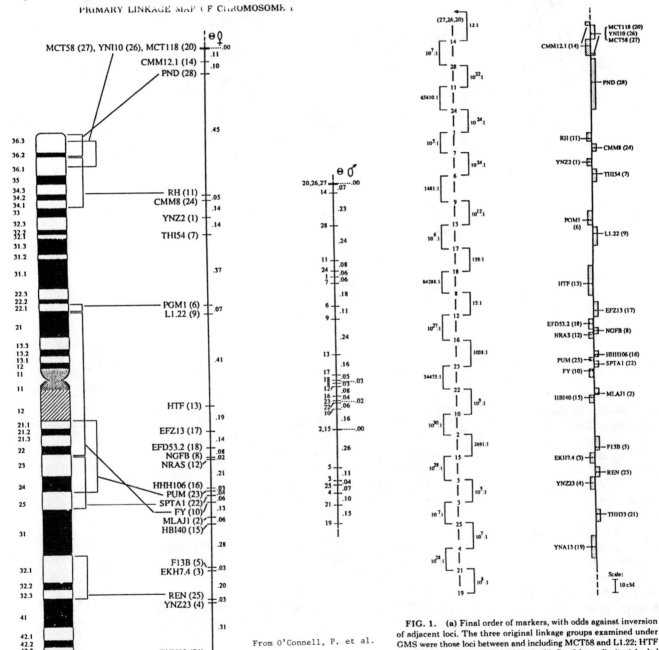

From O'Connell, P. et al.

Genomics 4:12-20, 1989

FIG. 1. (a) Final order of markers, with odds against inversion of adjacent loci. The three original linkage groups examined under GMS were those loci between and including MCT58 and L1.22; HTF and HBI40; and F13B and YNA13. (b) Confidence limits (shaded bars) of maximum likelihood location for each of 28 loci on the map of chromosome 1. Genetic distances between loci are derived from the sex-averaged recombination fractions in Table 4.

FIG. 2. Sex-specific genetic maps of chromosome 1, with physical locations of selected markers indicated on the karyogram. Maps are scaled in genetic distance (centimorgans), under the assumption of a variable sex ratio in each interval. θ = recombination fraction.

TABLE 1

Polymorphic Loci on Chromosome 1

Locus no.	Probe (locus)	Enzyme	Allele size (kb)	Allele frequencies	Observed heterozygosity	Number indiv. typed
1	pYNZ2 (D1S57)	TaqI	VNTR >8 alleles 1.0–3.0		0.65	736
2	pMLAJ1 (D1S61)	HinfI	VNTR >8 alleles 1.5–3.0		0.68	481
3	pEKH7.4 (D1S65)	TaqI	A1 5.0 A2 3.8	0.47 0.53	0.53	655
4	pYNZ23 (D1S58)	MspI	A1 5.0 A2 4.5	0.48 0.52	0.39	710
5	FXIIIB (F13B)	FXIIIB	A1 A2 A3 A4	0.73 0.13 0.14 0.00	0.44	564
6	PGM1 (PGM1)	PGM1	1$^+$ 1$^-$ 2$^+$ 2$^-$	0.63 0.12 0.20 0.04	0.58	548
7	pTHI54 (D1S62)	PvuII	A1 6.0 A2 5.0	0.51 0.49	0.45	624
8	N8C6 (NGFB)	BglII	A1 6.0 A2 4.1, 1.9	0.22 0.78	0.35	565
9	pL1.22 (D1S2)	BglII	A1 10.0 A2 7.0	0.81 0.19	0.36	661
10	Duffy (FY)	Duffy	Fy^a Fy^b	0.45 0.55	0.48	613
11	Rhesus (RH)	Rhesus	DCe dce DcE Dce DCE dCe dcE dCE	0.39 0.46 0.12 0.01 0.01 0.00 0.01 0.01	0.60	623
12	pMCR3 (NRAS)	EcoRI	A1 6.0 A2 4.8	0.72 0.28	0.36	686
13	pAP8 (HTF)	MspI	A1 3.2 A2 2.6	0.29 0.71	0.37	384
14	pCMM12.1 (D1S76)	TaqI	VNTR 4 alleles 3.5–4.5		0.58	436
15	pHBI40 (D1S66)	MspI	A1 8.0 A2 4.4	0.73 0.27	0.37	766
16	pHHH106 (D1S67)	MspI	A1 2.3 A2 2.0	0.46 0.54	0.51	711
17	pEFZ13 (D1S64)	MspI	A1 5.0 A2 3.3	0.66 0.34	0.53	554
18	pEFD53.2 (D1S73)	TaqI	A1 6.5 A2 4.5	0.58 0.42	0.42	628
19	cYNA13 (D1S74)	MspI	VNTR >10 alleles 2.0–8.0		0.96	678
20	pMCT118 (D1S80)	PvuII	VNTR 5 alleles 1.5–1.8		0.62	429
21	pTHH33 (D1S81)	RsaI	VNTR 6 alleles 4.0–7.0		0.76	543
22	p3021 (SPTA1)	MspI	A1 12.0 A2 11.5	0.61 0.39	0.37	675
23	pMUC10 (PUM)	RsaI	VNTR >10 alleles 3.0–8.0		0.79	734
24	pCMM8 (D1S79)	EcoRI	VNTR 4 alleles 4.0–9.0		0.59	710
	pCMM8.1 (D1S63)	MspI	A1 1.6 A2 1.0	0.74 0.26	0.38	647
25	pHRnES1.9 (REN)	HindIII	A1 8.7 A2 6.2	0.70 0.30	0.39	661
	pHRnX3.6	TaqI	A1 10.0 A2 9.0	0.28 0.72	0.41	185
26	pYNI10 (D1Z2)	TaqI	Midisatellite >8 alleles 1.0–12.0		0.80	229
27	pMCT58 (D1S77)	PvuII	VNTR 6 alleles 1.5–1.8		0.67	479
28	pJA110 (PND)	BglI	A1 10.0 A2 6.0	0.87 0.13	0.24	517

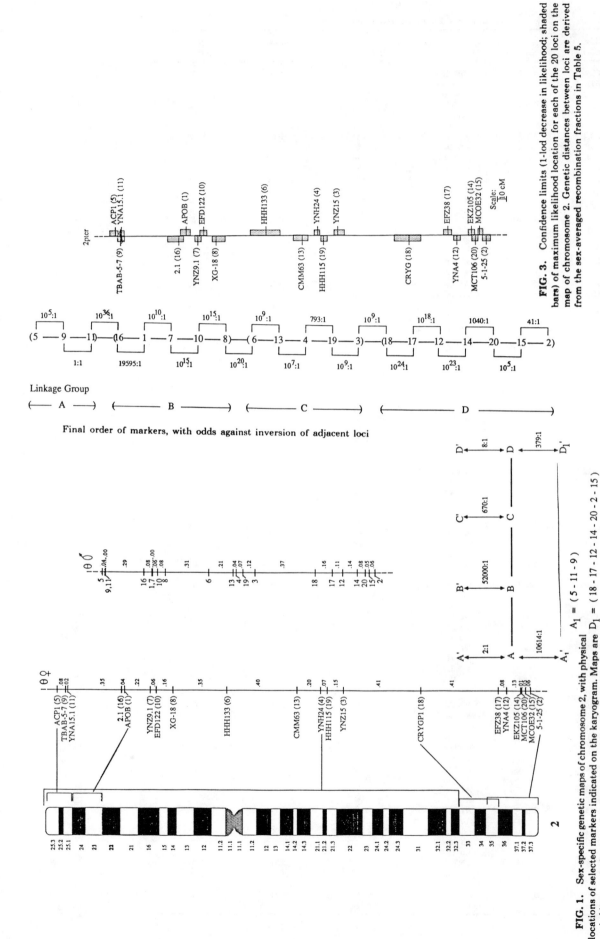

FIG. 3. Confidence limits (1-lod decrease in likelihood; shaded bars) of maximum likelihood location for each of the 20 loci on the map of chromosome 2. Genetic distances between loci are derived from the **sex-averaged recombination fractions in Table 5.**

From O'Connell, P. et al. Genomics 5:738–745, 1989
O'Connell, P. et al. Genomics, in press

Linkage Group

Final order of markers, with odds against inversion of adjacent loci

$A_1 = (5 - 11 - 9)$
$D_1 = (18 - 17 - 12 - 14 - 20 - 2 - 15)$

FIG. 1. Sex-specific genetic maps of chromosome 2, with physical locations of selected markers indicated on the karyogram. Maps are scaled in genetic distance determined by the Haldane mapping function (centimorgans), under the assumption of a variable sex ratio in each interval. θ = recombination fraction.

TABLE 1

Polymorphic Loci on Chromosome 2

Locus no.	Probe (locus)	Enzyme	Allele size (kb)	Allele frequency	Observed heterozygosity	No. indiv. typed
1	pB23 (APOB1)	XbaI	8.6 5.0	0.45 0.55	0.73	210
	pB8 (APOB1)	MspI	>2.5 <2.5	0.14 0.86	0.27	634
2	p5-1-25 (D2S3)	PstI	VNTR 4 alleles 2.0–2.8		0.55	406
3	pYNZ15 (D2S43)	TaqI	1.8 1.0	0.47 0.53	0.57	648
4	pYNH24 (D2S44)	MspI	VNTR >20 alleles 1.3–6.0		0.91	615
5	ACP1 (ACP1)	ACP1	ACP-1*A ACP-1*B ACP-1*C	0.33 0.67 0.00	0.37	601
6	pHHH133 (D2S45)	MspI	1.4 1.3	0.29 0.71	0.36	551
7	pYNZ9.1 (D2S46)	TaqI	1.1 1.0	0.56 0.44	0.59	588
8	pXG-18 (D2S6)	TaqI	5.5 4.6	0.45 0.55	0.56	648
9	pTBAB-5-7 (D2S47)	PvuII	VNTR 4 alleles 3.0–6.0		0.55	637
10	pEFD122 (D2S48)	MspI	3.2 3.0	0.44 0.56	0.44	741
11	pYNA15.1 (D2S49)	MspI	7.0 6.0	0.65 0.35	0.48	661
12	cYNA4 (D2S50)	MspI	VNTR >7 alleles 3.0–6.0		0.79	690
13	pCMM63 (D2S51)	MspI	7.0 4.5 4.3	0.07 0.91 0.02	0.15	751
14	pEKZ105 (D2S52)	RsaI	3.0 2.6	0.51 0.49	0.47	653
15	cMCOE32 (D2S53)	TaqI	VNTR >5 alleles 3.0–6.0		0.74	715
16	p2.1 (D2S*XX* 70)	HindIII	4.0 3.5	0.32 0.68	0.40	609
17	pEFZ38 (D2S60)	TaqI	4.0 3.5	0.74 0.26	0.46	617
18	p5G1(a) (CRYGP1)	TaqI	3.5 3.3	0.68 0.32	0.45	643
	p5G1(b) (CRYGP1)	TaqI	2.2 1.1	0.33 0.67	0.47	643
	p5G1(c) (CRYGP1)	TaqI	2.0 1.4, 0.6	0.17 0.83	0.30	648
19	pHHH115 (D2S54)	MspI	4.8 4.6 4.5	0.16 0.84 0.01	0.28	639
	pHHH115 (D2S54)	MspI	2.4 1.7	0.42 0.58	0.49	564
20	pMCT106 (D2S61)	PvuII	1.7 1.6 1.5	0.06 0.80 0.14	0.32	734

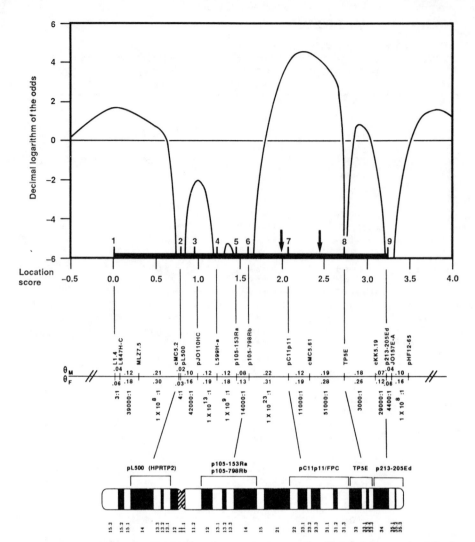

Fig. 1. Genetic linkage map of 16 DNA markers on chromosome 5, with location scores (top) for the FPC-GS gene on the female map. Two arrows on the abscissa indicate the 95% confidence boundaries for the location of the FPC-GS locus. The genetic map (center), also scaled to female genetic disease (in Morgans), indicates estimates of recombination distance (θ) between markers for males (above line) and females (below line).

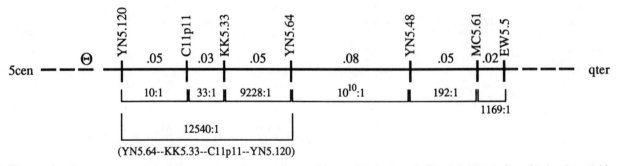

Figure I Map of markers on a short region of chromosome 5. Odds against inversion of adjacent loci are indicated in brackets. Odds are 12,540:1 against reversing the orientation of the subgroup containing the proximal four loci.

From Nakamura, Y. et al. Am. J. Hum. Genet. 43:53'   4, 1988
    Leppert, M. et al. Science 238:1411–1413, 1987

A PRIMARY GENETIC LINKAGE MAP OF CHROMOSOME 5

PROBE(LOCUS)	ENZYME	ALLELE SIZE (KB)	ALLELE FREQUENCIES	OBSERVED HETERO- ZYGOSITY	NUMBER OF INDIVIDUALS TYPED
1 L1.4 (D5S4)	EcoRI	0.7 0.6	0.80 0.20	0.31	346
2 pC11p11	TaqI	4.2 2.7	0.17 0.83	0.32	681
3 pL500 (HPRTP2)	MspI	3.6 1.3	0.23 0.77	0.42	460
4 p105-153Ra	MspI	8.0 5.0	0.40 0.60	0.53	771
5 p213-205Ed	MspI	6.0 3.9 3.8	0.43 0.35 0.22	0.60	721
6 IM4 (D5S6)	BamHI	11.0 9.6 7.6	0.36 0.54 0.11	0.68	264
7 TP5E	TaqI	13.0 5.0	0.77 0.23	0.40	597
8 pJO110HC	MspI	8.7 7.2 6.9	0.69 0.30 0.00	0.38	751
9 p105-798Rb	MspI	14.0 2.3	0.57 0.43	0.58	471
10 L565RI-b	MspI	6.2 4.6	0.44 0.56	0.44	192
11 L599H-a	TaqI	17.0 14.0 10.0	0.32 0.16 0.52	0.67	748
12 pHF12-65 (D5S2)	MspI	4.1 3.7	0.87 0.13	0.23	621
13 PP8C	TaqI	5.0 4.8	0.31 0.69	0.42	539
14 JO157E-A	MspI	4.0 3.5	0.38 0.63	0.52	754
15 L647H-C	MspI	9.0 7.0 5.5 4.4	0.07 0.29 0.33 0.31	0.74	273
16 Kell	Kel		0.02 0.98	0.04	630

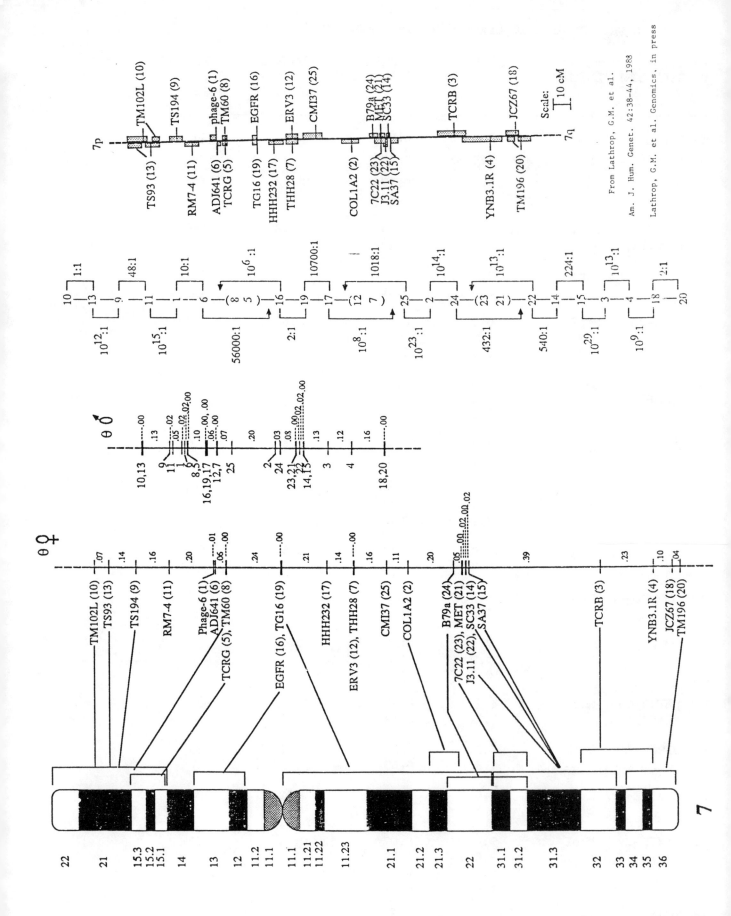

From Lathrop, G.M. et al.

Am. J. Hum. Genet. 42:38-44, 1983

Lathrop, G.M. et al. Genomics, in press

5.140

TABLE 1

Polymorphic Loci on Chromosome 7

Locus no.	Probe (locus)	Enzyme	Allele size (kb)	Allele frequency	Observed heterozygosity	Number ind. typed (No. of meioses from informative matings)
1	phage-6 (D7S11)	HindIII	9.8 7.8	0.23 0.77	0.35	387 (163)
2	pNJ-3 (COL1A2)	EcoRI	13.0 9.5	0.66 0.34	0.45	624 (295)
3	pJ2 (TCRB)	BglII	10.0 9.5	0.55 0.45	0.55	639 (332)
4	pYNB3.1R (D7S372)	RsaI	3.9 2.4	0.29 0.71	0.46	672 (333)
5	PST/BAM (TCRG)	PvuII	14.7 12.6 10.5	0.20 0.59 0.21	0.57	665 (403)
6	ΦADJ641 (D7S369)	TaqI	2.0 1.0	0.20 0.80	0.34	672 (261)
7	pTHH28 (D7S371)	MspI	3.4 2.3	0.29 0.71	0.43	453 (183)
8	pTM60 (D7S132)	TaqI	5.3 5.0	0.37 0.63	0.47	648 (368)
9	ΦTS194 (D7S150)	TaqI	1.8 1.5	0.79 0.21	0.42	654 (335)
10	pTM102L (D7S135)	TaqI	12.0 8.5	0.37 0.63	0.49	668 (360)
11	pRM7-4 (D7S370)	MspI	5.5 3.2	0.38 0.62	0.49	545 (293)
12	pHP1.7 (ERV3)	MspI	3.3 2.8	0.55 0.45	0.37	713 (330)
13	pTS93 (D7S149)	PstI	4.4 4.2	0.28 0.72	0.38	579 (288)
14	pSC33 (D7S126)	HindIII	4.3 4.0 3.6	0.20 0.14 *od 6*	0.51	543 (193)
15	pSA37 (D7S125)	PstI	7.0 5.0	0.45 0.55	0.38	716 (299)
16	pE7 (EGFR)	PstI	14.0 13.0	0.83 0.17	0.26	416 (129)
17	pHHH232 (D7S395)	PvuII	7.5 4.2	0.71 0.29	0.54	314 (180)
18	pJCZ67 (D7S396)	RsaI	VNTR 6 alleles 3.0–6.0		0.76	666 (349)
19	pTG16 (D7S129)	EcoRI	2.1 1.0	0.85 0.15	0.21	409 (98)
20	pTM196 (D7S392)	PstI	2.8 2.5	0.85 0.15	0.24	591 (193)
21	pMetH (MET)	TaqI	7.5 4.0	0.51 0.49	0.64	432
	pMetH (MET)	MspI	4.8 2.3 1.7	0.05 0.53 0.42	0.62	185
	pMetD (MET)	TaqI	5.0 4.3	0.80 0.20	0.31	659
	pMetHOS5 (MET)	TaqI	2.9 1.3	0.34 0.66	0.47	663
					0.70 (HAP)	(347)
22	pJ3.11 (D7S8)	MspI	4.0 1.6	0.38 0.70	0.54	639
	pJ3.11 (D7S8)	TaqI	6.0 3.1	0.96 0.04	0.08	587
					0.54 (HAP)	(414)
23	p7C22 (D7S16)	EcoRI	7.0 5.0	0.80 0.20	0.35	507 (187)
24	pB79a (D7S13)	HindIII	8.1 4.3	0.16 0.84	0.24	628
	pB79a (D7S13)	MspI	11.6 8.4	0.31 0.69	0.41	654
					0.41 (HAP)	(308)
25	pCMI37 (D7S368)	RsaI	VNTR 4 alleles 2.0–5.0		0.68	675 (535)

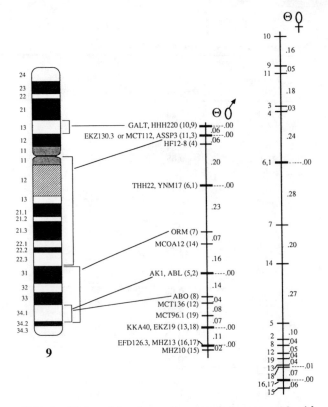

FIG. 2. Male and female genetic maps of chromosome 9, with physical locations of selected markers indicated on the karyogram. Scale is in genetic distance (morgans); recombination fractions (θ) are indicated in the intervals.

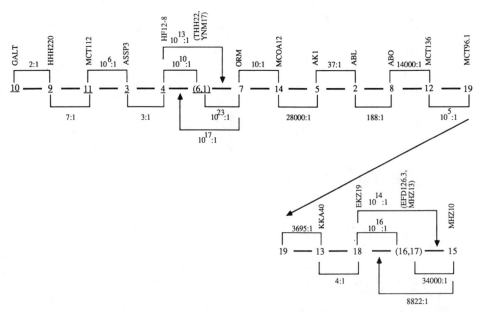

FIG. 1. Final locus order and odds against some alternatives. Underlined loci indicate the first linkage group as determined by GMS; the remaining loci except for AK1 (5) were included in the second linkage group. Odds against inversion or other placement of adjacent loci are indicated on the brackets.

Polymorphic Markers for Chromosome 9

Locus number	Probe (locus)	Enzyme	Allele size (kb)	Allele frequencies	Observed heterozygosity	Number indiv. typed
1	pYNM17 (D9S6)	*Taq*I	7.5 5.0	0.62 0.38	0.47	580
2	pAb1K2 (ABL)	*Taq*I	7.6 6.7 5.0	0.09 0.91 0.01	0.17	647
3	pAS-1 (ASSP3)	*Hind*III	10.1 5.3	0.74 0.26	0.43	607
4	pHF12-8 (D9S1)	*Taq*I	3.6 3.2	0.40 0.60	0.36	664
5	Adenylate kinase 1 (AK1)		A1 A2	0.95 0.05	0.10	626
6	pTHH22 (D9S12)	*Taq*I	12.0 8.0	0.09 0.91	0.16	710
7	Orosomucoid (ORM)		F S	0.58 0.42	0.58	619
8	ABO		O A_2 A_2 B	0.69 0.17 0.07 0.07	0.44	553
9	pHHH220 (D9S18)	*Taq*I	4.3 3.0 2.3	0.78 0.21 0.01	0.32	614
10	Galactose 1-*P*-uridyltransferase (GALT)		*N *D *LA	0.93 0.06 0.01	0.14	637
11	pMCT112 (D9S15) or pEKZ130.3 (D9S9)	*Msp*I	6.0 4.9	0.71 0.29	0.37	759
12	pMCT136* (D9S10)	*Pst*1	2.2 2.0	0.46 0.54	0.50	765
13	pKKA40* (D9S31)	*Pvu*II	1.8 1.7 1.5	0.33 0.58 0.09	0.44	610
14	pMCOA12* (D9S16)	*Taq*I	11.0 8.0 4.5 + 3.5	0.58 0.18 0.24	0.49	626
15	pMHZ10* (D9S11)	*Hinf*I	1.0–1.7 10 alleles		0.70	616
16	pEFD126.3* (D9S7)	*Hinf*I	1.7–2.6 16 alleles		0.75	617
17	pMHZ13* (D9S13)	*Pst*I	1.6–2.3 3 alleles		0.53	359
18	pEKZ19 (D9S17)	*Taq*I	3.2 2.5	0.31 0.69	0.40	731
19	pMCT96.1* (D9S14)	*Rsa*I	4.7 3.7	0.76 0.24	0.35	543

Note. Asterisks indicate VNTR marker.

From Lathrop, M. et al. Genomics 3:361-366, 1988

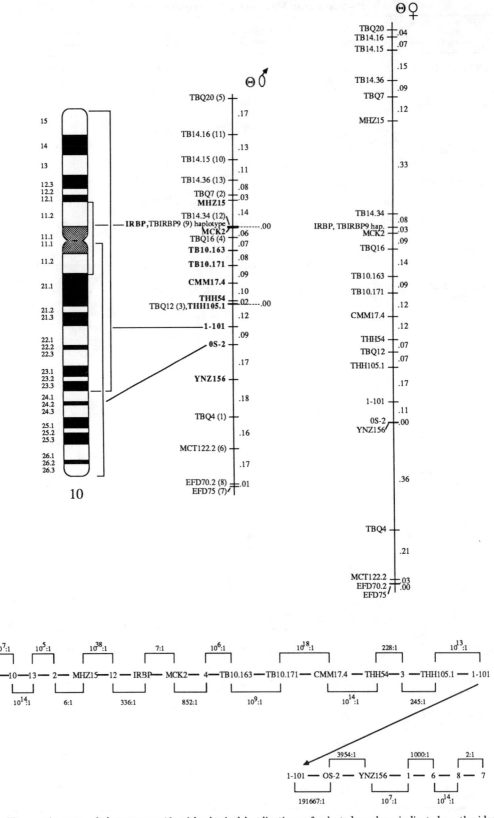

FIG. 1. Sex-specific genetic maps of chromosome 10, with physical localizations of selected markers indicated on the ideogram. Loci previously mapped are in boldface on the male map; numbers in parentheses on the male map identify the new loci for the depiction of odds against inversions of loci (below). Maps are scaled in genetic distance (morgans), under the assumption of a variable effect of sex in each interval. θ, recombination fraction.

Characterization of 13 New Polymorphic Markers on Chromsome 10

Locus No.	Probe (locus)	Enzyme	Allele size (kb)	Allele frequency	Observed heterozygosity (%)	Number of individuals typed
1	cTBQ4 (D10S27)	*Msp*I	4.8 3.4	0.45 0.55	50	725
2	cTBQ7 (D10S28)	*Taq*I	2.0–7.0 >20 alleles (VNTR marker)		92	488
3	cTBQ12 (D10S29)	*Taq*I	9.0 8.0	0.35 0.65	48	640
4	cTBQ16 (D10S30)	*Msp*I	2.5 2.0	0.68 0.32	47	602
5	cTBQ20 (D10S31)	*Msp*I	1.8 1.7	0.40 0.60	45	664
6	pMCT122.2 (D10S36)	*Taq*I	7.0 6.5	0.35 0.65	27	722
7	pEFD75 (D10S25)	*Taq*I	2.0–3.5 7 alleles (VNTR marker)		60	607
8	pEFD70.2 (D10S26)	*Pvu*II	1.7–3.3 4 alleles (VNTR marker)		52	634
9	cTBIRBP9 (IRBP)	*Taq*I	5.5 3.0	0.63 0.37	41	605
10	cTB14.15 (D10S32)	*Taq*I	3.6 3.5 2.8	0.15 0.51 0.34	64	527
11	cTB14.16 (D10S33)	*Taq*I	3.7–5.0 5 alleles (VNTR marker)		52	493
12	cTB14.34 (D10S34)	*Taq*I (1) (2)	7.5 7.0 6.5 6.0	0.40 0.60 0.44 0.56	49 46	582 574
13	cTB14.36 (10S35)	*Msp*I	4.4 3.1	0.51 0.49	55	511

From Nakamura, Y., et al. Genomics 3:389-392, 1988
      Lathrop, M., et al. Genomics 2:157-164, 1988

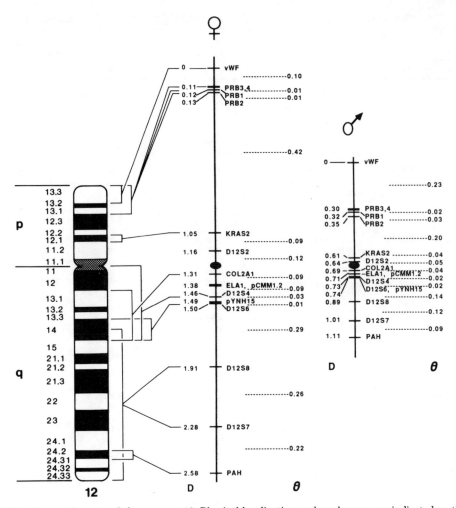

FIG. 5. Male and female genetic maps of chromosome 12. Physical localizations, where known, are indicated on the ideogram. Genetic distance (*D*) is measured cumulatively in Morgans from the VWF locus.

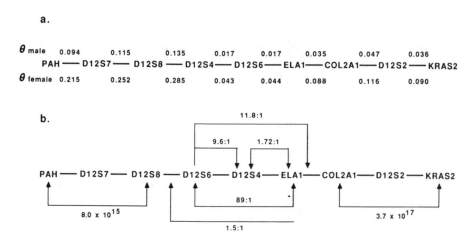

FIG. 2. (a) Trial map of nine loci, with recombination estimates in each interval based on multilocus analysis. (b) Maximum likelihood map of nine loci, showing odds against alternative orders.

O'Connell, P., et al. Genomics 1:93-102, 1987

Polymorphic Loci in Chromosome 12

Locus	Probe	Enzyme	Allele size (kb)	Allele frequency	Observed heterozygosity	Number of individuals typed
KRAS2	p640	*Taq*I	5.7	0.76	0.32	400
			3.3	0.24		
D12S2	p12-16	*Eco*RI	9.7	0.89	0.22	396
			8.5	0.11		
ELA1	pXP13	*Taq*I	4.3	0.82	0.26	394
			3.7	0.18		
D12S4	p9F11	*Taq*I	8.0	0.64	0.59	391
			3.0	0.36		
PRB1	pPRP	*Eco*RI-1	6.5	0.35	0.66	473
			6.3	0.48		
			6.1	0.07		
			6.3–6.3	0.01		
			6.5–6.1	0.04		
			6.3–6.1	0.01		
			—	0.06		
PRB2	pPRP	*Eco*RI-2	4.6	0.11	0.33	494
			4.3	0.74		
			4.2	0.10		
			4.0	0.04		
			3.8	0.01		
PRB3	pPRP	*Eco*RI-3	4.6	0.02	0.14	526
			4.3	0.03		
			4.1	0.93		
			4.0	0.02		
PRB4	pPRP	*Eco*RI-4	3.6	0.25	0.37	517
			3.5	0.63		
			3.3	0.13		
COL2A1	cosHcol2A	*Hin*dIII	14.0	0.56	0.70	285
			7.0	0.44		
VWF	pλSV2	*Bgl*II	9.7	0.64	0.40	374
			7.1	0.36		
D12S7	pDL32B	*Taq*I	6.5	0.08	0.66	496
			5.1	0.29		
			4.1–2.7	0.55		
			2.7–2.4	0.07		
D12S6	p1-7	*Msp*I	4.4	0.33	0.63	429
			3.6	0.67		
	p1-11	*Eco*RI	4.9	0.89		
			3.7	0.11		
D12S8	p7G11	*Msp*I	6.0	0.81	0.62	447
			4.3	0.19		
		*Taq*I-1	8.0	0.96		
			5.0	0.04		
		*Taq*I-2	4.0	0.60		
			3.0	0.40		
PAH	pPH72	*Msp*I	18.0	0.38	0.65	386
			16.0	0.63		
		*Hin*dIII	3.3	0.17		
			3.2	0.63		
			3.1	0.20		
—	pYNH15	*Msp*I	4.0	0.17	0.52	598
			3.2	0.63		
			2.6	0.20		
—	pCMM1.2	*Taq*I	3.0	0.15	0.28	669
			2.8	0.85		
—	p7A9	*Taq*I	6.4	0.13	0.22	397
			6.2	0.88		
—	p9F4	*Eco*RV	11.4	0.23	0.36	393
			5.5	0.77		

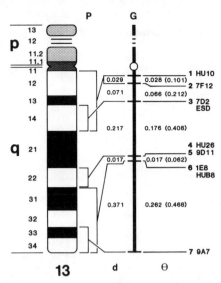

FIG. 4.—Map of chromosome 13 showing locations of seven loci based on genetic (G) as well as on physical (P) mapping data. Distances (d) between loci, in Morgans, are derived from recombination (θ) in males through Haldane's mapping function. The corresponding recombination values in females, under the assumption of a constant map ratio of 3.89, appear *in parentheses*.

SUMMARY OF GENETIC DATA ON CHROMOSOME 13

GENETIC SYSTEM		ALLELES (kb)	F	H	SUBJECTS TESTED Gpt	Pt	Off	SEGREGATION p	χ^2	HW χ^2
pHU10 *Xmn*I	1 ...	8.0	0.69	0.33	49	58	131	0.46	1.08	0.05
	2 ...	7.0	0.31							
pHU10 *Eco*RI	1 ...	13.0	0.83	0.41	39	32	111	0.53	0.32	2.08
	2 ...	9.6	0.17							
p7F12 *Msp*I	1 ...	4.3 2.1	0.48	0.55	97	60	241	0.47	1.20	0.70
	2 ...	3.4 2.1 0.9	0.52							
p7F12 *Taq*I	1 ...	6.9	0.27	0.32	66	60	154	0.47	0.43	0.51
	2 ...	5.9 1.0	0.73							
p7F12 *Bcl*I	1 ...	1.6	0.29	0.51	60	35	134	0.46	0.66	1.31
	2 ...	1.4 0.2	0.71							
p7D2 *Taq*I	1 ...	12.0	0.21	0.20	36	60	90	0.48	0.10	0.74
	2 ...	11.0	0.79							
Esterase D	1		0.88	0.22	89	55	208	0.53	0.22	0.02
	2		0.12							
pHU26 *Bgl*II	1 ...	9.6	0.82	0.12	23	58	61	0.47	0.15	nm
	2 ...	7.8	0.18							
p9D11 *Msp*I	1 ...	15.0	0.61	0.57	97	60	240	0.49	0.10	nm
	2 ...	10.5	0.36					0.52	0.24	
	3 ...	11.0	0.02							
p9D11 *Taq*I	1 ...	7.6 3.7	0.18	0.30	67	60	135	0.50	0.01	0.38
	2 ...	5.6 3.7	0.82							
pHUB8 *Eco*RI	1 ...	13.1	0.23	0.31	60	58	142	0.45	1.58	1.96
	2 ...	4.7	0.77							
pHUB8 *Hind*III	1 ...	10.8	0.84	0.16	37	58	70	0.51	0.05	0.41
	2 ...	8.5	0.16							
p1E8 *Msp*I	1 ...	10.1	0.51	0.55	94	60	232	0.51	0.18	0.80
	2 ...	7.4	0.49							
p9A7 *Msp*I	1 ...	4.5 1.4	0.36	0.42	72	60	195	0.49	0.18	0.53
	2 ...	4.5 1.0	0.64							
p9A7 *Hind*III	1 ...	3.2	0.62	0.43	79	58	184	0.53	0.99	0.34
	2 ...	0.9	0.38							

NOTE: For each system, we report the fragment sizes, in kilobases, corresponding to each allele; the gene frequencies (F); the observed heterozygosity (H); the numbers of grandparents (Gpt), parents (Pt), and offspring (Off) tested; segregation ratios (p) and a test of their departures from 0.5; and a test of departure from Hardy-Weinberg (HW) expectations. In two instances, the latter is not meaningful (nm) because of very low proportions in some genotype classes.

Leppert, M., et al. Am. J. Hum. Genet. 39:425–437, 1986

Locus no.	Probe (locus)	Enzyme	Allele size (kb)	Allele frequencies	Observed heterozygosity	Number typed
1	pAW101 (D14S1)	*Eco*RI	15.0–25.0 >10 alleles VNTR marker		0.64	365
2	p3.4 (IGHC)	*Bgl*II	2.9–3.7 >5 alleles VNTR marker		0.83	530
3	pMLJ14 (D14S13)	*Rsa*I	4.0–8.0 >20 alleles VNTR marker		0.94	718
4	pTHH37 (D14S16)	*Taq*I	3.0 2.3 2.1 VNTR marker	0.59 0.02 0.39	0.48	635
5	pHHH208 (D14S19)	*Bam*HI	6.5 5.8 4.0	0.45 0.47 0.08	0.61	623
6	pEFZ18.2 (D14S17)	*Taq*I	4.5 3.5	0.32 0.68	0.51	755
7	PI (α_1-antitrypsin)	PI	M1 M2 M3 S	0.67 0.14 0.11 0.08	0.59	605
8	cKKA39 (D14S23)	*Rsa*I	3.0–6.0 >10 alleles VNTR marker		0.83	666
9	pCMM62 (D14S21)	*Hind*III	4.0 2.0	0.77 0.23	0.37	676
10	pMHZ9 (D14S18)	*Taq*I	4.8 4.0	0.91 0.09	0.17	789
11	pMCOC12 (D14S20)	*Msp*I	1.0–1.7 6 alleles VNTR marker		0.66	570

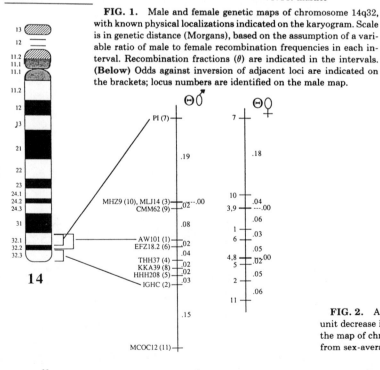

FIG. 1. Male and female genetic maps of chromosome 14q32, with known physical localizations indicated on the karyogram. Scale is in genetic distance (Morgans), based on the assumption of a variable ratio of male to female recombination frequencies in each interval. Recombination fractions (θ) are indicated in the intervals. **(Below)** Odds against inversion of adjacent loci are indicated on the brackets; locus numbers are identified on the male map.

FIG. 2. Approximate confidence limits, in shaded areas (1-lod-unit decrease in likelihood), for placement of each marker locus on the map of chromosome 14. Distances between markers are derived from sex-averaged recombination fractions.

From Nakamura, Y., et al. Genomics 4:76–81, 1989

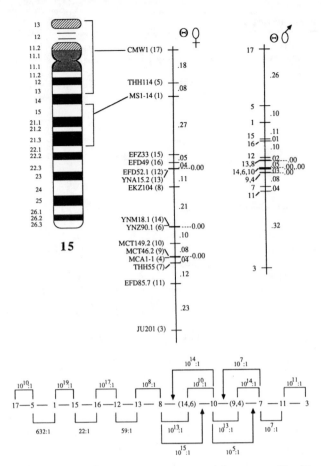

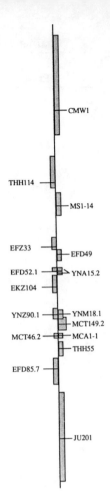

FIG. 1. Male and female genetic maps of chromosome 15, with physical localizations of selected markers indicated on the karyogram. Scale is in genetic distance (morgans) calculated under the variable–ratio model; recombination fractions (θ) are indicated in the intervals. (Below) Odds against inversion or other placement of adjacent loci are indicated on the brackets; locus numbers are identified on the female map.

FIG. 2. Approximate confidence limits, in shaded areas (one-lod-unit decrease in likelihood), for placement of each marker locus on the map of chromosome 15. Distances between markers are scaled in morgans and derived from female recombination frequencies under the constant–ratio model.

TABLE 1

Characterization of Polymorphic Loci on Chromosome 15

Locus No.	Probe (locus)	Enzyme	Allele size (kb)	Allele frequencies	Observed heterozygosity	No. of indiv. typed
1	pMS1-14 (D15S1)	*Msp*I	12.0 4.5	0.43 0.57	0.64	574
2	pDP151 (D15S2)	*Eco*RI	1.0 2.0	0.22 0.78	0.33	265
3	pJU201 (D15S3)	*Eco*RI	1.9 1.8	0.43 0.57	0.47	575
4	pMCA1-1 (D15S33)	*Pvu*II	5.7 5.2	0.77 0.23	0.47	321
5	pTHH114 (D15S25)	*Rsa*I	2.6 2.3	0.62 0.38	0.44	641
6	pYNZ90.1 (D15S28)	*Bam*HI	6.0 5.8	0.75 0.25	0.34	720
7	pTHH55 (D15S27)	*Msp*I	4.6 3.3	0.33 0.67	0.45	583
8	EKZ104 (D15S30)	*Msp*I	0.1 0.2	0.46 0.54	0.51	555
9	pMCT46.2 (D15S26)	*Pvu*II	5.9 5.3	0.88 0.12	0.25	661
10	pMCT149.2 (D15S34)	*Msp*I	0.1 0.2	0.50 0.50	0.24	741
11	pEFD85.7 (D15S37)	*Eco*RI	4.0 2.3	0.49 0.51	0.47	739
12	pEFD52.1 (D15S44)	*Taq*I	2.3 1.5	0.75 0.25	0.32	750
13	pYNA15.2 (D15S36)	*Taq*I	2.3 1.8	0.75 0.25	0.31	625
14	pYNM18.1 (D15S35)	*Taq*I	4.8 4.4 4.2	0.48 0.52 0.00	0.36	713
15	pEFZ33 (D15S45)	*Hind*III	4.0 3.7	0.28 0.72	0.38	601
16	pEFD49.2 (D15S38)	*Taq*I	4.4 2.4	0.49 0.51	0.57	274
	pEFD49.3 (D15S29)	*Msp*I	3.0 2.1 2.0	0.38 0.61 0.01	0.37	733
	Combined				0.51	
17	pCMW1 (D15S24)	*Taq*I	1.5–2.5 6 alleles (VNTR marker)		0.52	476

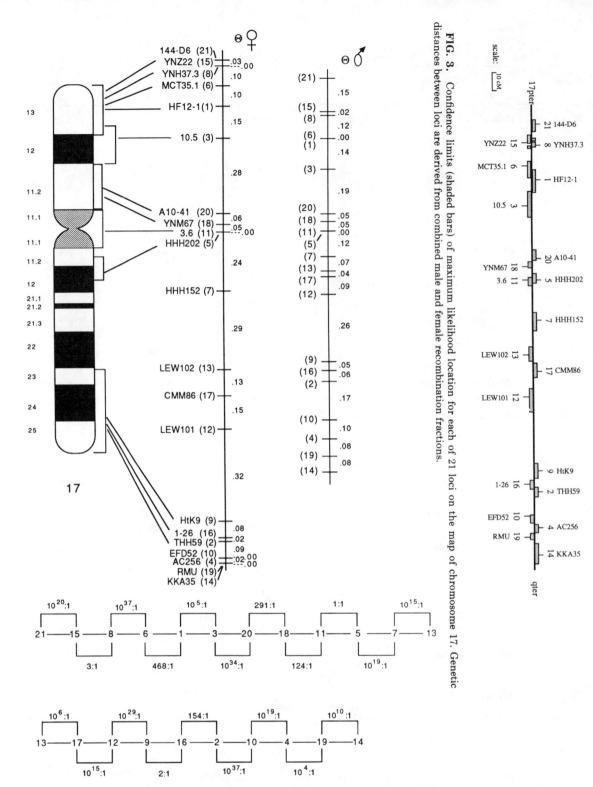

FIG. 3. Confidence limits (shaded bars) of maximum likelihood location for each of 21 loci on the map of chromosome 17. Genetic distances between loci are derived from combined male and female recombination fractions.

FIG. 2. Sex-specific genetic maps of chromosome 17, with physical localizations of selected markers indicated on the karyogram. Maps are scaled in centimorgans (cM), under the assumption of a variable sex ratio in each interval. θ = recombination fraction. **Below:** Odds against inversion of adjacent loci on the final map.

From Nakamura, Y. et al. Genomics 2:302–309, 1988

Polymorphic Loci on Chromosome 17

Locus No.	Probe (locus)	Enzyme	Allele size (KB)	Allele frequencies	Observed heterozygosity	Number typed
1	pHF12-1 (D17S1)	MspI	2.9 / 2.1	0.78 / 0.22	0.40	343
2	pTHH59 (D17S4)	TaqI	VNTR 8 alleles 3.0–4.0		0.71	669
3	p10.5 (MYH2)	HindIII	5.3 / 4.9	0.26 / 0.74	0.38	626
4	pAC256 (D17S79)	PvuII	VNTR 6 alleles 3.0–7.0		0.73	646
5	pHHH202 (D17S33)	RsaI	2.5 / 1.9	0.55 / 0.45	0.48	690
6	pMCT35.1 (D17S31)	MspI	2.4 / 1.8	0.75 / 0.25	0.43	735
7	pHHH152 (D17S32)	BamHI	10.5 / 9.6	0.39 / 0.61	0.45	572
8	pYNH37.3 (D17S28)	TaqI	VNTR 8 alleles 2.0–4.0		0.64	520
9	pHtK9 (TK1)	TaqI	4.3 / 1.3	0.54 / 0.46	0.60	588
10	cEFD52 (D17S26)	PvuII	VNTR >10 alleles 3.0–10.0		0.83	594
11	p3.6 (D17Z1)	EcoRI	2.3 / 1.6	0.72 / 0.28	0.45	608
12	pLEW101 (D17S40)	MspI	15.0 / 7.0	0.66 / 0.34	0.43	486
13	pLEW102 (D17S41)	TaqI	8.0 / 5.5	0.34 / 0.66	0.47	627
14	pKKA35 (D17S75)	MspI	2.0 / 1.8 / 1.6	0.73 / 0.20 / 0.08	0.37	420
15	pYNZ22 (D17S30)	RsaI	VNTR >10 alleles 1.3–2.3		0.86	717
16	c1-26 (D17S20)	TaqI	7.6 / 4.9	0.75 / 0.25	0.41	720
17	pCMM86 (D17S74)	HinfI	VNTR >20 alleles 1.0–3.5		0.91	604
18	pYNM67 (D17S29)	TaqI	3.8 / 3.2	0.69 / 0.31	0.41	593
	1pYNM67	RsaI	3.0 / 1.3	0.09 / 0.91	0.18	544
	2pYNM67	RsaI	1.8 / 0.8	0.18 / 0.82	0.31	490
	Haplotype				0.57	
19	pRMU3 (D17S24)	TaqI	VNTR 6 alleles 3.2–3.8		0.65	651
	pRMU1 (D17S27)	PstI	1.0 / 2.0	0.72 / 0.28	0.38	493
	haplotype				0.62	
20	pA10-41 (D17S71)	PvuII	2.7 / 2.6	0.17 / 0.83	0.29	658
	pA10-41	MspI	2.4 / 1.9	0.47 / 0.53	0.56	647
	Haplotype				0.57	
21	p144-D6 (D17S34)	RsaI	VNTR 14 alleles 1.7–5.3		0.84	715

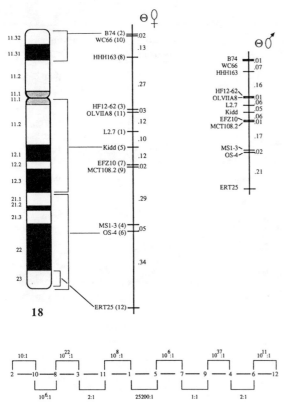

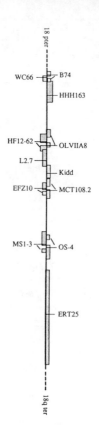

FIG. 1. Sex-specific genetic maps of chromosome 18, with physical localizations of selected markers indicated on the ideogram. Maps are scaled in genetic distance (morgans), under the assumption of a constant ratio of female/male crossingover occurrence in each interval. θ, recombination fraction. (Below) Odds against inversion of adjacent loci.

FIG. 2. Confidence limits (1-loci decrease in likelihood in shaded bars) of maximum likelihood location for each of 12 loci on the map of chromosome 19. Genetic distances (scaled in morgans) between loci are derived from female recombination fractions. Locus numbers are identified in Table 1 and Fig. 1.

Polymorphic Loci on Chromosome 18

Locus No.	Probe (locus)	Enzyme	Allele size (kb)	Allele frequency	Observed heterozygosity	Number of samples typed
1	pL2.7 (D18S6)	*Pst*I	10.0 7.5	0.76 0.24	0.42	441
2	pB74 (D18S3)	*Msp*I	4.8 1.7	0.51 0.49	0.40	625
3	pHF12-62 (D18S1)	*Taq*I	6.0 2.0	0.47 0.53	0.57	574
4	pMS1-3 (D18S19)	*Pst*I	1.0 2.0	0.54 0.46	0.48	596
5	JK (Kidd)	SEROL	JK(a) JK(b)	0.51 0.49	0.47	572
6	OS-4 (D18S5)	*Taq*I	7.6 6.0	0.81 0.19	0.33	633
7	pEFZ10 (D18S20)	*Pvu*II	VNTR 5 alleles 2.3–6.0 kb		0.70	569
8	pHHH163 (D18S21)	*Pvu*II	4.7 2.6	0.44 0.56	0.37	635
9	pMCT108.2 (D18S24)	*Taq*I	3.2 2.3	0.88 0.12	0.21	668
10	WC66 (D18S25)	*Bam*HI	18.5 7.8	0.40 0.60	0.56	550
11	OLVIIA8 (D18S7)	*Msp*I	4.2 3.8 3.1	0.67 0.32 0.01	0.42	702
12	pERT25 (D18S11)	*Hinf*I	VNTR 8 alleles 4.0–8.0 kb		0.96	135

From O'Connell, P. et al. Genomics 3:367–372, 1988

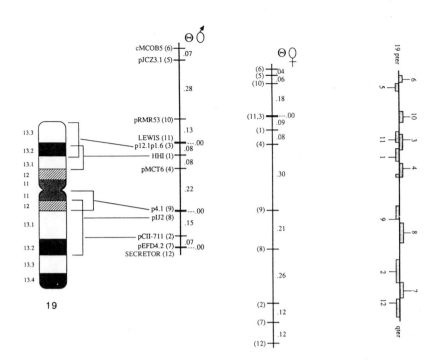

FIG. 1. Sex-specific genetic maps of chromosome 19, with physical localizations of selected markers indicated on the karyogram. Maps are scaled in centimorgans, under the assumption of a variable sex ratio in each interval. θ = recombination fraction. (**Below**) Odds against alternate orders of loci.

FIG. 2. Confidence limits (shaded bars) of maximum likelihood location for each of 12 loci on the map of chromosome 19. Scaled genetic distances between loci are derived from combined male and female recombination fractions. Locus numbers are identified in Table 1 and Fig. 1.

From Nakamura, Y. et al. Genomics 3:67-71, 1988

Polymorphic Loci on Chromosome 19

Locus no.	(Locus) Probe	Enzyme	Allele size (KB)	Allele frequencies	Observed heterozygosity	Number type
1	pHHI (LDLR)	*Pvu*II	19.0 16.0 + 2.6	0.72 0.28	0.40	673
2	pCII-711 (APOC2)	*Taq*I	3.8 3.5	0.47 0.43	0.51	653
3	p12.1p1.6 (INSR)	*Bgl*II	23.4 20.0 + 3.4	0.77 0.23	0.31	604
4	pMCT6 (D19S24)	*Bam*HI	15.0 12.0 9.0 8.5	0.12 0.22 0.59 0.07	0.67	301
		*Bam*HI	2.3 2.0	0.22 0.78	0.35	540
5	pJCZ3.1 (D19S20)	*Hinf*I	VNTR marker (>10 alleles) 1.0–4.0		0.79	706
6	cMCOB5 (D19S21)	*Pst*I	VNTR marker (8 alleles) 3.0–6.0		0.73	686
7	pEFD4.2 (D19S22)	*Pvu*II	VNTR marker 2.4 2.2	0.62 0.38	0.44	644
8	pIJ2 (D19S9)	*Eco*RI	9.0 5.0	0.13 0.87	0.25	427
9	p4.1 (D19S7)	*Msp*I	8.2 7.8	0.37 0.63	0.51	604
10	pRMR53 (D19S26)	*Bam*HI	8.0 7.0	0.85 0.15	0.24	510
	Protein polymorphisms					
11	Lewis				0.68	108
12	Secretor				0.58	276

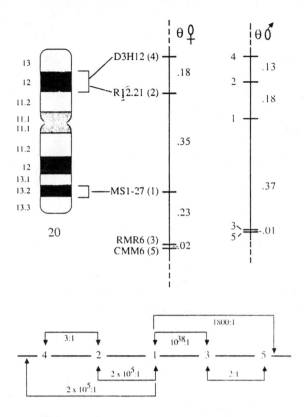

FIG. 1. (Top) Male and female genetic maps of chromosome 20, with known physical localizations indicated on the idiogram. Maps are scaled in centimorgans using the Haldane mapping function, under the assumption of a variable ratio of female/male genetic distance in each interval. θ, recombination fraction. (Bottom) Odds against alternative placements of individual loci on the maximum-likelihood map.

Characteristics of Five Polymorphic Loci on Chromosome 20

Locus no.	Probe (locus)	Enzyme	Allele size (kb)	Allele frequency	Observed heterozygosity	No. of individuals typed (No. of meioses from informative matings)
1	pMS1-27 (D20S4)	*Msp*I	6.5	0.59	50%	695
			1.5	0.40		(414)
			1.3	0.01		
2	pR12.21 (D20S5)	*Msp*I	3.8	0.31	46%	387
			3.0	0.69		(193)
3	pRMR6 (D20S20)	*Taq*I	2.5	0.27	40%	688
			2.3	0.73		(320)
4	pD3H12 (D20S6)	*Taq*I	13.1	0.52	45%	630
			8.6 + 4.5	0.48		(330)
5	pCMM6 (D20S19)	*Pst*I	1.5–10.0		90%	695
			>20 alleles			(766)
			VNTR marker			

From Nakamura, Y. et al. Genomics, in press.

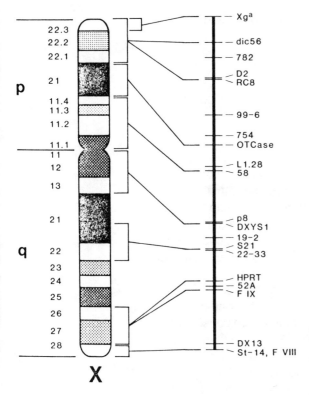

Chromosome X [1]. The Xg^a locus codes for a red blood cell antigen, HPRT for hypoxanthine phosphoribosyl transferase, FVIII and
FIX for blood clotting factors 8 and 9, and OTCase for ornithine
carbamoyl transferase. Others are arbitrary loci mapped with DNA
probes. The total length of the genetic map is approximately 185
recombination units.

From Drayna, D. & White, R. Science 230:753-758, 1985

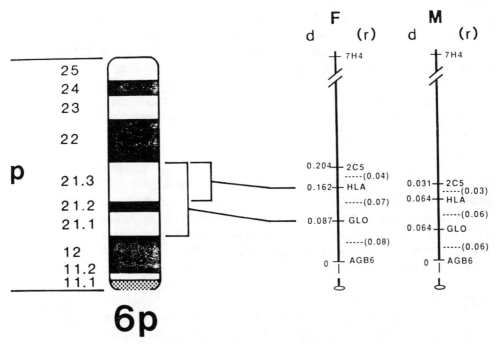

Chromosome 6p [2]. The Eighth Intenational Workshop on Human Gene
Mapping (HGM8) [9] has designated markers 2C5 and 7H4 as D6S8 and
D6S7, respectively; HLA is a haplotyped locus coding for histocompatibility antigens HLA-DRalpha, HLA-DQ alpha, and HLA-B; GLO
is the locus coding for glyoxylase I. AgB6 has not been assigned
a workshop symbol. Marker 7H4, unlinked to the others in pedigree
studies, was placed distal to the linkage group on the basis of
deletion mapping data.

From Leach, R., et al. Proc. Natl. Acad. Sci. USA 83:3909-3913, 1986

An On Average 6 cM RFLP Linkage Map of the Human Genome

Helen Donis-Keller, Cynthia Helms

Genetics Department, Washington University School of Medicine, St. Louis, MO 63110

The first published genetic linkage map of the human genome was comprised of 403 markers (393 RFLPs) and was estimated to link more than 95% of the genome with an average marker spacing of about 10 cM [2]. 147 new markers have been added and the map now contains 550 loci (see Table I and [3]), and includes 82 genes (see Table II). The map was generated from linkage analysis of genotypic data collected from three-generation pedigrees. The pedigree resource was made available, as cell lines, from the NIGMS Human Genetic Mutant Cell Repository and Ray White (University of Utah) (10 and 7 pedigrees, respectively), and as DNA from the CEPH resource (Centre d'Etude du Polymorphisme Humain). In addition, published genotypic data from 126 loci was made available for analysis from the database (Version 2) maintained by the CEPH organization. The chromosome maps were constructed using the linkage analysis program package CRI-MAP (current version 2.3, Philip Green [4]). A similar program package, MAPMAKER [8] was also utilized in the construction of the first genome map. Two point linkage analysis was used to arrange loci into LOD 4 linkage groups, and each linkage group was assigned to a chromosome based on physical localization of at least 1 marker per group. Linkage maps for each chromosome were then constructed by multilocus analysis [7]. In the CRI-MAP programs a pair of linked loci is first chosen as the nucleus of the map. Each remaining locus, in turn, is then placed in each possible position with respect to the nucleus, and the maximum likelihood computed for each locus order. An order is excluded if is is at least 100 fold less likely than another. A locus is added to the map when all locations except one are excluded and the process repeated with the remaining loci (often the odds for uniquely placed markers was 1000:1, e.g. chromosome 5; or 10,000:1, e.g. chromosome 16). Following construction of this map, any remaining loci (i.e., those not uniquely placed) are placed in their most likely positions. Physical localization of 95 loci in this map to the cytogenetic map of metaphase chromosomes provides reference points between these two representations of the human genome.

Chromosomes 7 and 16 are the most densely mapped with an average marker spacing of 2.4 cM and 3.0 cM, respectively, compared to the overall genome estimated average marker spacing of 6 cM, based on a sex-averaged genome size of 3300 cM. The chromosome 7 linkage map contains 71 loci and spans a distance of 173 cM in males, 275 cM in females, and a sex averaged distance of 219 cM. Physical localizations to metaphase chromosome preparations of 8 markers incorporated into the genetic linkage map help correlate this genetic map to the physical representation [3]. Recent efforts are aimed at defining the boundaries of the genetic maps by incorporating cloned telomeric regions. Preliminary evidence closely links a chromosome 7q telomere sequence (HTY146) to the distal markers on the chromosome 7 map [9][3]. The chromosome 16 map containing 45 markers spans 195 cM in females and 115 cM in males, with a sex averaged spacing of 149 cM [6]. While the maps in general appear to reflect Haldane's rule (i.e., that the heterogametic sex shows less recombination than the homogametic sex, [5]), in some cases significantly increased recombination in males has been observed especially in distal regions of chromosomes. Particularly striking is the region near the alpha globin locus on chromosome 16p in which the recombination fraction between the markers 3'HVR (573 meioses) and 0327 (276 meioses) are 0.01 in females and 0.14 in males (Chi-square = 9.67, p < 0.01) [6]. Substantially updated maps for chromosome 10 [1], chromosome 16 [6], and chromosome 5 [11][10] are shown.

We have defined a subset of 251 RFLP markers from the map that can be used for efficient systematic searches of the genome for linkage to inherited disorders for which adequate pedigree resources exist. The markers were chosen based on their informativeness and with the goal of a marker spacing of no more than 10 cM (see Table III). For chromosomes 7 and 16, as few as 10-20 informative meioses would be needed to achieve a LOD 3.0 in a search of a simple dominantly inherited disorder. The majority of probes listed were originally isolated at Collaborative Research, Inc. (CRI prefix) and may be obtained directly from them. All other probes are available from the ATCC or by contacting us for information.

The Maps: The following graphic representations of our linkage analysis results show male, female, and sex-average maps (in which the recombination fraction was not allowed to vary between male and female) for each chromosome. Recombination fractions were converted to cM using the Kosambi mapping function and are indicated to the right of the maps. Loci that are not placed uniquely (with odds of at least 100:1) are shown on the sex average maps with the most likely intervals indicated by a thin line to the side of the marker label. Genotype data incorporated from the V2 CEPH database is boxed; a dashed box indicates data also collected by other CEPH collaborators (total of 21 loci). Physically localized markers incorporated into the linkage maps are shown to the right of the chromosome ideograms (most citations for these physical localizations can be found in [1, 2, 3, 6, 10, 11] and the New Haven Gene Mapping Library Chromosome Plots Number 4, HGM 9.5, 1988).

Literature Citations

1. Bowden, D. W., T. C. Gravius, P. Green, K. Falls, D. Wurster-Hill, W. Noll, H. Muller-Kahle and H. Donis-Keller. A Genetic Linkage Map of 32 Loci on Human Chromosome 10. Genom. 5: In Press.

2. Donis-Keller, H., P. Green, C. Helms, S. Cartinhour, B. Weiffenbach, K. Stephens, T. P. Keith, D. W. Bowden, D. R. Smith, E. S. Lander, D. Botstein, G. Akots, K. S. Rediker, T. Gravius, V. A. Brown, M. B. Rising, C. Parker, J. A. Powers, D. E. Watt, E. R. Kauffman, A. Bricker, P. Phipps, H. Muller-Kahle, T. R. Fulton, S. Ng, J. W. Schumm, J. C. Braman, R. G. Knowlton, D. F. Barker, S. M. Crooks, S. E. Lincoln, M. J. Daly and J. Abrahamson. A Genetic Linkage Map of the Human Genome. Cell. 51(October 23): 319-337, 1987.

3. Donis-Keller, H., C. Helms, P. Green, H. Riethman, S. Ramachandra, K. Falls, D. W. Bowden, B. Weiffenbach, T. Keith, K. Stephens, L. A. Cannizzaro, T. B. Shows, G. D. Stewart and M. Van Keuren. A human genome linkage map with more than 500 RFLP loci and average marker spacing of 6 centiMorgans. Cytogenet. Cell Genet. : In Press.

4. Green, P. CRIMAP: a multilocus linkage analysis program package. : In Preparation.

5. Haldane, J. B. S. Sex ratio and unisexual sterility in hybrid animals. J. Genet. 12: 101-109, 1922.

6. Keith, T. P., P. Green, S. T. Reeders, V. A. Brown, P. Phipps, A. Bricker, K. Falls, K. Rediker, J. A. Powers, C. Hogan, C. Nelson, R. Knowlton and H. Donis-Keller. Genetic Linkage Map of 45 DNA Markers on Human Chromosome 16. Proc. Natl. Acad. Sci. USA. : In Press.

7. Lander, E. S. and P. Green. Construction of Multilocus GeneticLinkage Maps in Humans. Proc. Natl. Acad. Sci. USA. 84: 2363-2367, 1987.

8. Lander, E. S., P. Green, J. Abrahamson, A. Barlow, M. Daly, S. Lincoln and L. Newburg. MAPMAKER: an interactive computer package for constructing primary genetic linkage maps of experimental and natural populations. Genom. 1: 174-181, 1987.

9. Riethman, H. C., R. K. Moyzis, J. Meyne, D. T. Burke and M. V. Olson. Cloning human telomeric DNA fragments into Saccharomyces Cerevisiae using a yeast artificial-chromosome vector. Proc. Natl. Acad. Sci. USA. 86 (August):6240-6244, 1989.

10. Weiffenbach, B., K. Falls, P. Green, N. Shute, T. Keith and H. Donis-Keller. A Genetic Linkage Map of Human Chromosome 5 with 53 RFLP Loci. : In Preparation.

11. Weiffenbach, B., K. Falls, P. Green, N. Shute, T. Keith and H. Donis-Keller. Genetic Map of Chromosome 5 with 53 RFLP Loci. Cytogenet. Cell Genet. : In Press.

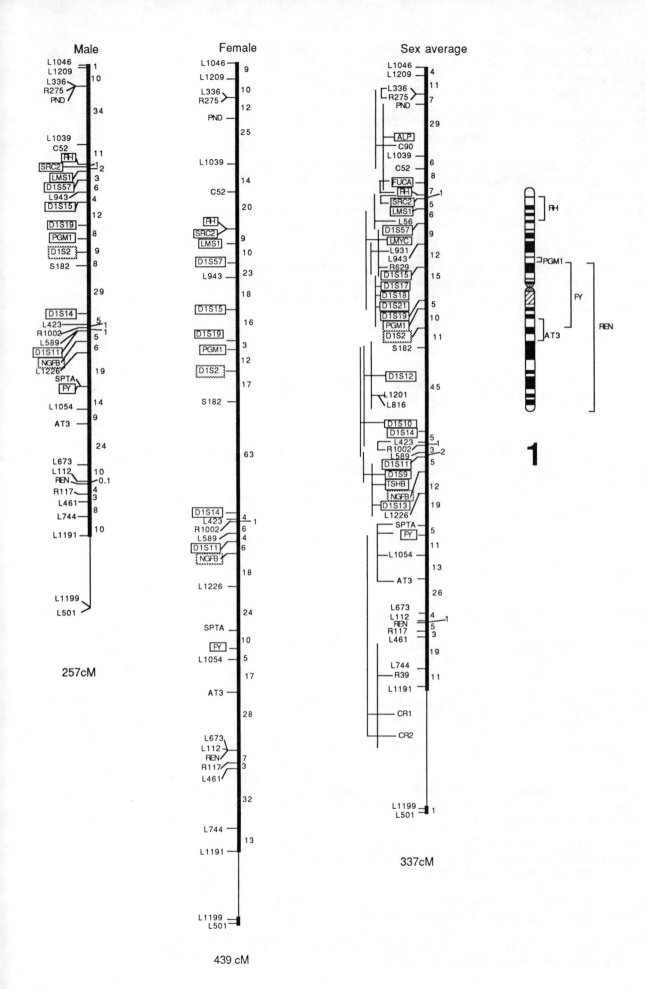

Male

L1046
L1209
L336
R275
PND
34
L1039
C52
11
RH
2
SRC2
LMS1
3
D1S57
6
L943
4
D1S15
12
D1S19
8
PGM1
D1S2
9
S182
8
29
D1S14
5
L423
1
R1002
1
L589
5
D1S11
6
NGFB
L1226
19
SPTA
FY
L1054
14
AT3
9
24
L673
L112
10
REN
0.1
R117
4
L461
3
L744
8
L1191
10
L1199
L501

257cM

Female

L1046
9
L1209
10
L336
R275
12
PND
25
L1039
14
C52
20
RH
9
SRC2
LMS1
10
D1S57
L943
23
18
D1S15
16
D1S19
3
PGM1
12
D1S2
17
S182
63
D1S14
4
L423
1
R1002
6
L589
4
D1S11
6
NGFB
18
L1226
24
SPTA
FY
10
L1054
5
17
AT3
28
L673
L112
REN
7
R117
3
L461
32
L744
13
L1191
L1199
L501

439 cM

Sex average

L1046
4
L1209
11
L336
R275
7
PND
29
ALP
C90
6
L1039
8
C52
FUCA
7
RH
1
SRC2
5
LMS1
6
L56
D1S57
9
LMYC
L931
12
L943
R629
D1S15
15
D1S17
D1S18
D1S21
5
D1S19
10
PGM1
11
D1S2
S182
D1S12
45
L1201
L816
D1S10
D1S14
L423
5
R1002
1
L589
3
2
D1S11
5
D1S9
TSHB
12
NGFB
D1S13
19
L1226
SPTA
5
FY
11
L1054
13
AT3
26
L673
L112
4
REN
1
R117
5
L461
3
19
L744
R39
11
L1191
CR1
CR2
L1199
1
L501

337cM

RH

PGM1

FY

AT3

REN

1

5.160

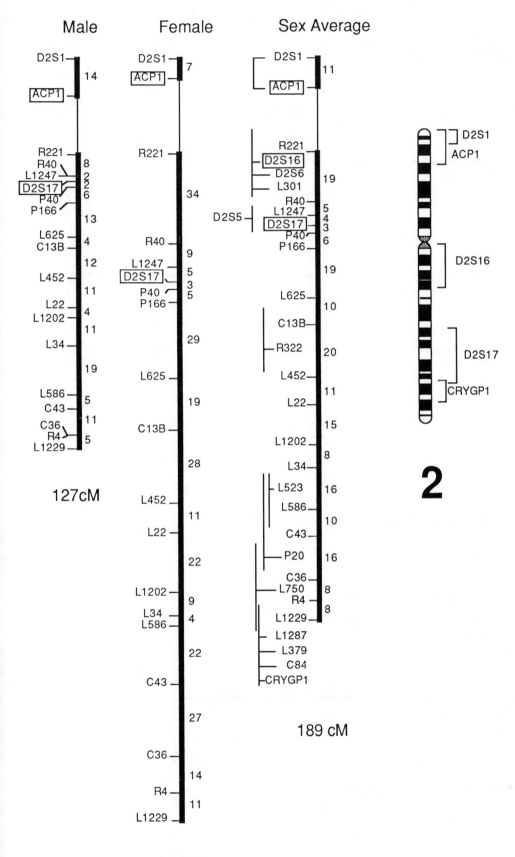

Male

D2S1
ACP1
14

R221
8
R40
L1247
2
D2S17
2
P40
6
P166
13
L625
4
C13B
12
L452
11
L22
4
L1202
11
L34
19
L586
5
C43
11
C36
R4
5
L1229

127cM

Female

D2S1
7
ACP1

R221

34

R40
9
L1247
5
D2S17
3
P40
5
P166

29

L625

19

C13B

28

L452
11
L22

22

L1202
9
L34
4
L586

22

C43

27

C36
14
R4
11
L1229

259cM

Sex Average

D2S1
11
ACP1

R221
D2S16
D2S6
19
L301
R40
5
L1247
4
D2S5
D2S17
3
P40
6
P166

19

L625
10
C13B
R322
20
L452
11
L22
15
L1202
8
L34
L523
16
L586
10
C43
P20
16
C36
L750
8
R4
L1229
8
L1287
L379
C84
CRYGP1

189 cM

D2S1
ACP1

D2S16

D2S17

CRYGP1

2

5.161

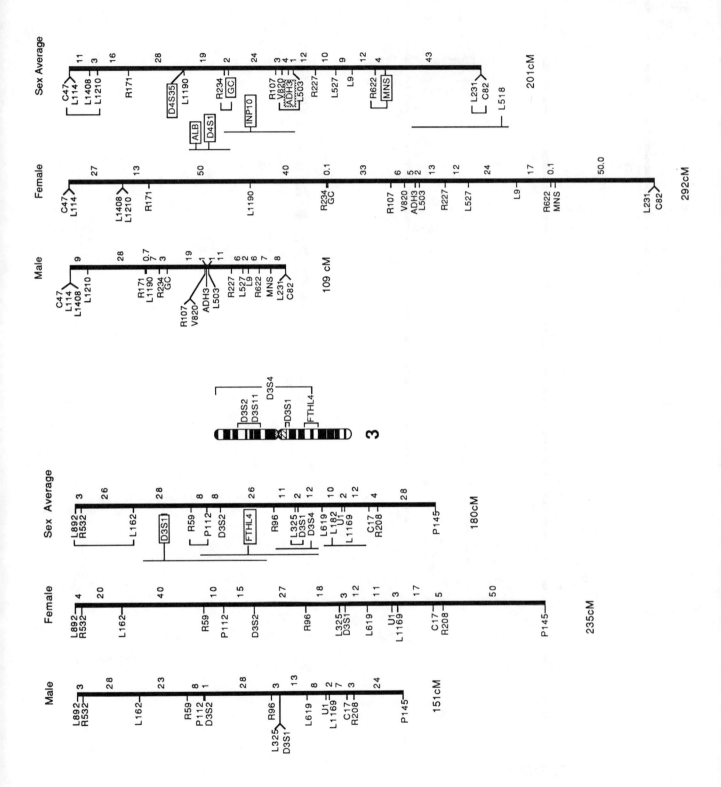

5.162

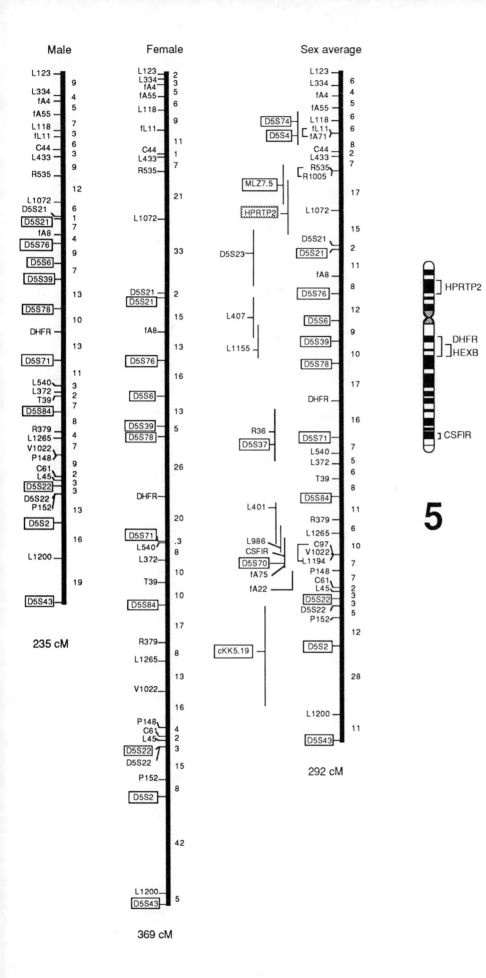

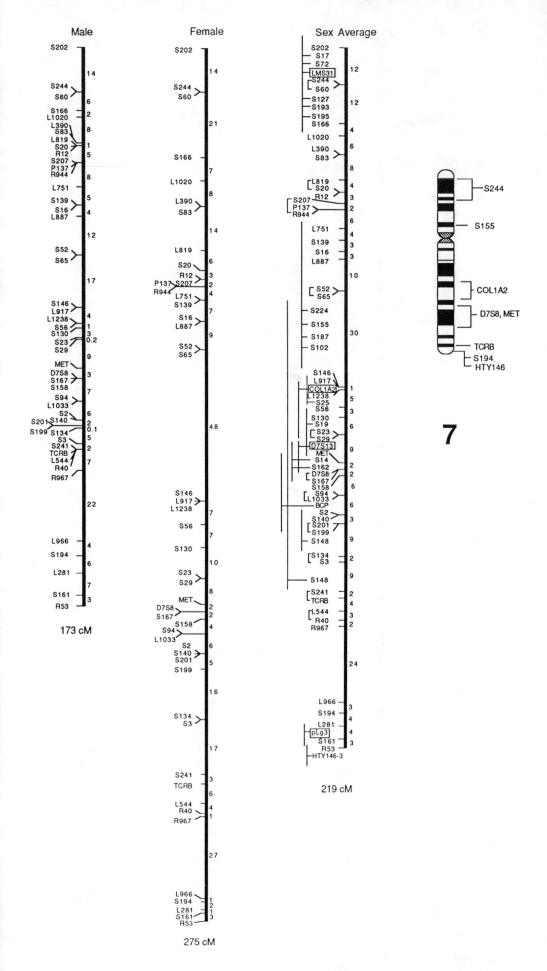

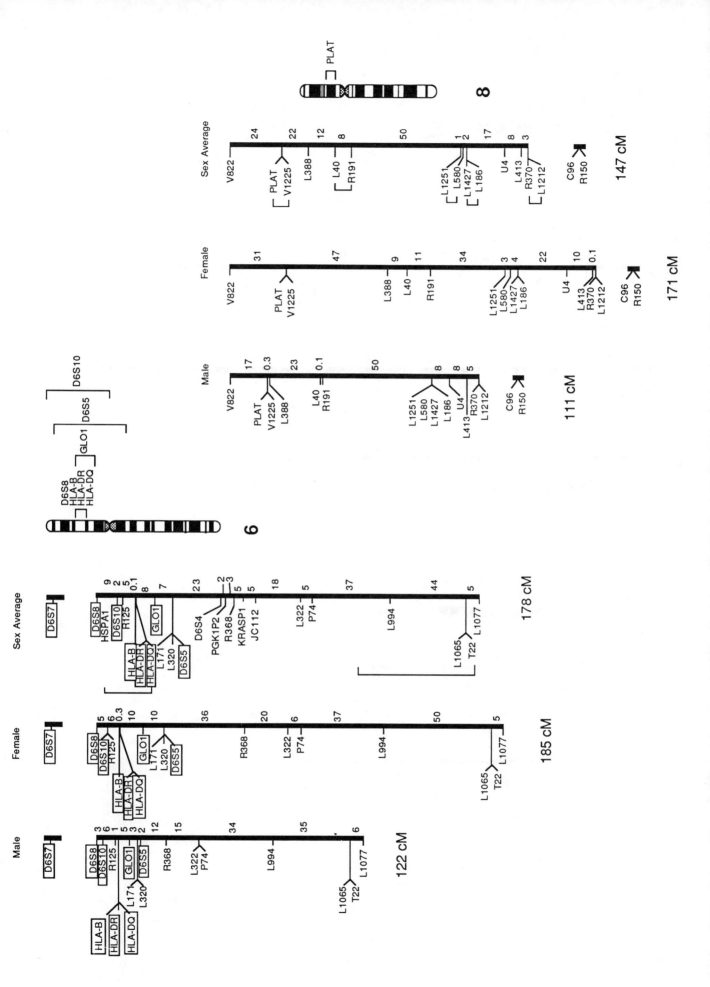

5.165

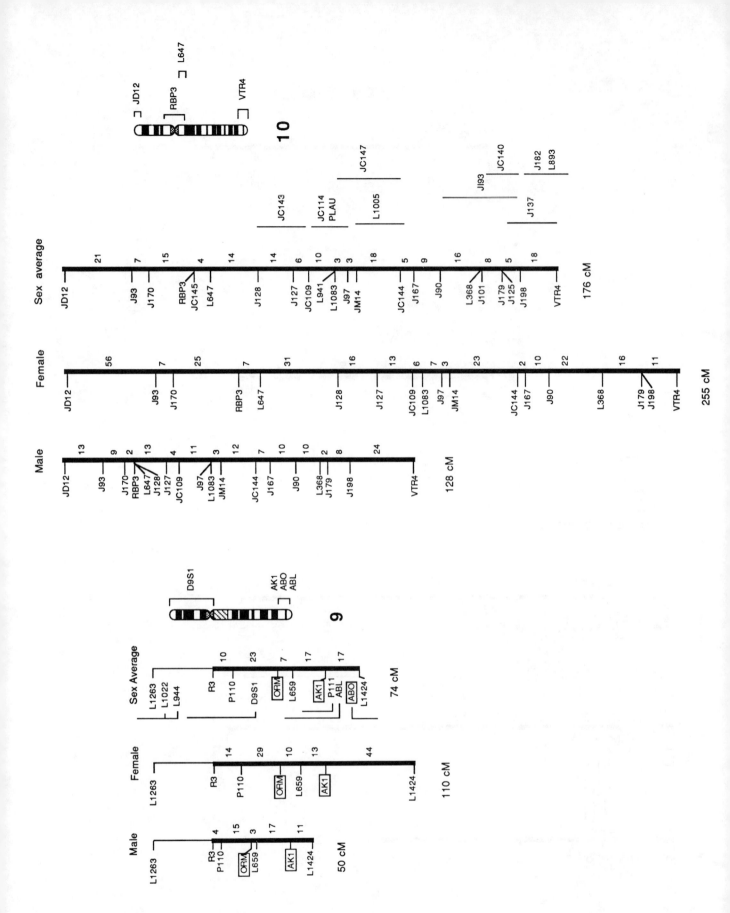

5.166

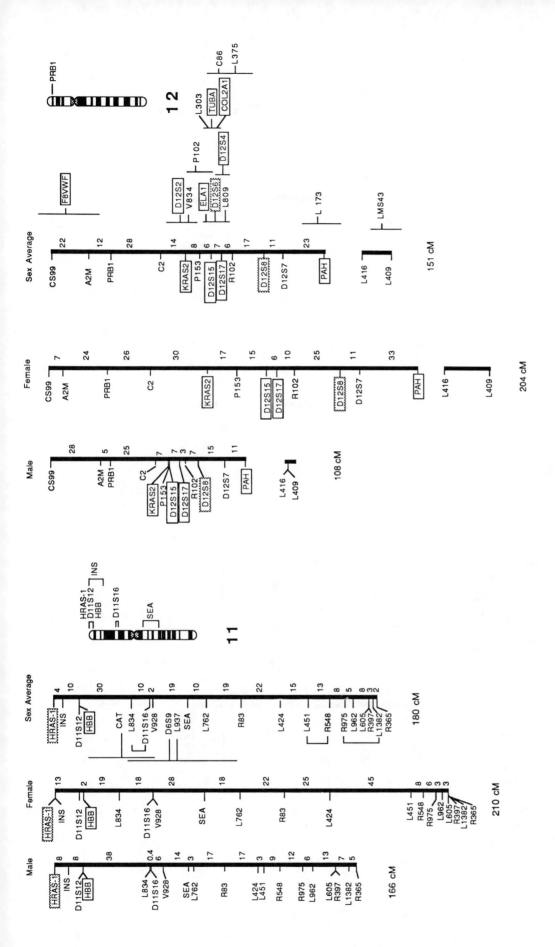

5.168

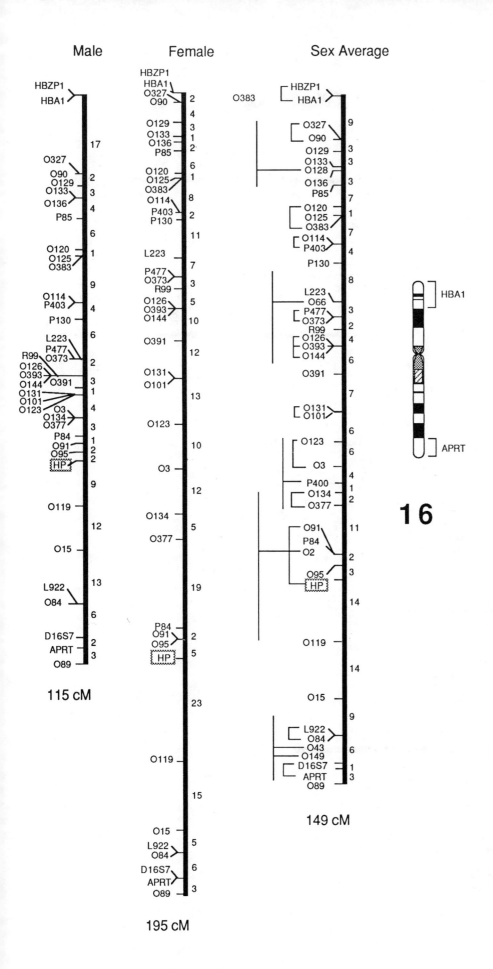

Male

HBZP1
HBA1

17

O327
O90
O129
O133
O136
P85

O120
O125
O383

O114
P403
P130

L223
P477
R99
O373
O126
O393
O144 O391
O131
O101
O123 O3
O134
O377
P84
O91
O95
HP

O119

O15

L922
O84

D16S7
APRT
O89

115 cM

Female

HBZP1
HBA1
O327
O90

O129
O133
O136
P85

O120
O125
O383
O114
P403
P130

L223

P477
O373
R99
O126
O393
O144

O391

O131
O101

O123

O3

O134

O377

O119

P84
O91
O95
HP

O119

O15
L922
O84
D16S7
APRT
O89

195 cM

Sex Average

O383

HBZP1
HBA1

O327
O90
O129
O133
O128
O136
P85

O120
O125
O383
O114
P403

P130

L223
O66
P477
O373
R99
O126
O393
O144

O391

O131
O101

O123

O3

P400
O134
O377

O91
P84
O2

O95
HP

O119

O15

L922
O84
O43
O149
D16S7
APRT
O89

149 cM

HBA1

APRT

16

5.169

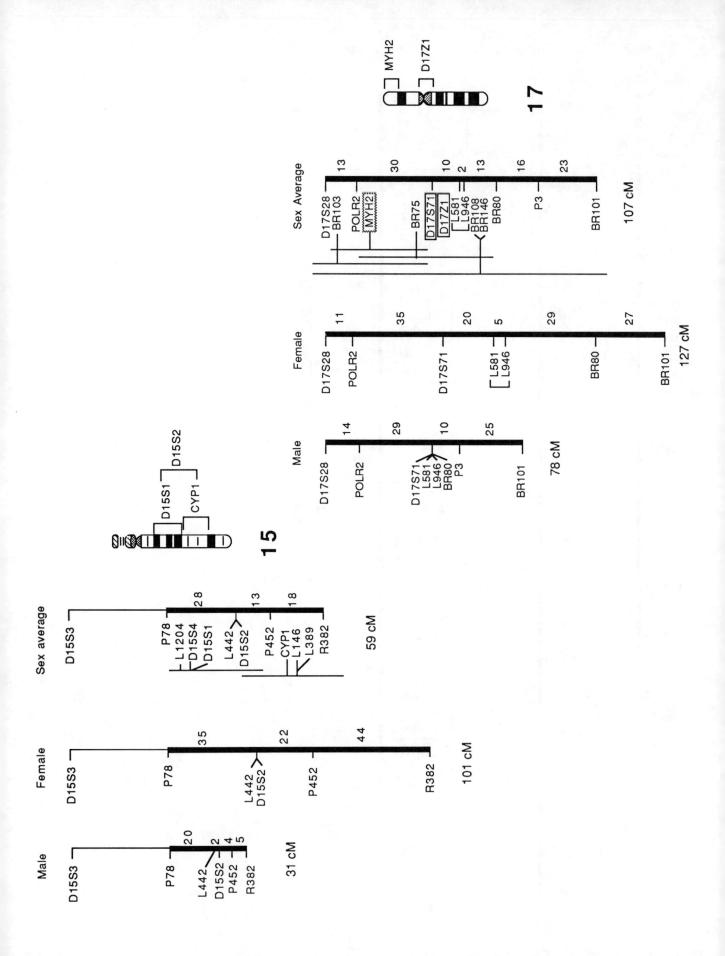

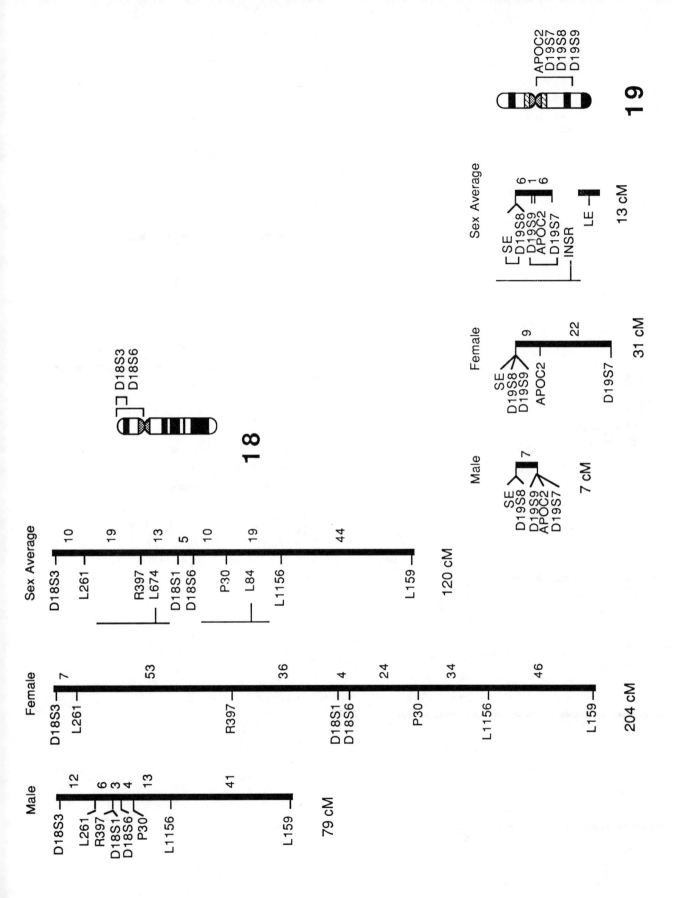

5.171

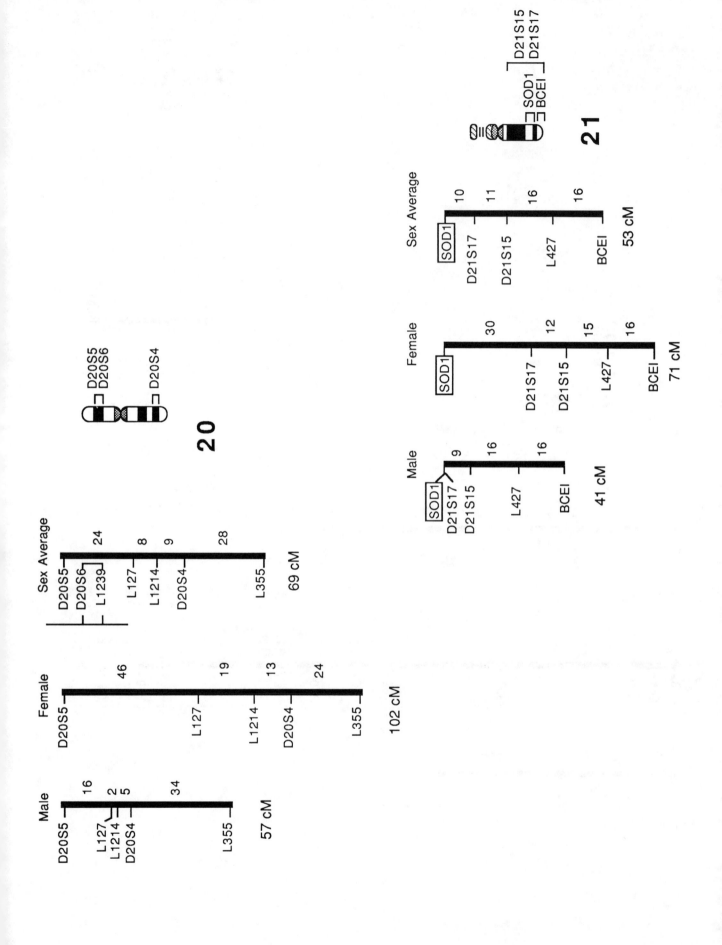

5.172

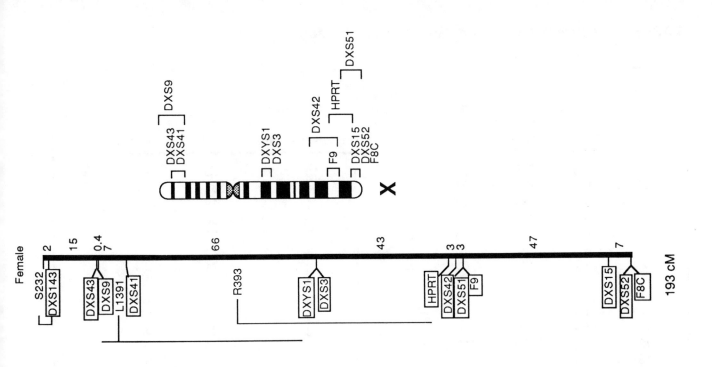

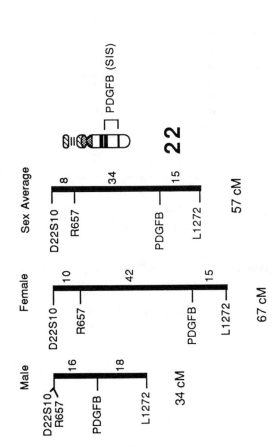

TABLE I. Current Combined Genome RFLP Map

chr.	% genome	sex avg. length [1]	RFLPs [2]	physically localized RFLPs [3]		CEPH V2 [4]		total loci [5]	avg spacing (cM)
				genes	other	genes	other		
1	7.5	249	28	7	1	10	14	58	4.3
2	6.7	221	24	2	3	1	2	32	6.9
3	5.8	192	14	0	3	1	1	19	10.1
4	5.4	177	18	3	0	9	2	30	5.9
5	5.3	175	33	3	4	1	15	53	3.3
6	5.1	167	13	2	0	3	5	23	7.3
7	5.2	174	61	4	2	3	4	71	2.4
8	5.1	167	17	1	0	0	0	18	9.3
9	4.4	147	8	1	1	3	0	13	11.3
10	4.9	162	29	2	1	0	0	32	5.7
11	4.2	140	16	4	1	3	4	26	5.7
12	5.3	175	13	3	3	7	7	30	5.8
13	3.4	112	2	1	3	1	9	13	8.6
14	3.4	112	5	0	1	0	0	6	14.0
15	3.7	122	7	1	4	0	0	12	10.2
16	4.2	138	40	4	1	1	0	45	3.1
17	4.2	139	7	2	1	1	1	11	12.6
18	3.8	125	7	0	3	0	0	10	12.5
19	3.7	124	0	2	3	2	0	7	17.7
20	3.8	126	4	0	3	0	0	7	18.0
21	2.1	69	2	2	4	1	0	8	8.6
22	2.3	76	2	1	1	0	0	4	19.0
X,Y	5.1	167	6	0	1	1	14	22	7.3
TOTALS:		3300	356	44	45	48	78	550	6.0

1. Estimated genetic lengths are based on chiasmata in male meioses (Morton et al, (1982) Hum. Genet. 62, 266-270) extrapolated to a sex-averaged genome size of 3300 cM (e. g., Chr 1 = 195 cM x 3300 cM / 2600 cM = 249 cM.)

2. Includes published and recently developed RFLPs (originally isolated at CRI). Probes having less than 20 informative meioses are not included.

3. Physically localized probes for RFLP loci obtained from other researchers or ATCC.

4. CEPH database Version 2 RFLP loci designated published.

5. The totals represent the number of individual loci (each of the twenty-one loci duplicated in the CEPH database are counted only once).

Table II. Genes Incorporated Into Human Genetic Linkage Map

Locus	Probe*	Gene name	Physical location	CEPH collab.
AT3	pAT3c	antithrombin III	1q23	1
ALP	ALP	alkaline phosphatase, liver/bone/kidney	1p36.1-p34	38
CR1	pCR1-1	complement component (3b/4b) receptor 1	1q32	1
CR2	pCR2-1.6	complement component (3d/ Epstein Barr virus) receptor 2	1q32	1
FGR	Src2B8	Gardner-Rasheed feline sarcoma viral (v-fgr) oncogene homolog	1p36.2-P36.1	38
FUCA1	FUCA	fucosidase, alpha-L-1, tissue	1p34	38
FY	---	Duffy blood group	1q22-q23	20
MYCL	LMYC	avian myelocytomatosis viral oncogene, lung derived	1p32	38
NGFB	phbetaN8C6	nerve growth factor receptor, beta polypeptide	1p22.1 or p13	1
	N8C6	"	1p22.1 or p13	38
PGM1	---	phosphoglucomutase1	1p22.1	20
PND	pJA119	pronatriodilatin	1p36	1
REN	pHRnES1.9	renin	1q32 or 1q42	1
RH	---	Rhesus blood group	1p34-p36.2	20
SPTA1	pHalphaSp5	spectrin, alpha, erythrocytic 1	1q22-q25	1
TSHB	TSHB	thyroid stimulating hormone, beta polypeptide	1p22	38
ACP1	---	acid phosphatase 1, soluble	2p23 or p25	20
CRYGP1	p5G1	crystallin, gamma polypeptide pseudogene 1	2q33-q35	1
FTHL4	FTH	ferritin, heavy polypeptide-like 4	3q21-q23	20
ADH3	pADH74	alcohol dehydrogenase (class 1) gamma polypeptide	4q21-q25	16
	pADH73	"	4q21-q25	1
ALB	F47-B44	albumin	4q11-q13	16
EGF	EGF121	epidermal growth factor	4q25	16
FBB/FBG	H1B2-H1G2	fibrinogen, B beta polypeptide	4q31	16
GC	---	group-specific component (vitamin D binding protein)	4q12-q13	20
INP10	giFN31-7	protein 10 from gamma-interferon induced cell line	4q21	16
IL2	IL2-5	interleukin 2	4q26-q27	16
MNS	---	MNS blood group	4q28-q31	20
MT2P1	pHM6	metallothionein 2 pseudogene 1 (processed)	4p11-q21	1
	MT11	"	4p11-q21	16
RAF1P1	RAF2p52	murine leukemia viral (v-raf-1) oncogene pseudogene 1	4p16.1	1

Locus	Probe*	Gene name	Physical location*	CEPH collab.
CSF1R	Cvfms	colony stimulating factor 1 receptor [(v-fms)oncogene homolog]	5q33.2-q33.3	1
DHFR	cHB203	dihydrofolate reductase	5q11.1-q13.3	1
HPRTP2	pL500	hypoxanthine phosphoribosyltransferase pseudogene 2	5p14-p13	20
	plambda500	"	5p14-p13	1
GLO1	---	glyoxylase I	6p21.3-p21.1	20
HLA-B	pDP001	major histocompatibility complex, class I	6p-21.3	20
HLA-DQA	DCH1	major histocompatibility complex, class II, alpha	6p21.3	20
HLA-DRA	DRH7	major histocompatibility complex, class II, alpha	6p21.3	20
	DRH7	"	6p21.3	20
HSPA1	pH2.3	heat shock 70 kD protein 1	6p22-p21.3	1
KRAS1P	pJ819	Kirsten rat sarcoma 1 viral (v-Ki-ras 1) oncogene homolog, processed pseudogene	6p12-11	1
PGK1P2	pGK4	phosphoglycerate kinase 1, pseudogene 2	6p21-q12	2
BCP	JHN-blue	blue cone pigment	7q22-qter	1
COL1A2	NJ3	collagen type I, alpha 2	7q21.3-q22.1	20,5
MET	pMETH	met proto-oncogene	7q31	20,5
	pMETH	"	7q31	20,1,5
	pMETD	"	7q31	20,1,5
	pHOS6	"	7q31	20
TCRB	VB8	T cell receptor, beta polypeptide	7q35	11
	VB11	"	7q35	11
	pJ2	"	7q35	11,20,5
	pT10	"	7q35	1
PLAT	pCGE217	plasminogen activator, tissue	8p12	1
ABL	pablK2	Abelson murine leukemia viral (v-abl) oncogene homolog	9q34	1
ABO	---	ABO blood group	9q34.1-q34.2	20
AK1	---	adenylate kinase 1	9q34.1-q34.2	20
ORM1	---	orosomucoid 1	9q31-qter	20
PLAU	pCGE194	plasminogen activator, urokinase	10q24-qter	1
RBP3	H.4 IRBP	retinol-binding protein 3, interstitial	10p11.2-q11.2	1
CAT	pINT-800	catalase	11p13	1
HBB	2JW151	hemoglobin, beta	11p15.5	20
	1JW151	"	11p15.5	20
	JW102	"	11p15.5	20

Locus	Probe*	Gene name	Physical location	CEPH collab.
HRAS1	pTBB-2	Harvey rat sarcoma 1 viral (v-Ha-ras-1)oncogene homolog	11p15.5	20
	pUCEJ6.6	"	11p15.5	1
INS	pINS-310	insulin	11p15.5	20
	phins 310	"	11p15.5	1
SEA	clone 3	S13 avian erythroblastosis oncogene homolog	11q13	1
A2M	pha2ml	alpha-2-macroglobulin	12	1
COL2A1	cosHcol1	collagen , type II, alpha 1	12q13.1-q14.3	20
ELA1	pXP13	elastase 1	12	20
F8VWF	pISV2	coagulation factor VIII VWF (von Willebrand factor)	12p12.2-pter	20
KRAS2	p640	Kirsten rat sarcoma viral (v-Ki-ras2) oncogene homolog	12p12.1	20
PAH	pPH72	phenylalanine hydroxylase	12q22-q24.2	20
	pPH72	"	12q22-q24.2	20
PRB1	pPRPII2.2RP	proline rich protein BstNI subfamily 1	12p13.2	1
TPI1	phPI-450-3'	triosephosphate isomerase 1	12p13	1
TUBAL1	TUBA	tubulin, alpha-like 1	12q	11
ESD	--	esterase D/ formylglutathione hydrolase	13q14.1-q14.2	20
TUBBP2	21beta3'UT	tubulin, beta polypeptide pseudogene 2	13	1
CYP1	pHUAP/B2.2	cytochrome P450, subfamily I (aromatic compound inducible)	15q22-q24	1
APRT	3'HVR, 5'HVR	adenine phosphoribosyltransferase	16q24	1
HBA1	--	hemoglobin, alpha 1	16p13	1
HP	hp2alpha	haptoglobin	16q22.1	20
		"	16q22.1	1
HBZP1	pBRZ	hemoglobin, zeta pseudogene1	16p13.3	1
MYH2	p10-5	myosin, heavy polypeptide 2, skeletal muscle, adult	17p13.1	19
	p10-3	"	17p13.1	1
POLR2	pHRp5.5	polymerase (RNA) II (DNA directed) large polypeptide	17p13.1-p12	1
APOC2	pCII-711	apolipoprotein C-II	19q12-13.2	1
INSR	pH1R/P12-1	insulin receptor	19p13.3-p13.2	1
LE	--	Lewis blood group	19	20
SE	--	ABH secretion	19q12-q13.2	20
BCEI	pS2	estrogen-inducible sequence, expressed in breast cancer	21q22.3	1
SOD1	--	superoxide dismutase 1, soluble	21q22.1	20
	pS61-10	"	21q22.1	1

Locus	Probe*	Gene name	Physical location	CEPH collab.
PDGFB	pSM-1	platelet derived growth factor beta polypeptide (simian sarcoma viral (v-sis) oncogene homolog	22q12.3-q13.1	1
F8C	--	coagulation factor VIIIc, procoagulant component(hemophilia A)	Xq28	20
F9	F9	coagulation factor IX (plasma thromboplastic component (hemophilia B)	Xq26.3-q27.2	20,3,4
HPRT	pL500	hypoxanthine phosphoribosyltransferase	Xq26	20

Explanatory Notes

* polymorphic protein and serological markers are shown as ---

Locus symbols, gene names, and physical localizations are from the New Haven Human Gene Mapping Library Chromosome Plots, Number 4, HGM9.5.

CEPH collaborators contributing published data on genes to the CEPH version 2 database used to construct our maps:

ID#	CEPH collaborators	
1	H. Donis-Keller	Washington University, St. Louis, MO (formerly CRI)
2	C.E.P.H.	College De France, Paris, France
3	J. L. Mandel	Institute De Chimie Biologique, Strasbourg, France
4	K. Davies	John Radcliffe Hospital, Headington, Oxford, UK
5	R. Williamson	St. Mary's Hospital Medical School, London, UK
11	R.A. Gatti	UCLA School of Medicine, Los Angeles, CA
16	J. Murray	University of Iowa Hospitals, Iowa City, IA
19	C. Schwartz	Greenwood Genetic Center, Greenwood, SC
20	R. White	Howard Hughes Medical Institute, Salt Lake City, UT
38	N. C. Dracopoli	MIT Center for Cancer Research, Cambridge, MA

Table III. Chromosome Subsets: RFLP Mapping Panel

Chr.	Probe, Enz.	Locus	Het.	Spacing (cM)*	Chr.	Probe, Enz.	Locus	Het.	Spacing (cM)*
1	CRI-L1046, T, M	D1S49	.74		2	CRI-L586, T, M	D2S39	.40	16
	CRI-L1209, T	D1S50	.64	04		CRI-C43, E	D2S23	.21	10
	CRI-LA336, R	D1S47	.76	11		CRI-C36, E	D2S24	.33	16
	pJA119, T	PND	.17	07		CRI-L1229, T	D2S35	.57	16
	CRI-L1039, M	D1S71	.40	29					
	CRI-C52, E	D1S56	.57	06	3	CRI-L892, T	D3S17	.71	
	Src2B8, R	FGR	.51	16		CRI-R532, M	D3S22	.62	03
	pYNZ2, T	D1S57	.55	11		CRI-L162, T	D3S18	.19	26
	CRI-L943, T	D1S41	.36	09		CRI-R59, M	D3S12	.67	28
	1-11B, Bg	D1S15	.60	12		CRI-P112, M	D3S23	.40	08
	4-03, S	D1S19	.46	15		pHF12-32, M	D3S2	.38	08
	L1.22, Bg	D1S2	.48	15		CRI-R96, M	D3S13	.64	26
	CRI-S182, M	D1S38	.43	11		HS3, H	D3S1	.40	13
	6-02, T	D1S14	.50	45		CRI-L619, T	D3S16	.40	12
	CRI-L589, E	D1S55	.52	09		CRI-L1169, M	D3S20	.74	12
	1-18, Bc	D1S11	.44	02		CRI-C17, M	D3S47	.69	12
	CRI-L1226, M	D1S36	.67	17		CRI-R208, Bg, M	D3S14	.76	04
	pHalphaSp5, Hc	SPTA	.29	19		CRI-P145, T	D3S26	.48	28
	CRI-L1054, M	D1S42	.36	16					
	pAT3c, P	AT3	.43	13	4	CRI-C47, B	D4S117	.33	
	CRI-L673, M	D1S53	.40	26		CRI-L114, H	D4S110	.29	00
	CRI-L112, B	D1S52	.45	04		CRI-L1408, M	D4S107	.50	11
	CRI-L461, T	D1S70	.50	09		CRI-R171, Bg	D4S104	.31	19
	CRI-L744, R	D1S48	.52	19		CRI-L1190, M, T	D4S103	.67	28
	CRI-L1191, M	D1S51	.40	11		CRI-R234, Bg, He	D4S105	.57	19
	CRI-L1199, M	D1S68	.90	UL		CRI-R107, M	D4S108	.50	26
						CRI-L503, T	D4S100	.48	08
2	L2.30, Bg	D2S1	.69			CRI-R227, E, T	D4S101	.81	12
	CRI-R221, T	D2S27	.40	UL		CRI-L527, T	D4S120	.45	10
	CRI-R40, M, Bg	D2S28	.38	19		CRI-L9, H	D4S109	.40	09
	CRI-L1247, M	D2S34	.55	05		CRI-R622, T	D4S121	.21	12
	CRI-P166, M	D2S30	.67	13		CRI-L518, T	D4S130	.83	UL
	CRI-L625, M, T	D2S38	.62	19					
	CRI-C13B,	D2S25	.67	10	5	CRI-L123, M	D5S48	.71	
	CRI-L452, Bg	D2S41	.55	20		CRI-L334, T	D5S73	.64	06
	CRI-L22, T	D2S21	.79	11		CRI-L118, P E	D5S47	.40	15
	CRI-L34, M	D2S20	.40	23		CRI-C44, M	D5S56	.74	14

Enzyme Abbreviations: BamHI = B, BclII = Bc, BglII = Bg, DraI = D, EcoRI = E, EcoRV = EV, HaeIII = He, HincII = Hc, HinfI = Hn, HindIII = H, MspI = M, PvuII = P, RsaI = Pv, SacI = R, TaqI = S, T. UL means unlinked to the main linkage group.

5.179

Chr.	Probe, Enz.	Locus	Het.	Spacing
5	CRI-R535, M	D5S60	.45	09
	CRI-L1072, T	D5S69	.36	17
	J0110HC, M	D5S21	.36	15
	L599H-a, T	D5S76	.67	21
	M4, B	D5S6	.60	12
	p105-153Ra, M	D5S39	.53	09
	p105-798Rb, M	D5S78	.58	10
	cHB203, M	DHFR	.31	17
	CRI-L540, P	D5S50	.67	23
	pc11p11, T	D5S71	.32	16
	CRI-L372, M	D5S49	.64	12
	cMC5.61, M	D5S84	.55	14
	CRI-R379, M	D5S58	.55	11
	CRI-V1022, M	D5S54	.55	16
	CRI-P148, T	D5S72	.57	07
	CRI-L45, M	D5S61	.76	09
	p213-205eD, M	D5S22	.60	06
	pHf12-65, M	D5S2	.23	17
	CRI-L1200, T	D5S62	.40	28
	LMS8, Hn	D5S43	1.0	11
6	p7H4, E	D6S7	.30	UL
	p2C5, M	D6S8	.37	06
	CRI-R125, H	D6S28	.40	06
	CRI-L171, T, M	D6S19	.55	20
	CRI-R368, M	D6S23	.52	23
	CRI-L322, M	D6S26	.48	18
	CRI-P74, P	D6S27	.36	04
	CRI-L994, T	D6S33	.31	31
	CRI-L1065, R	D6S21	.74	44
	CRI-L1077, T	D6S22	.50	05
7	CRI-S202, P	D7S108	.24	
	CRI-S244, T	D7S112	.41	12
	CRI-L1020, H, T	D7S62	.76	16
	CRI-S83, M	D7S86	.52	06
	CRI-S20, M	D7S77	.41	12

Chr.	Probe, Enz.	Locus	Het.	Spacing
7	CRI-P137, T	D7S65	.45	05
	CRI-L751, M	D7S57	.65	06
	CRI-L887, M	D7S59	.55	10
	CRI-S65, H	D7S84	.26	10
	CRI-L917, H, Hc	D7S15	.71	30
	CRI-S23, T	D7S78	.48	15
	pMETH, M, T	MET	.60	09
	CRI-S94, T	D7S87	.52	10
	CRI-S140, M	D7S93	.46	06
	CRI-S3, H	D7S72	.39	14
	CRI-S241, E	D7S111	.59	09
	CRI-R967, M	D7S70	.40	11
	CRI-S194, T	D7S104	.80	27
	CRI-R53, Bg	D7S68	.40	11
8	CRI-V822, E	D8S23	.57	24
	CRI-V1225, E	D8S22	.69	22
	CRI-L388, T	D8S33	.40	12
	CRI-L40, Bg	D8S35	.40	08
	CRI-R191, M, T	D8S26	.74	50
	CRI-L1251, M	D8S29	.62	03
	CRI-L186, T	D8S34	.57	25
	CRI-L413, P	D8S32	.64	UL
	CRI-R150, H	D8S27	.50	
9	CRI-L1263, T	D9S19	.64	
	CRI-R3, E	D9S20	.50	UL
	CRI-P110, E	D9S22	.45	10
	CRI-L659, T	D9S26	.45	30
	CRI-L1424, H	D9S23	.43	34
10	CRI-JD12, Bg	D10S63	.40	21
	CRI-J93, M, E	D10S41	.60	07
	CRI-J170, P	D10S49	.64	15
	H,4IRBP, M, Bg	IRBP	.62	
	CRI-CS76, T	D10S65	.21	00
	CRI-CT19, M	D10S66	.26	00

Enzyme Abbreviations: BamHI = B, BclII = Bc, BglII = Bg, DraI = D, EcoRI = E, EcoRV = EV, HaeIII = He, HincII = Hc, HinfI = Hn, HindIII = H, MspI = M, PvuII = M, RsaI = Pv, RsaI = R, SacI = S, TaqI = T. UL means unlinked to the main linkage group.

Chr.	Probe, Enz.	Locus	Het.	Spacing
10	CRI-L647, T	D10S11	.50	04
	CRI-J128, T	D10S46	.17	14
	CRI-J127, H	D10S45	.31	14
	CRI-JC109, M	D10S54	.62	06
	CRI-L1083, M	D10S7	.55	10
	CRI-JM14, H	D10S62	.62	06
	CRI-JC144, M	D10S58	.62	18
	CRI-J90, M	D10S40	.52	14
	CRI-L368, H	D10S12	.74	16
	CRI-J179, Hc, M	D10S50	.50	08
	CRI-J198, P	D10S53	.57	05
	VTR41, E	D10S6	.55	18
11	pTBB-2, T	HRAS	.58	
	pINS-310, Pv	INS	.60	04
	1JW151, H	HBB	.46	10
	CRI-L834, M	D11S134	.45	30
	CRI-V928, M	D11S142	.48	12
	clone 3, H	SEA	.33	19
	CRI-L762, T	D11s141	.62	10
	CRI-R83, H, M, T	D11S137	.71	19
	CRI-L424, E, H	D11S132	.57	22
	CRI-L451, E, M	D11S133	.36	15
	CRI-R548, P	D11S138	.21	13
	CRI-R975, T	D11S131	.50	08
	CRI-L605, M	D11S127	.71	13
	CRI-R365, Bg	D11S129	.48	05
12	pha2ml, T	A2M	.24	22
	pPRP112.2RP, E	PRB1	.83	12
	CRI-C2, P, E	D12S20	.71	28
	p640, T	KRAS2	.32	14
	CRI-P153, M	D12S23	.45	08
	pCMM1.2, T	D12S15	.28	06
	pYNH15, M	D12S17	.52	07
	CRI-R102, M	D12S22	.60	06
	p7G11, T	D12S8	.52	17
12	pDL32B, T	D12S7	.69	11
	CRI-L416, E	D12S26	.52	UL
13	p9F4, EV	D13S?	.36	43
	p7F12, M	D13S1	.55	17
	21beta3'UT, M	TUBBP2	.50	20
	CRI-V1134, H	D13S56	.60	08
	CRI-R214, T	D13S55	.40	19
	p1E8, M	D13S4	.60	UL
	p9A7, H	D13S3	.44	
14	CRI-L436, T	D14S26	.52	04
	CRI-L329, T	D14S25	.48	21
	CRI-C70, M	D14S24	.71	UL
	CRI-L1013, Bg	D14S27	.43	13
	CRI-L1113, P	D14S28	.48	04
	pAW101, E	D14S1	.74	
15	PJU201, Bg	D15S3	.31	UL
	CRI-P78, B	D15S40	.36	28
	CRI-L442, M	D15S48	.60	00
	pDP151, E	D15S2	.40	13
	CRI-P452, M	D15S51	.24	18
	CRI-R382, M	D15S52	.62	
16	3'HVR, M	HBA	.93	09
	CRI-090, E,D	D16S45	.74	06
	CRI-0136, Hc	D16S60	.83	06
	CRI-0120, B	D16S51	.64	10
	CRI-0114, E	D16S49	.43	08
	CRI-L223, T	D16S76	.48	12
	CRI-R99, H, M	D16S75	.64	05
	CRI-0391,Hc, T	D16S67	.67	10
	CRI-0101, H	D16S48	.45	07
	CRI-03, P	D16S39	.52	06
	CRI-0377, Bg, T, Hc	D16S65	.76	07
	CRI-091, M, T	D16S46	.64	11

Enzyme Abbreviations: BamHI = B, BclII = Bc, BglII = Bg, DraI = D, EcoRI = E, EcoRV = EV, HaeIII = He, HincII = Hc, HinfI = Hn, HindIII = H, MspI = M, PvuII = P, RsaI = R, SacI = S, TaqI = T. UL means unlinked to the main linkage group.

Chr.	Probe, Enz.	Locus	Het.	Spacing
16	CRI-095, E	D16S47	.50	02
	CRI-0119, T	D16S50	.55	17
	CRI-015, T	D16S40	.62	14
	CRI-084, H, P	D16S43	.83	09
	CRI-089, Bg	D16S44	.62	10
17	pYNH37.3, T	D17S28	.29	
	pHRp5.5, Bg, H	POLR2	.55	13
	pABL10-41, Pv	D17S71	.29	30
	CRI-L946, M	D17S36	.67	12
	CRI-BR80, E	D17S231	.45	13
	CRI-pP3-1, M	D17S35	.52	16
	CRI-BR101, Bg	D17S232	.29	23
18	B74, M	D18S3	.48	
	CRI-L261, M	D18S16	.52	10
	CRI-R397, P	D18S12	.57	19
	pHF12-62, T	D18S1	.57	13
	L2.7, P	D18S6	.48	05
	CRI-P30, Bg, T	D18S13	.52	10
	CRI-L1156, M	D18S14	.57	19
	CRI-pL159-1, P	D18S17	.74	44
	CRI-pL159-2, P	D18S17	.74	00
19	p17.1, T	D19S8	.38	
	pC11-711, T	APOC2	.38	07
	p4.1, M	D19S7	.52	06
20	pR12-21, M	D20S5	.45	
	CRI-L127, M	D20S17	.43	24
	CRI-L1214, Bg	D20S16	.98	08
	pMB1-27, M	D20S4	.43	09
	CRI-L355, Bg	D20S15	.74	28
21	pS61-10, M	SOD1	.12	
	pGSH8, Bg	D21S17	.40	10
	GMG21S1, M	D21S15	.50	11

Chr.	Probe, Enz.	Locus	Het.	Spacing
21	CRI-L427, R	D21S112	.95	16
	pS2, B	BCE1	.17	16
22	22.C.1, T	D22S10	.40	08
	CRI-R657, T	D22S17	.45	34
	pSM-1, H	PDGFB	.40	15
	CRI-L1272, Bg	D22S18		
X	CRI-S232, E	DXS278	.90	17
	pD2, Pv	DXS43	.91	07
	p99-6, P	DXS41	.50	66
	p19-2, M	DXS3	.31	46
	52A, T	DXS51	.86	47
	DX13, Bg	DXS15	.95	07
	St14-1, T	DXS52	.46	

Enzyme Abbreviations: BamHI = B, BclI = Bc, BglII = Bg, DraI = D, EcoRI = E, EcoRV = EV, HaeIII = He, HincII = Hc, HinfI = Hn, HindIII = H, MspI = M, PvuII = M, RsaI = R, SacI = S, TaqI = T. UL means unlinked to the main linkage group.

Chromosome 21
Genetic Linkage Map of the Human (Homo sapiens sapiens)
2N=46
August, 1989
James F. Gusella, Jonathon L. Haines, Rudolph E. Tanzi
Neurogenetics Laboratory
Massachusetts General Hospital
and Harvard Medical School
Boston, MA 02114

Two genetic maps are presented (Figures 1 and 2). Figure 1 presents the map
of Tanzi et al. (1,3) including more recent updated information. Figure 2
presents the work of Warren et al. (2). Figure 1 is based on the Venezuelan
reference set, while Figure 2 is based on the CEPH reference set. These maps
are consistent in order, as well as in findings of excess female recombination
only in the centromeric region. CentiMorgan estimates differ across maps for
three possible reasons: differences in the underlying populations, sampling
error, and differences in methods of error checking the data.

TABLE 1.

Markers on Human Chromosome 21

Locus Designation	Probe name	Reference
D21S1	pPW228C	4
D21S3	pPW231F	4
D21S4	pPW233F	4
D21S8	pPW245D	4
D21S11	pPW236B	4
D21S12	pPW267C	4
D21S13	pGSM21	5
D21S15	pGSE8	5
D21S16	pGSE9	5
D21S17	pGSH8	5
D21S19	pGSB3	6
D21S23	pPW244D	4
D21S52	pPW511-1H	4
D21S53	pPW512	4
D21S54	pPW513-5H	4
D21S55	pPW518-1R	4
D21S56	pPW520	4
D21S58	pPW524-5P	4
D21S59	pPW552-3H	4
D21S110	p21-4U	10
D21S111	pCWpcq21	7
D21S112	CRI-L427	15
APP	pHL124	8
SOD1	SOD 4A	9
ETS2	H33ets-2	11
CD18	p3.1.1	16
BCEI	pS2	12
PFKL	pG-PFKL3.3	13
COL6A1	ML18	14

References

1). Tanzi, R.E. et al, Genomics. 3: 129-136, 1988.
2). Warren, A.C. et al, Genomics. 4: 579-591, 1989.
3). Tanzi, R.E. et al, in preparation.
4). Watkins, P.C. et al, Cytogenet Cell Genet. 40: 773-774, 1985.
5). Stewert, G.D. et al, NAR. 13: 4125-4132, 1985.
6). Stewert, G.D., et al, NAR. 13: 7168, 1985.
7). Warren, A.C. et al, Science. 237: 652-654, 1987.
8). Tanzi, R.E. et al, Nature. 329: 156-157, 1987.
9). Hallewell, R.A. et al, NAR. 13: 2017-2034, 1985.
10). Stewert, G.D. et al, Cytogenet Cell Genet. 46: 492, 1987.
11). Creau-Goldberg, N. et al, Hum Genet. 76: 396-398, 1987.
12). Masiakowski, P. et al, NAR. 10: 7895-7903, 1982.
13). Levanon, D. et al, Biochem Biophys Res Comm. 141: 374-380, 1986.
14). Weil, D. et al, Am J Hum Genet. 42: 435-445, 1988.
15). Green, P. et al, Cytogenet Cell Genet. 46: 623, 1987.
16). Corbi, A.L. et al, J Exp Med. 167: 1597-1607, 1988.

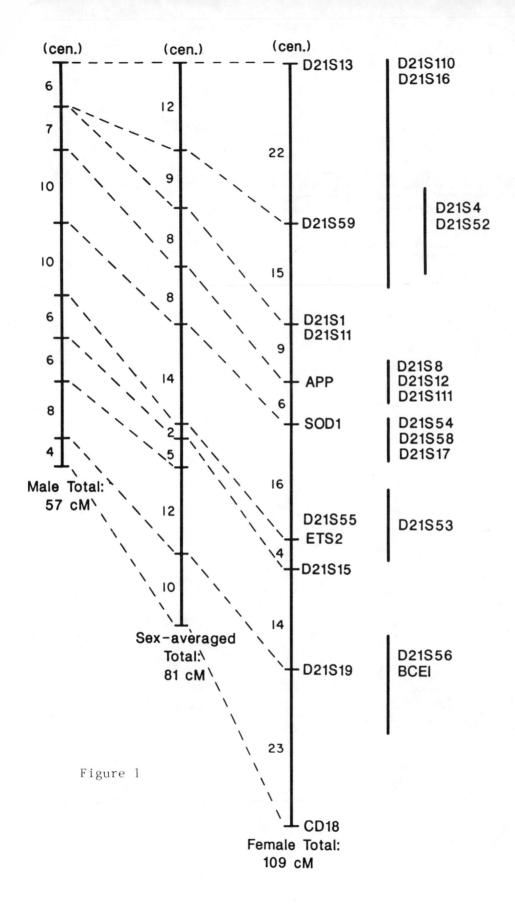

Figure 1

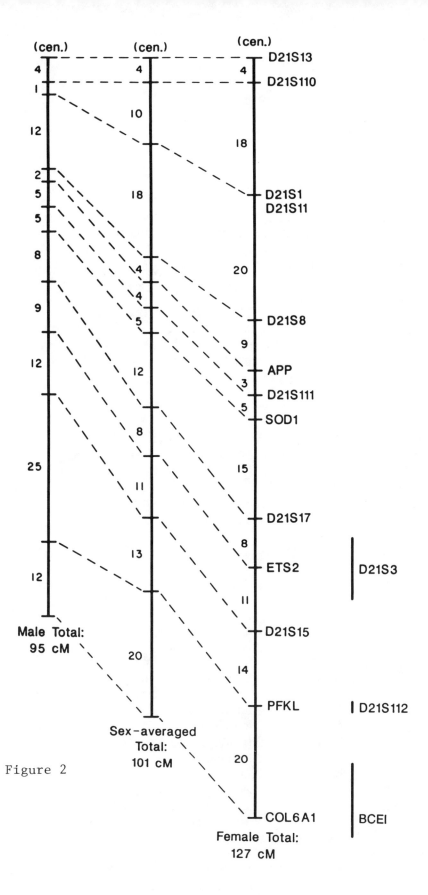

Figure 2

Male Total:
95 cM

Sex-averaged
Total:
101 cM

Female Total:
127 cM

5.185

Chromosome Rearrangements in Human Neoplasia

October 1989

Clara D Bloomfield and Joy L Frestedt
Department of Medicine
Roswell Park Memorial Institute
666 Elm Street
Buffalo, NY 14263

Chromosome abnormalities in malignant cells are acquired, clonal, and confined to the neoplastic tissue. The following table lists recurring structural as well as numerical chromosome abnormalities. The structural abnormalities are based primarily on the report of the committee on structural chromosome changes in neoplasia for the Human Gene Mapping 10 (HGM10) Workshop (1). The numerical abnormalities are based primarily on the report of the committee on chromosome rearrangements in neoplasia and on fragile sites for the HGM8 Workshop (2), and the previous genetic maps 1987 (3). Only abnormalities associated with specific neoplasms are reported. These abnormalities are listed by chromosome number and region or band number. The diversity of chromosome changes in neoplasia is enormous. Numerous reports have been published but few correlate specific neoplasms with chromosome abnormalities. Since some citations may be controversial, this report follows the policies outlined in HGM10 (1) for inclusion in this genetic map.

References

1) Trent JM, Kaneko Y, and Mitelman F (HGM10) Cytogenet Cell Genet 51:533-62, 1989
2) Berger R, Bloomfield CD, and Sutherland GR (HGM8) Cytogenet Cell Genet 40:490-535, 1985
3) Bloomfield CD. In: Genetic Maps 1987. O'Brien SJ, ed. Cold Spring Harbor, New York: Cold Spring Harbor Laboratory, Volume 4:605-609, 1987.
4) Hoyle, CF, Boughton BJ, and Hayhoe FGJ Cancer Genet Cytogenet 34:29-32, 1988
5) Suciu S, Weh H, Kuse R, et al. Cancer Genet Cytogenet 33:19-23, 1988
6) Dohner H, Arthur DC, Ball ED, et al. Blood 74 (in press)

Notes to tables

Abbreviations used in disease names

AC	Adenocarcinoma	MB	Medulloblastoma
ALL	Acute lymphoblastic leukemia, subclassified L1 through L3	MDS	Myelodysplastic syndrome
		MEL	Malignant melanoma
AML	Acute myeloid leukemia, subclassified M1 through M7	MH	Malignant histiocytosis
		ML	Malignant lymphoma
ATL	Adult T-cell leukemia/lymphoma	MM	Multiple myeloma
B-	B Cell	MN	Meningioma
BL	Burkitt's lymphoma	MPD	Myeloproliferative disorder
CLD	Chronic lymphoproliferative disorder	MT	Mesothelioma
CLL	Chronic lymphocytic leukemia	NB	Neuroblastoma
CML	Chronic myelocytic leukemia	PA	Pleomorphic adenoma
CMML	Chronic myelomonocytic leukemia	PCL	Plasma cell leukemia
ES	Ewing's sarcoma	PLL	Prolymphocytic leukemia
GCT	Germ cell tumor	PV	Polycythemia vera
GL	Glioma	RB	Retinoblastoma
HCL	Hairy cell leukemia	RMS	Rhabdomyosarcoma
LI	Lipoma	SCC	Small cell carcinoma
LIS	Liposarcoma	SS	Synovial sarcoma
LM	Leiomyoma	T-	T Cell
LMS	Leiomyosarcoma	WT	Wilms' tumor

Region or Band	Type of abberation	Disease	References
Chromosome 1			
1p36	t(1;1)(p36;p11-12)	ML	1
1p36	t(1;3)(p36;q21)	MDS, AML	1,2
1p36	t(1;17)(p36;q21)	AML-M3	1,2
1p36-p32	deletions	ML	1
1p36-p32	deletions, translocations	NB,GL	1,2
1p34-p32	t(1;14)(p32-34;q11)	T-ALL	1
1p32	t(1;11)(p32;q23)	ALL	1
1p22-p11	deletions, translocations	MEL	1,2
1p13	deletions	ML	1
1p13-p12	deletions, translocations	LMS intestinal	1
1p13-p11	deletions, translocations	MT pleural,AC breast	1
1p12-11	t(1;1)(p36;p11-12)	ML	1
1p11	t(1;17)(p11 or q11;q11 or p11)	CML, MPD, AML, ML	1,2
1p11	t(1;7)(p11;p11)	MPD, MDS, AML	1,2
1p11-q11	i(1q)	ML, AC uterus	1
1p11-q11	i(1q), translocations	WT	1
1q11-q12	t(1;6)(q11-12;q15-21)	MEL	1
1q11-q12	translocations	MEL	1
1q12-q31	duplications	BL, ALL-L3, ML	1
1q21	del(1)(q21)	ML	1
1q21	t(1;11)(q21;q23)	AML	1
1q21-q23	translocations	AC breast	1
1q21-q23	deletions, translocations	AC bladder, AC uterus	1
1q21-q25	t(1;14)(q21-25;q32)	ML	1,2
1q21-q32	dup(1)(q21q32)	MPD	1
1q23	t(1;19)(q23;p13)	ALL	1,2
1q23-q25	t(1;6)(q23-25;p21-25)	MPD	1
1q32	del(1)(q32)	ML	1
1q42	del(1)(q42)	ML	1
1q42	t(1;14)(q42;q32)	B-ML	1
Chromosome 2			
2p23	t(2;5)(p23;q35)	MH, T-ML	1
2p22-p13	t(2;3)(p13-22;q26-29)	AML	1
2p21	t(2;11)(p21;q23)	MDS, AML	1,2
2p21	del(2)(p21)	ML	1
2p13	t(2;14)(p13;q32)	B-CLL	1,2
2p12	t(2;8)(p12;q24)	BL, ALL-L3	1,2
2q21-q23	t(2;3)(q21-23;q27)	ML	1
2q31	del(2)(q31)	AML, CLL	1
2q32	del(2)(q32)	ML	1
2q35-q37	t(2;13)(q35-37;q14)	RMS	1,2
Chromosome 3			
3p23-p14	del(3)(p14p23)	SCC lung, AC lung	1,2
3p21	del(3)(p21)	ML	1
3p21	t(3;8)(p21;q12)	PA	1,2
3p21	t(3;14)(p21;q32)	ML	1,2
3p21-p14	deletions, translocations	AC kidney	1
3p21-p13	del(3)(p13-21)	AC ovarian	1,2
3p13-p11	deletions, translocations	AC breast	1

Region or Band	Type of abberation	Disease	References
3q11-q13	deletions, translocations	AC breast	1
3q21	t(1;3)(p36;q21)	MDS, AML	See chrom. 1
3q21	t(3;5)(q21;q31)	AML	1,2
3q21	ins(3;3)(q26;q21q26) inv(3)(q21q26) t(3;3)(q21;q26)	AML, MDS, MPD	1,2
3q24-q25	t(3;5)(q24-25;q32-34)	AML	1,2
3q26	ins(3;3)(q26;q21q26) inv(3)(q21q26) t(3;3)(q21;q26)	AML, MDS, MPD	1,2
3q26	t(3;21)(q26;q22)	CML, MDS	1
3q26-q29	t(2;3)(p13-22;q26-29)	AML	See chrom. 2
3q27	t(2;3)(q21-23;q27)	ML	See chrom. 2
3q27-q28	t(3;12)(q27-28;q13-14)	LI	1
Trisomy 3	+3	MM,ATL	2

Chromosome 4

Region or Band	Type of abberation	Disease	References
4p13-p14	del(4)(p13-14)	ML	1
4q21	t(4;11)(q21;q23)	ALL, AML	1,2
Trisomy 4	+4	AML	4,5

Chromosome 5

Region or Band	Type of abberation	Disease	References
5p13	del(5)(p13)	ML	1
5p11-q11	i(5p)	AC bladder	1,2
5q11-q35	del(5)(q11-35)	MDS, AML	1,2
5q31	t(3;5)(q21;q31)	AML	See chrom. 3
5q32-q34	t(3;5)(q24-25;q32-34)	AML	See chrom. 3
5q35	t(2;5)(p23;q35)	MH, T-ML	See chrom. 2
Trisomy 5	+5	MM	2
Monosomy 5	-5	MDS, AML	2,3

Chromosome 6

Region or Band	Type of abberation	Disease	References
6p25-p21	t(1;6)(q23-25;p21-25)	MPD	See chrom. 1
6p23	t(6;9)(p23;q34)	AML	1,2
6p23-p22	translocations	LI	1
6p23-p21	deletions	T-ML	1,2
6p21	t(6;14)(p21;q32)	CLD, ML	1
6p11-q11	i(6p), translocations	MEL	1,2
6p11-q11	i(6p)	ML, ALL, RB	1
6q11-q27	deletions, translocations	MEL	1,2
6q14-q27	deletions	ML, ALL, ATL, HCL, PLL	1,2
6q15-q21	t(1;6)(q11-12;q15-21)	MEL	See chrom. 1
6q15-q23	deletions	AC ovarian	1,2
6q15-q27	deletions, translocations	GB	1
6q27	t(6;11)(q27;q23)	AML-M5	1,2

Chromosome 7

Region or Band	Type of abberation	Disease	References
7p15	t(7;11)(p15;p15)	AML, MPD	1,2
7p14-p13	del(7)(p13-14)	ML	1
7p13-p11	dic(7;9)(p11-13;p11)	ALL	1
7p11	t(1;7)(p11;p11)	MPD, MDS, AML	See chrom. 1
7p11-q11	i(7q)	MDS, AML, ML, ALL	1

Region or Band	Type of abberation	Disease	References
7q11	translocations	MEL	1
7q21-q31	del(7)(q21-31)	LM uterus	1
7q22	del(7)(q22)	AC prostate	1
7q22-q34	deletions, translocations	GL	1
7q22-q36	deletions	AML, MDS	1,2
7q32	del(7)(q32)	ML	1
7q35	t(7;11)(q35;p13-14)	T-ALL	1
Monosomy 7	-7	AML,MDS	2,3
Trisomy 7	+7	AC colon, MM	2

Chromosome 8

Region or Band	Type of abberation	Disease	References
8p23-p21	deletions, translocations	GL	1
8p11	t(8;9)(p11;q34)	MPD	1,2
8p11	t(8;16)(p11;p13)	AML-M5	1,2
8p11-q11	i(8q)	T-CLL, T-PLL	1
8q12	t(3;8)(p21;q12)	PA	See chrom. 3
8q22	t(8;14)(q22;q32)	ML	1,2
8q22	t(8;21)(q22;q22)	AML-M2	1,2
8q24	t(2;8)(p12;q24)	BL, ALL-L3	See chrom. 2
8q24	t(8;14)(q24;q11)	T-ALL	1,2
8q24	t(8;14)(q24;q32)	BL, ALL-L3, ML	1,2
8q24	t(8;22)(q24;q11)	BL, ALL-L3	1,2
Trisomy 8	+8	AML, CML, AC colon	2,3

Chromosome 9

Region or Band	Type of abberation	Disease	References
9p24-p13	deletions, translocations	GL	1
9p22-p13	t(9;12)(p13-22;q13-15)	PA	1
9p22-p21	t(9;11)(p21-22;q23)	AML-M4, M5	1,2
9p21	del(9)(p21)	ALL	1,2
9p13	del(9)(p13)	ML	1
9p13-q11	dic(9;12)(p11-13;p11-12)	ALL	1
9p12-p11	t(9;12)(p11-12;p12)	ALL	1
9p11	dic(7;9)(p11-13;p11)	ALL	See chrom. 7
9p11-q11	i(9q)	ALL	1
9q11-q32	deletions	AML	1,2
9q34	t(6;9)(p23;q34)	AML	See chrom. 6
9q34	t(8;9)(p11;q34)	MPD	See chrom. 8
9q34	t(9;22)(q34;q11)	CML, ALL, AML, T-ML	1,2
Trisomy 9	+9	MM	2

Chromosome 10

Region or Band	Type of abberation	Disease	References
10p15-p11	t(10;11)(p11-15;q23) ins(10;11)(p11;q23q24)	AML-M5	1
10p14	t(10;11)(p14;q13-14)	AML-M4, M5	1,2
10p11	t(10;14)(p11;q32)	ML	1
10q22-q23	del(10)(q22-23)	ML	1
10q24	t(10;14)(q24;q11)	T-ALL	1,2
10q24	del(10)(q24)	AC prostate	1,2

Region or Band	Type of abberation	Disease	References
Chromosome 11			
11p15	t(7;11)(p15;p15)	AML, MPD	See chrom. 7
11p15	t(11;20)(p15;q11)	AML	1
11p14-p13	t(7;11)(q35;p13-14)	T-ALL	See chrom. 7
11p13	deletions, translocations	WT	1,2
11p13	t(11;14)(p13;q11)	T-ALL	1,2
11q13	t(11;14)(q13;q32)	ML, B-CLL, B-PLL, MM/PCL	1,2
11q13-q14	t(10;11)(p14;q13-14)	AML-M4, M5	See chrom. 10
11q13-q23	duplications	ML	1
11q13-q25	duplications	ML	1
11q14	del(11)(q14)	MDS	1,2
11q21	t(11;14)(q21;q32)	ML	1
11q23	del(11)(q23)	ML	1
11q23	t(1;11)(p32;q23)	ALL	See chrom. 1
11q23	t(1;11)(q21;q23)	AML	See chrom. 1
11q23	t(2;11)(p21;q23)	MDS, AML	See chrom. 2
11q23	t(4;11)(q21;q23)	ALL, AML	See chrom. 4
11q23	t(6;11)(q27;q23)	AML-M5	See chrom. 6
11q23	t(9;11)(p21-22;q23)	AML-M4, M5	See chrom. 9
11q23	t(10;11)(p11-15;q23) ins(10;11)(p11;q23q24)	AML-M5	See chrom. 10
11q23	t(11;14)(q23;q32)	ML	1,2
11q23	t(11;17)(q23;q25)	AML-M4, M5	1,2
11q23	t(11;19)(q23;p13)	ALL, AML	1,2
11q24	t(11;22)(q24;q12)	ES	1,2
Trisomy 11	+11	MM	2
Chromosome 12			
12p13-p12	t(12;17)(p12-13;q12)	ALL	1
12p13-p11	del(12)(p11-13), translocations	ALL, CMML, AML	1,2
12p12	t(9;12)(p11-12;p12)	ALL	See chrom. 9
12p12-p11	dic(9;12)(p11-13;p11-12)	ALL	See chrom. 9
12p12-p11	del(12)(p11-12)	ML	1,2
12p11-q11	i(12p)	GCT	1,2
12q13	t(12;16)(q13;p11)	LIS myxoid	1,2
12q13-q14	t(3;12)(q27-28;q13-14)	LI	See chrom. 3
12q13-q14	translocations	LI	1
12q13-q15	translocations	PA	1
12q13-q15	t(9;12)(p13-22;q13-15)	PA	See chrom. 9
12q13-q15	t(12;14)(q13-15;q23-24)	LM uterus	1
12q13-q22	dup(12)(q13q21-22)	ML	1,2
12q22	del(12)(q22)	ML	1
Trisomy 12	+12	B-CLL,ML,AC colon	2,3
Chromosome 13			
13q12-q22	del(13)(q12-22)	MPD, MDS	1,2
13q12-q13	deletions, translocation	LI	1
13q14	del(13)(q14q14)	RB	1,2
13q14	t(2;13)(q35-37;q14)	RMS	See chrom. 2
13q22	del(13)(q22)	ML	1
Trisomy 13	+13	AML	6

Region or Band	Type of abberation	Disease	References
Chromosome 14			
14q11	t(1;14)(p32-34;q11)	T-ALL	See chrom. 1
14q11	t(8;14)(q24;q11)	T-ALL	See chrom. 8
14q11	t(10;14)(q24;q11)	T-ALL	See chrom. 10
14q11	t(11;14)(p13;q11)	T-ALL	See chrom. 11
14q11	inv(14)(q11q32)	T-CLL, T-PLL, ATL	1,2
14q22-q24	del(14)(q22-24)	B-CLL,ML	1,2
14q23-q24	t(12;14)(q13-15;q23-24)	LM uterus	See chrom. 12
14q32	inv(14)(q11q32)	ATL, T-CLL, T-PLL	1
14q32	t(1;14)(q21-25;q32)	ML	See chrom. 1
14q32	t(1;14)(q42;q32)	B-ML	See chrom. 1
14q32	t(2;14)(p13;q32)	B-CLL	See chrom. 2
14q32	t(3;14)(p21;q32)	ML	See chrom. 3
14q32	t(6;14)(p21;q32)	ML, CLD	See chrom. 6
14q32	t(8;14)(q22;q32)	ML	See chrom. 8
14q32	t(8;14)(q24;q32)	BL, ALL-L3, ML	See chrom. 8
14q32	t(10;14)(p11;q32)	ML	See chrom. 10
14q32	t(11;14)(q13;q32)	ML, B-CLL, MM/PCL, B-PLL	See chrom. 11
14q32	t(11;14)(q21;q32)	ML	See chrom. 11
14q32	t(11;14)(q23;q32)	ML	See chrom. 11
14q32	t(14;17)(q32;q23)	B-CLL	1,2
14q32	t(14;18)(q32;q21)	ML	1,2
14q32	t(14;19)(q32;q13)	B-CLL	1,2
14q32	t(14;22)(q32;q11)	ALL, CLD	1
Chromosome 15			
15q22	t(15;17)(q22;q11-12)	AML-M3	1,2
Chromosome 16			
16p13	inv(16)(p13q22) t(16;16)(p13;q22)	AML-M4EO	1,2
16p13	t(8;16)(p11;p13)	AML-M5	See chrom. 8
16p11	t(12;16)(q13;p11)	LIS myxoid	See chrom. 12
16q22	del(16)(q22)	AML-M4EO	1,2
16q22	inv(16)(p13q22) t(16;16)(p13;q22)	AML-M4EO	1,2
Chromosome 17			
17p11-q11	i(17q), translocations	AC colon	1,2
17p11-q11	i(17q)	CML, MDS, AML, ML, ALL, CLL, MB	1,2
17q11	t(12;17)(p12-13;q11)	ALL	See chrom. 12
17q11	t(1;17)(p11 or q11;q11 or p11)	CML,MPD,AML,ML	See chrom. 1
17q11-q12	t(15;17)(q22;q11-12)	AML-M3	See chrom. 15
17q21	t(1;17)(p36;q21)	AML-M3	See chrom. 1
17q23	t(14;17)(q32;q23)	B-CLL	See chrom. 14
17q25	t(11;17)(q23;q25)	AML-M4, M5	See chrom. 11

Region or Band	Type of abberation	Disease	References
Chromosome 18			
18p11-q11	i(18q)	ML	1
18q11	t(X;18)(p11;q11)	SS	1
18q21	t(14;18)(q32;q21)	ML	See chrom. 14
Chromosome 19			
19p13	t(1;19)(q23;p13)	ALL	See chrom. 1
19p13	t(11;19)(q23;p13)	AML, ALL	See chrom. 11
19q13	translocations	GL	1
19q13	t(14;19)(q32;q13)	B-CLL	See chrom. 14
Chromosome 20			
20q11	t(11;20)(p15;q11)	AML	See chrom. 11
20q11-q13	del(20)(q11q13) or del(20)(q11)	PV, MDS, AML	1,2
Chromosome 21			
21p11-q11	i(21q)	AML, MDS	1
21q22	t(3;21)(q26;q22)	CML, MDS	See chrom. 3
21q22	t(8;21)(q22;q22)	AML-M2	See chrom. 8
Trisomy 21	+21	ALL	2,3
Chromosome 22			
22q11	t(8;22)(q24;q11)	BL, ALL-L3	See chrom. 8
22q11	t(9;22)(q34;q11)	CML, ALL, AML, T-ML	See chrom. 9
22q11	t(14;22)(q32;q11)	ALL, CLD	See chrom. 14
22q11-q12	del(22)(q11-12)	ML	1,2
22q12	t(11;22)(q24;q12)	ES	See chrom. 11
22q12-q13	del(22)(q12-13)	MN	1,2
Monosomy 22	-22	MN	2,3
X Chromosome			
Xp11	t(X;18)(p11;q11)	SS	See chrom. 18
Xq13	idic(X)(q13)	MDS, AML	1

Human Fragile Sites
June, 1989
Grant R. Sutherland
Cytogenetics Unit
Adelaide Children's Hospital
North Adelaide   SA   5006
Australia

Symbol	Location	Type	References
FRA1A	1p36	Common, aphidicolin	16, 22, 56
FRA1B	1p32	Common, aphidicolin	9, 16
FRA1C	1p31.2	Common, aphidicolin	10, 54, 61
FRA1L	1p31	Common, aphidicolin	22, 34, 35, 60
FRA1D	1p22	Common, aphidicolin	9, 16, 19, 22
FRA1E	1p21.2	Common, aphidicolin	22, 54, 61
FRA1J	1q12	Common, 5-azacytidine	19, 60
FRA1F	1q21	Common, aphidicolin	19, 54, 61
FRA1G	1q25.1	Common, aphidicolin	16, 54, 61
FRA1K	1q31	Common, aphidicolin	22, 54
FRA1H	1q42	Common, 5-azacytidine	39, 51
FRA1I	1q44	Common, aphidicolin	9, 54, 61
FRA2C	2p24.2	Common, aphidicolin	22, 54, 61
FRA2D	2p16.2	Common, aphidicolin	54, 61
FRA2E	2p13	Common, aphidicolin	9, 16, 61
FRA2A	**2q11.2**	**Rare, folic acid**	**13, 24, 33, 46**
FRA2B	**2q13**	**Rare, folic acid**	**13, 24, 33, 46**
FRA2F	2q21.3	Common, aphidicolin	54, 22, 61
FRA2K	**2q22.3**	**Rare, folic acid**	**29**
FRA2G	2q31	Common, aphidicolin	9, 16
FRA2H	2q32.1	Common, aphidicolin	22, 54, 61
FRA2I	2q33	Common, aphidicolin	9, 10, 16
FRA2J	2q37.3	Common, aphidicolin	54, 61
FRA3A	3p24.2	Common, aphidicolin	16, 54, 61
FRA3B	3p14.2	Common, aphidicolin	4, 16, 28
FRA3D	3q25	Common, aphidicolin	14, 17, 34
FRA3C	3q27	Common, aphidicolin	16, 61
FRA4A	4p16.1	Common, aphidicolin	9, 19, 54, 61
FRA4D	4p15	Common, aphidicolin	19, 60
FRA4B	4q12	Common, BrdU	51
FRA4E	4q27	Common, unclassified	19, 60
FRA4C	4q31.1	Common, aphidicolin	54, 61
FRA5E	5p14	Common, aphidicolin	22, 60
FRA5A	5p13	Common, Brdu	51
FRA5B	5q15	Common, Brdu	51
FRA5D	5q15	Common, aphidicolin	17, 19, 21

Symbol	Location	Type	References
FRA5F	5q21	Common, aphidicolin	14, 34, 60
FRA5C	5q31.1	Common, aphidicolin	16, 54, 61
FRA6B	6p25.1	Common, aphidicolin	34, 54, 61
FRA6A	**6p23**	**Rare, folic acid**	**49**
FRA6C	6p22.2	Common, aphidicolin	54, 61
FRA6D	6q13	Common, BrdU	19, 51
FRA6G	6q15	Common, aphidicolin	14, 17, 34
FRA6F	6q21	Common, aphidicolin	17, 19, 60
FRA6E	6q26	Common, aphidicolin	9, 25
FRA7B	7p22	Common, aphidicolin	21, 61
FRA7C	7p14.2	Common, aphidicolin	22, 34, 54, 61
FRA7D	7p13	Common, aphidicolin	3, 16
FRA7A	**7p11.2**	**Rare, folic acid**	**31, 46**
FRA7J	7q11	Common, aphidicolin	9, 17, 19
FRA7E	7q21.2	Common, aphidicolin	54, 61
FRA7F	7q22	Common, aphidicolin	9, 16, 22, 34
FRA7G	7q31.2	Common, aphidicolin	3, 61
FRA7H	7q32.3	Common, aphidicolin	16, 61
FRA7I	7q36	Common, aphidicolin	21, 61
FRA8B	8q22.1	Common, aphidicolin	16, 61
FRA8A	**8q22.3**	**Rare, folic acid**	**33, 49**
FRA8C	8q24.1	Common, aphidicolin	21, 61
FRA8E	**8q24.1**	**Rare, distamycin A**	**30, 53**
FRA8D	8q24.3	Common, aphidicolin	21, 54, 61
FRA9A	**9p21**	**Rare, folic acid**	**18, 49**
FRA9C	9p21	Common, Brdu	51
FRA9F	9q12	Common, 5-azacytidine	19, 60
FRA9D	9q22.1	Common, aphidicolin	17, 54, 61
FRA9B	**9q32**	**Rare, folic acid**	**32, 51**
FRA9E	9q32	Common, aphidicolin	16, 61
FRA10C	10q21	Common, BrdU	51
FRA10D	10q22.1	Common, aphidicolin	34, 54, 61
FRA10A	**10q23.3 or 10q24.2**	**Rare, folic acid**	**12, 15, 44**
FRA10B	**10q25.2**	**Rare, BrdU**	**36, 48**
FRA10E	10q25.2	Common, aphidicolin	54, 61
FRA10F	10q26.1	Common, aphidicolin	17, 54, 61
FRA11C	11p15.1	Common, aphidicolin	21, 61
FRA11I	**11p15.1**	**Rare, distamycin A**	**20, 53**
FRA11D	11p14.2	Common, aphidicolin	9, 61
FRA11E	11p13	Common, aphidicolin	3, 16
FRA11H	11q13	Common, aphidicolin	17, 21, 60
FRA11A	**11q13.3**	**Rare, folic acid**	**33, 44**
FRA11F	11q14.2	Common, aphidicolin	9, 19, 54, 61
FRA11B	**11q23.3**	**Rare, folic acid**	**49, 58**
FRA11G	11q23.3	Common, aphidicolin	54, 61

Symbol	Location	Type	References
FRA12A	**12q13.1**	**Rare, folic acid**	**15, 33, 47**
FRA12B	12q21.3	Common, aphidicolin	22, 34, 61, 54
FRA12E	12q24	Common, aphidicolin	21, 60
FRA12C	**12q24.2**	**Rare, BrdU**	**51, 57**
FRA12D	**12q24.13**	**Rare, folic acid**	**1**
FRA13A	13q13.2	Common, aphidicolin	22, 54, 61
FRA13B	13q21	Common, BrdU	19, 51
FRA13C	13q21.2	Common, aphidicolin	34, 54, 61
FRA13D	13q32	Common, aphidicolin	14, 17, 19, 60
FRA14B	14q23	Common, aphidicolin	3, 21
FRA14C	14q24.1	Common, aphidicolin	9, 16, 54, 61
FRA15A	15q22	Common, aphidicolin	3
FRA16A	**16p13.11**	**Rare, folic acid**	**6, 11, 44, 51**
FRA16E	**16p12.1**	**Rare, distamycin A**	**20**
FRA16B	**16q22.1**	**Rare, distamycin A**	**5, 27, 37, 43**
FRA16C	16q22.1	Common, aphidicolin	9, 43, 54, 61
FRA16D	16q23.2	Common, aphidicolin	42, 46, 49, 61
FRA17A	**16p12**	**Rare, distamycin A**	**8, 50**
FRA17B	17q23.1	Common, aphidicolin	17, 54, 61
FRA18A	18q12.2	Common, aphidicolin	54, 61
FRA18B	18q21.3	Common, aphidicolin	22, 54, 61
FRA19B	**19p13**	**Rare, folic acid**	**7, 55**
FRA19A	19q13	Common, 5-azacytidine	19, 39, 51
FRA20B	20p12.2	Common, aphidicolin	54, 61
FRA20A	**20p11.23**	**Rare, folic acid**	**40, 44**
FRA22B	22q12.2	Common, aphidicolin	16, 19, 54, 61
FRA22A	**22q13**	**Rare, folic acid**	**21, 59**
FRAXB	Xp22.31	Common, aphidicolin	16, 61
FRAXC	Xq22.1	Common, aphidicolin	16, 61
FRAXD	Xq27.2	Common, aphidicolin	23, 45
FRAXA	**Xq27.3**	**Rare, folic acid**	**26, 52**

REFERENCES

1. AMAROSE, A.P., et al. 1987. *Hum. Genet.* 75:4-6.
2. ARVEILER, B., et al. 1988. *Am. J. Hum. Genet.* 42:380-389.
3. BARBI, G., et al. 1984. *Hum. Genet.* 68:290-294.
4. BROGGER, A., 1968. *Tidsskr. Nor. Laegeforen.* 88:1741-1747.
5. CALLEN, D.F., et al. 1988. *Genomics* 2:144-153.
6. CALLEN, D.F., et al. 1989. *Genomics* 4:348-354.
7. CHODIRKER, B.M., et al. 1987. *Clin. Genet.* 31:1-6.
8. COURT BROWN, W.M., et al. 1966. *Chromosome Studies on Adults. Eugenics Laboratory Memoirs*, Vol. XLII. Cambridge University Press, New York.

References

9. CRAIG-HOLMES, A.P., et al. 1987. *Hum. Genet.* 76:134–137.
10. DANIEL, A., et al. 1984. *Am. J. Med. Genet.* 18:483–491.
11. DAY, E.J., et al. 1967. *Lancet* ii:1307.
12. DUTRILLAUX, B., et al. 1985. *Ann. Genet. (Paris)* 28:161–163.
13. FORD, C.E. and K. MADAN, 1973. *Chromosome Identification*, Academic Press, New York and London, 98–103.
14. FUSTER, C., et al. 1989. *Hum. Genet.* 81:243–246.
15. GIRAUD, F., et al. 1976. *Hum. Genet.* 34:125–136.
16. GLOVER, T.W., et al. 1984. *Hum. Genet.* 67:136–142.
17. GREEN, R.J., et al. 1988. *Hum. Genet.* 81:9–12.
18. GUICHAOUA, M., et al. 1982. *J. Génét. Hum.* 30:183–187.
19. HECHT, F., et al. 1988. *Cancer Genet. Cytogenet.* 33:1–9.
20. HORI, T., et al. 1988. *Cancer Genet. Cytogenet.* 34:177–187.
21. KAHKONEN, M., et al. 1989. *Hum. Genet.* 82:3–8.
22. KUWANO, A. and T. KAJII. 1987. *Hum. Genet.* 75:75–78.
23. LEDBETTER, S.A. and D.H. LEDBETTER. 1988. *Am. J. Hum. Genet.* 42:694–702.
24. LEJEUNE, J., et al. 1968. *C.R. Acad. Sci. Paris* 266:24–26.
25. LEVERSHA, M.A., et al. 1981. *Lancet* i:49.
26. LUBS, H.A., 1969. *Am. J. Hum. Genet.* 21: 231–244.
27. MAGENIS, R.E., et al. 1970. *Science* 170:85–87.
28. MARKKANEN, A., et al. 1982. *Hereditas* 96:317–319.
29. MATTEI, M.G., et al. 1987. *Cytogenet. Cell Genet.* 46:658.
30. OCHI, H., et al. 1988. *Jap. J. Cancer Res.* 79:145–147.
31. PAVEY, S.M. and G.C. WEBB, 1982. Presented 6th Annual Scientific Meeting, Human Genetics Society of Australasia, Adelaide.
32. PETIT, P., et al. 1986. *Clin. Genet.* 29:96–100.
33. QUACK, B., et al. 1978. *J. Genet. Hum.* 26:55–67.
34. RAO, P.N., et al. 1988. *Hum. Genet.* 78:21–26.
35. RUDOLPH, B., et al. 1988. *Cancer Genet. Cytogenet.* 31:83–94.
36. SCHERES, J.M.J.C. and T.W.J. HUSTINX, 1980. *Am. J. Hum. Genet.* 32:628–629.
37. SCHMID, M., et al. 1980. *Cytogenet. Cell Genet.* 28:87–94.
38. SCHMIDT, A. 1986. *7th Int. Cong. Hum. Gen.* 106.
39. SCHMIDT, M., et al. 1985. *Hum. Genet.* 71:342–350.
40. SHABTAI, F. and I. HALBRECHT, 1981. *Clin. Genet.* 19:536.
41. SHABTAI, F., et al. 1982. *Hum. Genet.* 61:177–179.
42. SHABTAI, F., et al. 1983. *Hum. Genet.* 64:273–276.
43. SIMMERS, R.N., et al. 1987. *Science* 236:92–94.
44. SUTHERLAND, G.R., 1979. *Am. J. Hum. Genet.* 31:136–148.
45. SUTHERLAND, G.R. and E. BAKER, 1989. Submitted.
46. SUTHERLAND, G.R. and F. HECHT, 1985. *Fragile Sites on Human Chromosomes*, Oxford University Press, New York.
47. SUTHERLAND, G.R. and L. HINTON, 1981. *Hum. Genet.* 57: 217–219.
48. SUTHERLAND, G.R., et al. 1980. *Am. J. Hum. Genet.* 32:542–548.
49. SUTHERLAND, G.R., et al. 1983. *Am. J. Hum. Genet.* 35:432–437.
50. SUTHERLAND, G.R., et al. 1984. *Am. J. Hum. Genet.* 36:110–122.
51. SUTHERLAND, G.R., et al. 1985. *Hum. Genet.* 69:233–237.
52. SUTHERLAND, G.R., et al. 1987. *Cytogenet. Cell Genet.* 46:700.
53. TAKAHASHI, E., et al. 1988. *Hum. Genet.* 80:124–126.
54. TEDESCHI, B., et al. 1987. *Am. J. Med. Genet.* 27:471–482.
55. TOMMERUP, N., et al. 1985. *Clin. Genet.* 27:510–514.
56. VENTRUTO, V., et al. 1986. *Ann. Genet.* 29:59–61.
57. VOICULESCU, I., et al. 1988. *Hum. Genet.* 78:183–185.
58. VOULLAIRE, L.E., et al. 1987. *Hum. Genet.* 76:202–204.
59. WEBB, T. and A. THAKE, 1984. *Clin. Genet.* 26:125–128.
60. YUNIS, J.J., et al. 1987. *Oncogene* 1:59–69.
61. YUNIS, J.J. and A.L. SORENG, 1984. *Science* 226:1199–1204.

Homo sapiens   2N=46

October 1989

An abridged human gene map including:

(1) Oncogenes (Table I)
(2) Growth factors
(3) Cell surface receptors
(4) Endogenous retroviral sequences

Stephen J. O'Brien
Patricia A. Johnson
Laboratory of Viral Carcinogenesis
National Cancer Institute
Frederick, MD 21701-1013

Note: In the figure oncogenes are depicted to the right of each human chromosome. Gene designations, gene names, and regional locations of each of these are listed in Tables I and II.

Table I
Human Oncogene Loci

Gene Symbol	Gene Name	Chromosomal Position	References
ABL	Abelson murine leukemia viral (v-abl) oncogene homolog	9q34	132, 142, 162, 233, 234, 356, 391
ABLL	Abelson murine leukemia viral (v-abl) oncogene homolog-like	1q24-q25	180
AKT1	Murine thymoma viral (v-akt) oncogene homolog 1	14q32.32-q32.33	62, 146, 350, 362
ARAF1	Murine sarcoma 3611 viral (v-raf) oncogene homolog 1	Xp11.4-p11.23	10, 18, 152, 182, 291
ARAF2	Murine sarcoma 3611 viral (v-raf) oncogene homolog 2	7p14-q21	152
BLYM	Human B-lymphoma	1p32	236
CSF1R	Colony-stimulating factor 1 receptor, formerly McDonough feline sarcoma viral (v-fms) oncogene homolog. Previous symbol FMS	5q33-q35	131, 190, 298, 308, 374
EGFR	Epidermal growth factor receptor, formerly avian erythroblastic leukemia viral (v-erb-b) oncogene homolog	7p13-p12	74, 315, 338, 349, 404
ELK1	ELK1, member of ETS oncogene family	Xp22.1-p11	286
ELK2	ELK2, member of ETS oncogene family	14q32	286
ERBAL2	Avian erythroblastic leukemia viral (v-erb-a) oncogene homolog-like 2	19	229
ERBAL3	Avian erythroblastic leukemia-viral (v-erb-a) oncogene homolog-like 3	5	229
ERBA2L	Avian erythroblastic leukemia viral (v-erb-a) oncogene homolog 2-like	17q25	125
ERBB2	Avian erythroblastic leukemia viral (v-erb-b2) oncogene homolog 2 (neuro/glioblastoma derived oncogene homolog. Previous symbol NGL	17q11.2-q12	61, 107, 378, 400, 404
ERG	Avian erythroblastosis virus E26 (v-ets) oncogene related	21q22.3	231, 287

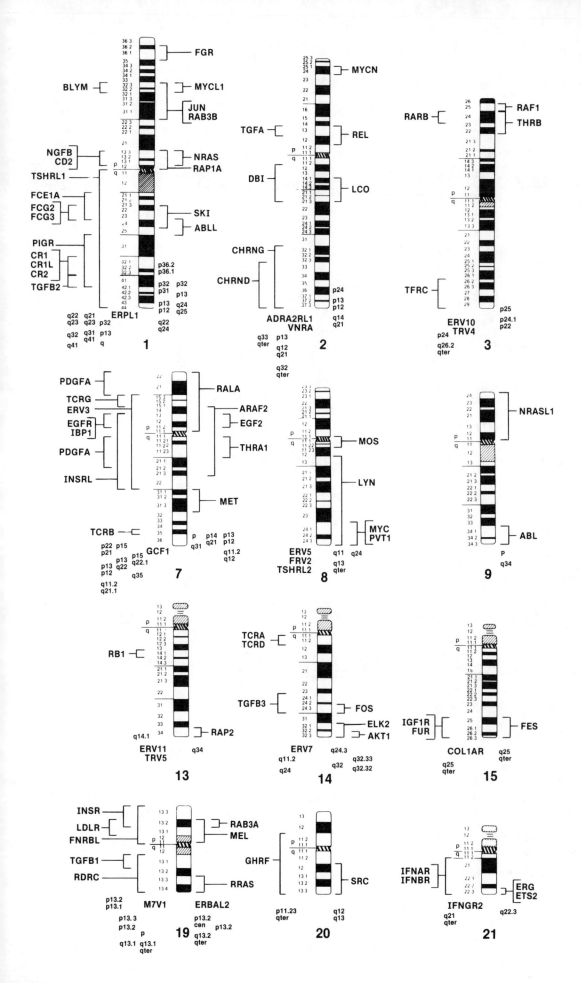

5.198

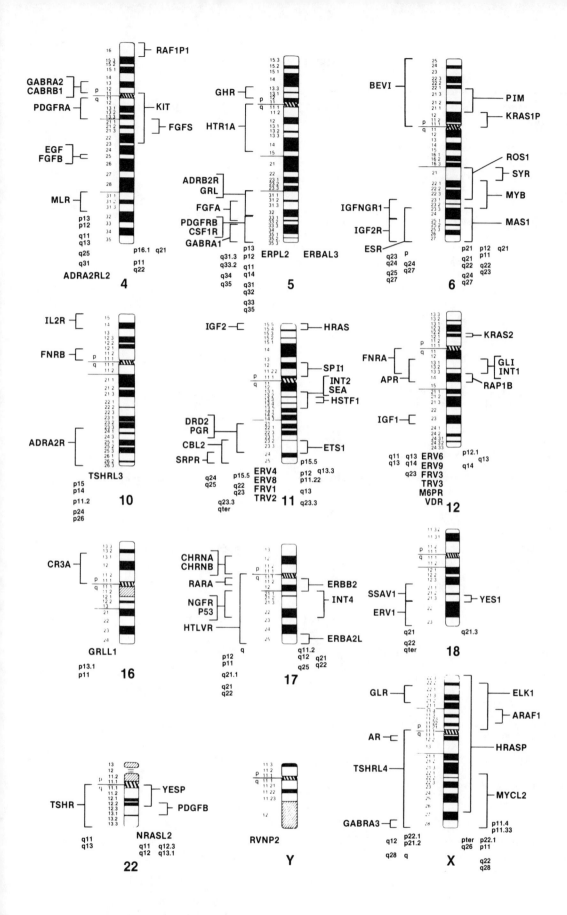

Table I continued

Gene Symbol	Gene Name	Chromosomal Position	References
ETS1	Avian erythroblastosis virus E26 (v-ets) oncogene homolog 1	11q23.3	77, 80, 309, 310, 344, 387a,b
ETS2	Avian erythroblastosis virus E26 (v-ets) oncogene homolog 2	21q22.3	26, 27, 88, 95, 310, 311, 387a
FES	Feline sarcoma viral (v-fes) oncogene; Fujinami avian sarcoma viral (v-fps) oncogene homolog	15q25-qter	69, 137, 141
FGF5	Fibroblast growth factor-like (fgf.5 oncogene)	4q21	254
FGR	Gardner-Rasheed feline sarcoma viral (v-fgr) oncogene homolog. Previous symbol SRC2	1p36.2-p36.1	257, 270, 368
FOS	Murine FBJ osteosarcoma viral (v-fos) oncogene homolog	14q24.3	12, 63, 92
GLI	Glioma-associated oncogene homolog (zinc finger protein)	12q13	6, 68, 169, 170
HRAS	Harvey rat sarcoma viral (v-Ha-ras) oncogene homolog. Previous symbol HRAS1	11p15.5	11, 29, 40, 60, 117, 122, 134, 189, 211, 216, 221, 222, 241, 393
HRASP	Harvey rat sarcoma viral (v-Ha-ras) oncogene homolog. Previous symbol HRAS2	Xpter-q26	230, 259
HSTF1	Heparan secretory transforming protein 1 (Kaposi sarcoma oncogene)	11q13.3	112, 139, 154, 382, 410
INT1	Murine mammary tumor virus integration site (v-int-1) oncogene homolog	12q13	7, 370, 379
INT2	Murine mammary tumor virus integration site (v-int-2) oncogene homolog	11q13	51, 139, 258, 410
INT4	Murine mammary tumor virus integration site (v-int-4) oncogene homolog	17q21-q22	296
JUN	Avian sarcoma virus 17 (v-jun) oncogene homolog	1p32-p31	136, 140
KIT	Hardy-Zuckerman 4 feline sarcoma viral (v-kit) oncogene homolog	4p11-q22	24, 48, 67, 406
KRAS1P	Kirsten rat sarcoma 1 viral (v-Ki-ras1) oncogene processed pseudogene	6p12-p11	217, 259, 278, 314
KRAS2	Kirsten rat sarcoma 2 viral (v-Ki-ras2) oncogene homolog	12p12.1	217, 259, 277, 313, 314, 381
LCO	Liver cancer oncogene	2q14-q21	265, 366
LYN	Yamaguchi sarcoma viral (v-yes-1) related oncogene homolog	8q13-qter	402
MAS1	MAS1 oncogene	6q24-q27	158, 283
MEL	Cell line NK14-derived transforming oncogene	19p13.2-cen	255, 348
MET	Met proto-oncogene	7q31	17, 33, 78, 100, 110, 171, 204, 251, 268, 269, 279, 318, 319, 334, 347, 359, 369, 376, 380, 383, 385, 386, 394, 416

Table I continued

Gene Symbol	Gene Name	Chromosomal Position	References
MOS	Moloney murine sarcoma viral (v-mos) oncogene homolog	8q11	52, 253, 273, 280
MYB	Avian myeloblastosis viral (v-myb) oncogene homolog	6q22-q23	69, 87, 137, 159, 217, 249, 413
MYC	Avian myelocytomatosis viral (v-myc) oncogene homolog. Previous symbol MYC1	8q24	70, 115, 219, 253, 315, 358
MYCL1	Avian myelocytomatosis viral (v-myc) oncogene homolog 1, lung carcinoma derived. Previous symbol MYCL.	1p32	85, 91, 217, 415
MYCL2	Avian myelocytomatosis viral (v-myc) oncogene homolog 2	Xq22-q28	240
MYCN	Avian myelocytomatosis viral (v-myc) related oncogene, neuroblastoma derived. Previous symbol NMYC	2p24	111, 165, 177, 325, 337
NRAS	Neuroblastoma RAS viral (v-ras) oncogene homolog	1p13	4. 217, 243, 263
NRASL1	Neuroblastoma RAS viral (v-ras) oncogene homolog-like 1	9p	224
NRASL2	Neuroblastoma RAS viral (v-ras) oncogene homolog-like 2	22	223
P53	Nuclear transformation antigen	17q21-q22; 17p	22, 187, 226
PDGFB	Platelet-derived growth factor beta polypeptide (simian sarcoma viral (v-sis) oncogene homolog). Previous symbol SIS.	22q12.3-q13.1	16, 71, 135, 163, 234, 303, 355
PIM	pim oncogene homolog	6p21	66, 248, 249
PVT1	Mouse pvt-1 oncogene homolog, MYC activator	8q24	127
RAB3A	RAB3A, member RAS oncogene family	19p13.2	305
RAB3B	RAB3B, member RAS oncogene family	1p32-p31	305
RAF1	Murine leukemia viral (v-raf-1) oncogene homolog 1	3p25	30, 116, 363
RAF1P1	Murine leukemia viral (v-raf-1) oncogene pseudogene 1. Previous symbol RAF2.	4p16.1	30, 64, 119, 210
RALA	Simian leukemia viral (v-ral) oncogene homolog A (ras related) Previous symbol RAL	7p	307
RAP1A	Member of RAS oncogene family (k-rev)	1p13-p12	305
RAP1B	Member of RAS oncogene family (k-rev)	12q14	305
RAP2	Member of RAS oncogene family (k-rev)	13q34	305
RB1	Human retinoblastoma	13q14.1	118, 242, 384
REL	Avian reticuloendotheliosis viral (v-rel) oncogene homolog	2p13-p12	44, 45
ROS1	Avian UR2 sarcoma virus oncogene (v-ros) homolog. Previous symbol ROS	6q21-q22	249, 283
RRAS	Related RAS viral (r-ras) oncogene homolog	19q13.3-qter	207, 323

Table I continued

Gene Symbol	Gene Name	Chromosomal Position	References
SEA	S13 avian erythroblastosis oncogene homolog	11q13	86, 258, 396, 397
SKI	Avian sarcoma viral (v-ski) oncogene homolog	1q22-q24	53
SPI1	Spleen focus forming virus (SFFV) proviral integration oncogene spi1	11p12-p11.22	373
SRC	Avian sarcoma viral (v-src) oncogene homolog. Previous symbol SRC1	20q12-q13	188, 270, 312
SYR	SRC/YES-related oncogene homolog	6q21	329, 411
THRA1	Thyroid hormone receptor, alpha 1 (avian erythroblastic leukemia viral (v-erb-a) oncogene homolog 1, formerly ERBA1)	7q11.2-q12	75, 83, 125, 332, 349, 365, 378, 400
THRB	Thyroid hormone receptor, beta (avian erythroblastic leukemia viral 1 (v-erb-a) oncogene homolog 2). Previous symbol ERBA2.	3p24.1-p22	83, 89, 125, 297, 332, 349, 412
YESP	Yamaguchi sarcoma viral (v-yes-1) oncogene homolog. Previous symbol YES2.	22q11-q12	330
YES1	Yamaguchi sarcoma viral (v-yes-1) oncogene homolog 1	18q21.3	266, 328, 409

Table II
Growth Factors, Cell Surface Receptors, and Endogenous Retroviral Sequences

	Gene Symbol	Gene Name	Chromosomal Position	References
Growth factors	EGF	Epidermal growth factor	4q25	20, 37, 178, 239, 244, 335
	EGFR	Epidermal growth factor receptor, formerly avian erythroblastic leukemia viral (v-erb-b) oncogene homolog. Previous symbol ERBB.	7p13-p12	74, 315, 338, 349, 404
	FGFA	Fibroblast growth factor, acidic (endothelial growth factor)	5q31.3-q33.2	160, 220, 341
	FGFB	Fibroblast growth factor, basic	4q25	48, 220, 336, 341
	FGF5	Fibroblast growth factor-like (fgf.5 oncogene)	4q21	254
	GCF1	Growth control factor 1	7	84
	GHRF	Growth hormone releasing factor	20p11.23-qter	215, 295, 340
	IBP1	Insulin-like growth factor binding protein	7p13-p12	3, 36
	IGF1	Insulin-like growth factor 1	12q23	107, 148, 191, 302, 320
	IGF1R	Insulin-like growth factor 1 receptor	15q25-qter	107, 372
	IGF2	Insulin-like growth factor 2	11p15.5	29, 143, 211, 238, 264, 403
	IGF2R	Insulin-like growth factor 2 receptor	6q25-q27	184
	NGFB	Nerve growth factor, beta polypeptide	1p13	4, 90, 111, 243, 263, 285, 414
	NGFR	Nerve growth factor receptor	17q21-q22	23, 153, 292, 378, 400

Table II continued

Gene Symbol	Gene Name	Chromosomal Position	References
PDGFA	Platelet-derived growth factor alpha polypeptide	7p22-p21 or 7q11.2-q21.1	25, 32, 353
PDGFB	Platelet-derived growth factor beta polypeptide (simian sarcoma viral (v-sis) oncogene homolog. Previous symbol SIS	22q12.3-q13.1	16, 71, 135, 163, 234, 303, 355
PDGFRA	Platelet-derived growth factor receptor alpha polypeptide	4q11-q13	81, 212
PDGFRB	Platelet-derived growth factor receptor, beta polypeptide. Previous symbol PDGFR.	5q33-q35	107, 298, 408
TGFA	Transforming growth factor, alpha	2p13	38, 107, 367
TGFB1	Transforming growth factor, beta 1. Previous symbol TGFB	19q13.1	14, 39, 109, 155, 323, 351
TGFB2	Transforming growth factor, beta 2	1q41	14
TGFB3	Transforming growth factor, beta 3	14q24	14, 360

	Gene Symbol	Gene Name	Chromosomal Position	References
Cell Surface Receptors	ADRA2R	Adrenergic, alpha-2-, receptor. Previous symbol ADRAR	10q24-q26	145, 175, 405
	ADRA2RL1	Adrenergic, alpha-2-, receptor-like 1. Previous symbol ADRARL1.	2	405
	ADRA2RL2	Adrenergic, alpha-2-, receptor-like 2. Previous symbol ADRARL2.	4	405
	ADRB2R	Adrenergic, beta-2-, receptor, surface. Previous symbol ADRBR.	5q31-q32	174, 333
	APR	Apolipoprotein receptor	12q13-q14	247
	AR	Androgen receptor (dihydrotestosterone receptor; testicular feminization. Previous symbol DHTR	Xq12	41, 182, 208, 225, 395
	CD2	Antigen CD2 (p50), sheep red blood cell receptor	1p13	42, 55, 326
	CHRNA	Cholinergic receptor, nicotinic, alpha polypeptide	17p12-p11	19
	CHRNB	Cholinergic receptor, nicotinic, beta polypeptide	17p12-p11	19
	CHRND	Cholinergic receptor, nicotinic, delta polypeptide. Previous symbol ACHRD	2q33-qter	19, 56, 206
	CHRNG	Cholinergic receptor, nicotinic, gamma polypeptide. Previous symbol ACHRG	2q32-qter	56, 206
	COL1AR	Collagen, type I, alpha, receptor	15	275
	CR1	Complement component (3b/4b) receptor 1	1q32	47, 294, 299, 390, 398
	CR1L	Complement component (3b/4b) receptor 1-like	1q32	398
	CR2	Complement component (3d/Epstein Barr virus) receptor 2	1q32	47, 256, 294, 299, 390
	CR3A	Complement component receptor 3, alpha; also known as CD11b (p170), macrophage antigen alpha polypeptide	16p13.1-p11	8, 59
	CSF1R	Colony-stimulating factor 1 receptor, formerly McDonough feline sarcoma viral (v-fms) oncogene homolog. Previous symbol FMS.	5q33-q35	131, 190, 298, 308, 374
	DBI	Diazepam binding inhibitor (? GABA receptor modulator)	2q12-q21	79

Table II continued

Gene Symbol	Gene Name	Chromosomal Position	References
DRD2	Dopamine receptor D2	11q22-q23	114, 128
EGFR	Epidermal growth factor receptor, formerly avian erythroblastic leukemia viral (v-erb-b) oncogene homolog. Previous symbol ERBB.	7p13-p12	74, 315, 338, 349, 404
ESR	Estrogen receptor	6q24-q27	57, 126, 419
FCE1A	Fc fragment of IgE, high affinity receptor for; alpha polypeptide	1q21-q23	361
FCG2	Fc fragment of IgG, low affinity II, receptor for (CD32). Previous symbol CDW32.	1q22-q23	133, 192, 271, 316
FCG3	Fc fragment of IgG, low affinity III, receptor for (CD16). Previous symbol CD16.	1q22-q23	133, 271
FNRA	Fibronectin receptor, alpha polypeptide	12q11-q13	346
FNRB	Fibronectin receptor, beta polypeptide	10p11.2	50, 124, 246, 399, 417
FNRBL	Fibronectin receptor, beta polypeptide-like	19p	399
FUR	Furin, membrane associated receptor protein	15q25-q26	300
GABRA1	Gamma-aminobutyric acid (GABA) A receptor, alpha 1	5q34-q35	46
GABRA2	Gamma-aminobutyric acid (GABA) A receptor, alpha 2	4p13-p12	46, 201
GABRA3	Gamma-aminobutyric acid (GABA) A receptor, alpha 3	Xq28	46, 21
GABRB1	Gamma-aminobutyric acid (GABA) A receptor, beta 1	4p13-p12	46, 281
GHR	Growth hormone receptor	5p13-p12	15
GLR	Glycine receptor	Xp22.1-p21.2	130, 342
GRL	Glucocorticoid receptor	5q31-q32	96, 105, 113, 121, 364
GRLL1	Glucocorticoid receptor-like 1	16	147
HTLVR	Human T cell leukemia virus (I and II) receptor	17q	345
HTR1A	5-hydroxytryptamine receptor 1A	5q11-q14	99, 120, 176
IFNAR	Interferon, alpha; receptor	21q21-qter	54, 232, 357, 418
IFNBR	Interferon, beta; receptor	21q21-qter	54, 232, 357, 418
IFNGR1	Interferon, gamma; receptor 1 (binding subunit). Previous symbol IFNGR.	6q23-q24	2, 31, 164, 274, 288
IFNGR2	Interferon, gamma; receptor 2 (confers antiviral resistance)	21	31, 101, 289
IGF1R	Insulin-like growth factor 1 receptor	15q25-qter	107, 372
IGF2R	Insulin-like growth factor 2 receptor	6q25-q27	184
IL2R	Interleukin 2 receptor	10p15-p14	5, 200
INSR	Insulin receptor	19p13.3-p13.2	39, 93, 331, 407
INSRL	Insulin receptor-like	7p13-q22	181
LDLR	Low density lipoprotein receptor (familial hypercholesterolemia)	19p13.2-p13.1	40, 106, 166, 202, 205, 209, 250, 331

Table II continued

Gene Symbol	Gene Name	Chromosomal Position	References
LNHR	Lymph node homing receptor	unassigned	1
MLR	Mineralocorticoid receptor (aldosterone receptor)	4q31	9, 48, 97, 235
M6PR	Mannose-6-phosphate receptor (cation dependent)	12	276, 301
M7V1	Baboon M7 virus receptor	19	43
NGFR	Nerve growth factor receptor	17q21-q22	23, 81, 153, 212, 292, 378, 400
PDGFRA	Platelet-derived growth factor receptor alpha polypeptide	4q11-q13	81, 212
PDGFRB	Platelet-derived growth factor receptor, beta polypeptide. Previous symbol PDGFR.	5q33-q35	107, 298, 408
PGR	Progesterone receptor	11q22-q23	185, 213, 252, 258, 304, 306
PIGR	Polymeric immunoglobulin receptor	1q31-q41	73
RARA	Retinoic acid receptor, alpha	17q21.1	35, 272
RARB	Retinoic acid receptor, beta	3p24	35, 214
RDRC	RD114 virus receptor	19q31.1-qter	166, 321
SRPR	Signal recognition particle receptor, "docking protein."	11q24-q25	406
TCRA	T cell receptor, alpha (V,D,J,C)	14q11.2	13, 58, 65, 72, 189, 203, 219
TCRB	T cell receptor, beta cluster	7q35	33, 100, 157, 167, 186, 237, 369, 371, 392, 416
TCRD	T cell receptor, delta (V,D,J,C)	14q11.2	28, 144, 156, 179, 267, 317
TCRG	T cell receptor, gamma cluster	7p15	49, 103, 104, 150, 151, 193, 194, 195, 196, 245, 282
TFRC	Transferrin receptor (p90)	3q26.2-qter	94, 123, 227, 284, 327, 375, 377
THRA1	Thyroid hormone receptor, alpha 1 (avian erythroblastic leukemia viral (v-erb-a) oncogene homolog 1, formerly ERBA1	7q11.2-q12	75, 83, 125, 332, 349, 365, 378, 400
THRB	Thyroid hormone receptor, beta (avian erythroblastic leukemia viral (v-erb-a) oncogene homolog 2. Previous symbol ERBA2.	3p24.1-p22	83, 89, 125, 297, 332, 349, 412
TSHR	Thyroid stimulating hormone receptor	22q11-q13	218, 228
TSHRL1	Thyroid stimulating hormone receptor-like 1	1q	218
TSHRL2	Thyroid stimulating hormone receptor-like 2	8	218
TSHRL3	Thyroid stimulating hormone receptor-like 3	10	218
TSHRL4	Thyroid stimulating hormone receptor-like 4	Xq	218
VDR	Vitamin D receptor	12	98
VNRA	Vitronectin receptor, alpha polypeptide	2	346

Table II continued

	Gene Symbol	Gene Name	Chromosomal Position	References
Endogenous retroviral sequences	BEVI	Baboon M7 endogenous retrovirus integration site	6p	43, 198, 199
	CBL2	Cas-Br-M (murine) ecotropic retroviral transforming sequence	11q23.3-qter	183, 389
	ERPL1	Endogenous retroviral pol gene-like sequence 1 (clone HLM2)	1	149
	ERPL2	Endogenous retroviral pol gene-like sequence 2	5	149
	ERV1	Endogenous retroviral sequence 1	18q22-qter	260, 293
	ERV3	Endogenous retroviral sequence 3	7p15-q22.1	261, 262
	ERV4	Endogenous retroviral sequence 4	11	352
	ERV5	Endogenous retroviral sequence 5	8	352
	ERV6	Endogenous retroviral sequence 6	12	352
	ERV7	Endogenous retroviral sequence 7	14	352
	ERV8	Endogenous retroviral sequence 8	11	352
	ERV9	Endogenous retroviral sequence 9	12	352
	ERV10	Endogenous retroviral sequence 10	3	352
	ERV11	Endogenous retroviral sequence 11	13	352
	FRV1	Full length retroviral sequence 1 (band F6, 14.5kb)	11	352
	FRV2	Full length retroviral sequence 2 (band F9, 11.2kb)	8	352
	FRV3	Full length retroviral sequence 3 (band F21, 3.5kb)	12	352
	RVNP2	Retroviral sequences NP2	Y	343
	SSAV1	Simian sarcoma-associated virus 1/gibbon ape leukemia virus-related endogenous retroviral element	18q21	34
	TRV2	Truncated retroviral sequence 2 (band T2, 21.0kb)	11	352
	TRV3	Truncated retroviral sequence 3 (band T3, 17.8kb)	12	352
	TRV4	Truncated retroviral sequence 4 (band T4, 14.9kb)	3	352
	TRV5	Truncated retroviral sequence 5	13	352

References

1. Ala-Kapee, M., et al. (HGM10) Cytogenet. Cell Genet. 51:948, 1989. (abs.)
2. Alcaide-Loridan, C., et al. (HGM10) Cytogenet. Cell Genet. 51:949, 1989. (abs.)
3. Alitalo, T., et al. (HGM10) Cytogenet. Cell Genet. 51:950, 1989. (abs.)
4. Anderson, L. A., et al. (HGM10) Cytogenet. Cell Genet. 51:951, 1989. (abs.)
5. Ardinger, R. H., et al. Nucleic Acids Res. 16:8201, 1988.
6. Arheden, K., et al. Hum. Genet. 82:1-2, 1989.
7. Arheden, K., et al. (HGM9) Cytogenet. Cell Genet. 47:86-87, 1988. (abs.)
8. Arnaout, M. A., et al. Proc. Natl. Acad. Sci. USA 85:2776-2780, 1988.
9. Arriza, J. L., et al. Science 237:268-275, 1987.
10. Avner, P., et al. Somatic Cell Mol. Genet. 13:267-272, 1987.
11. Bale, S. J., et al. Genet. Epidemiol. Suppl. 1:117-121, 1986.
12. Barker, P. E., et al. Proc. Natl. Acad. Sci. USA 81:5826-5830, 1984.
13. Barker, P. E., et al. (HGM8) Cytogenet. Cell Genet. 40:576-577, 1985. (abs.)
14. Barton, D. E., et al. Oncogene Res. 3:323-331, 1988.
15. Barton, D. E., et al. Cytogenet. Cell Genet., in press, 1989.
16. Bartram, C. R., et al. Blood 63:223-225, 1984.
17. Beaudet, A., et al. Am. J. Hum. Genet. 39:681-693, 1986.
18. Beck, T. W., et al. Nucleic Acids Res. 15:595-609, 1987.
19. Beeson, D., et al. (HGM10) Cytogenet. Cell Genet. 51:960, 1989. (abs.)
20. Bell, G. I., et al. Nucleic Acids Res. 14:8427-8446, 1986.
21. Bell, M. V., et al. (HGM10) Cytogenet. Cell Genet. 51:960, 1989. (abs.)
22. Benchimol, S., et al. Somatic Cell Mol. Genet. 11:505-510, 1985.
23. Bentley, K. L., et al. (HGM10) Cytogenet. Cell Genet. 51:961, 1989. (abs.)
24. Berdahl, L. D., et al. Nucleic Acids Res. 16:4740, 1988.
25. Betsholtz, C., et al. Nature 320:695-699, 1986.
26. Blanquet, V., et al. (HGM9) Cytogenet. Cell Genet. 46:582, 1987. (abs.)
27. Blanquet, V., et al. Ann. Genet. (Paris) 30:68-69, 1987.
28. Boehm, T., et al. EMBO J. 7:385-394, 1988.
29. Bonaiti-Pellie, C., et al. Genet. Epidemiol. Suppl. 1:129-134, 1986.
30. Bonner, T., et al. Science 223:71-74, 1984.
31. Bono, R., et al. (HGM9) Cytogenet. Cell Genet. 46:584, 1987. (abs.)
32. Bonthron, D. T., et al. Proc. Natl. Acad. Sci. USA 85:1492-1496, 1988.
33. Bowcock, A. M., et al. Am. J. Hum. Genet. 39:699-706, 1986.
34. Brack-Werner, R., et al. Genomics 4:68-75, 1989.
35. Brand, N., et al. Nature 332:850-853, 1988.
36. Brinkman, A., et al. EMBO J. 7:2417-2423, 1988.
37. Brissenden, J. E., et al. Nature 310:781-784, 1984.
38. Brissenden, J. E., et al. Cancer Res. 45:5593-5597, 1985.
39. Brook, J. D., et al. (HGM9) Cytogenet. Cell Genet. 46:587, 1987. (abs.)
40. Brook, J. D., et al. Cytogenet. Cell Genet. 41:30-37, 1986.
41. Brown, C. J., et al. Am. J. Hum. Genet. 44:264-269, 1989.
42. Brown, M. H., et al. Hum. Genet. 76:191-195, 1987.
43. Brown, S., et al. Cell 18:135-143, 1979.
44. Brownell, E., et al. Am. J. Hum. Genet. 39:194-202, 1986.
45. Brownell, E., et al. Oncogene 2:527-529, 1988.
46. Buckle, V. J., et al. (HGM10) Cytogenet. Cell Genet. 51:972, 1989. (abs.)
47. Buetow, K. H., et al. (HGM10) Cytogenet. Cell Genet. 51:972, 1989. (abs.)
48. Buetow, K. H., et al. (HGM10) Cytogenet. Cell Genet. 51:973, 1989. (abs.)
49. Buresi, C., et al. Immunogenetics 29:161-172, 1989.
50. Carson, N. L. and Simpson, N. E. (HGM9) Cytogenet. Cell Genet. 51:974, 1989. (abs.)
51. Casey, G., et al. Mol. Cell. Biol. 6:502-510, 1986.
52. Caubet, J.-F., et al. EMBO J. 4:2245-2248, 1985.
53. Chaganti, R. S. K., et al. Cytogenet. Cell Genet. 43:181-186, 1986.
54. Chany, C., et al. Proc. Natl. Acad. Sci. USA 72:3129-3133, 1975.
55. Clayton, L. K., et al. J. Immunol. 140:3617-3621, 1988.
56. Cohen-Haguenauer, O., et al. (HGM9) Cytogenet. Cell Genet. 46:595, 1987. (abs.)
57. Coleman, R. T., et al. Nucleic Acids Res. 16:7208, 1988.
58. Collins, M. K. L., et al. Nature 314:273-274, 1985.
59. Corbi, A. L., et al. J. Exp. Med. 167:1597-1607, 1988.
60. Couillin, P., et al. Genomics 4:7-11, 1989.

61. Coussens, L., et al. Science 230:1132-1139, 1985.
62. Cox, D. W., et al. (HGM9) Cytogenet. Cell Genet. 46:599, 1987. (abs.)
63. Cox, D. W., et al. (HGM10) Cytogenet. Cell Genet. 51:980, 1989. (abs.)
64. Cox, D. R., et al. Genomics 4:397-407, 1989.
65. Croce, C. M., et al. Science 227:1044-1047, 1985.
66. Cuypers, H. T., et al. Hum. Genet. 72:262-265, 1986.
67. d'Auriol, L., et al. Hum. Genet. 78:374-376, 1988.
68. Dal Cin, P., et al. (HGM10) Cytogenet. Cell Genet. 51:982, 1989. (abs.)
69. Dalla-Favera, R., et al. Proc. Natl. Acad. Sci. USA 79:4714-4717, 1982.
70. Dalla-Favera, R., et al. Proc. Natl. Acad. Sci. USA 79:7824-7827, 1982.
71. Dalla-Favera, R., et al. Science 218:686-688, 1982.
72. Davey, M. P., et al. Proc. Natl. Acad. Sci. USA 85:9287-9291, 1988.
73. Davidson, M. K., et al. Cytogenet. Cell Genet. 48:107-111, 1988.
74. Davies, R. L., et al. Proc. Natl. Acad. Sci. USA 77:4188-4192, 1980.
75. Dayton, A. I., et al. Proc. Natl. Acad. Sci. USA 81:4495-4499, 1984.
76. de Pagter-Holthuizen, P., et al. FEBS Lett. 214:259-264, 1987.
77. de Taisne, C., et al. Nature 310:581-583, 1984.
78. Dean, M., et al. Nature 318:385-388, 1985.
79. DeBernardi, M. A., et al. Proc. Natl. Acad. Sci. USA 85:6561-6565, 1988.
80. Diaz, M. O., et al. Science 231:265-267, 1986.
81. Disteche, C. M., et al. (HGM10) Cytogenet. Cell Genet. 51:990, 1989.
82. Dixon, R. A. F., et al. Nature 321:75-79, 1986.
83. Dobrovic, A., et al. Cancer Res. 48:682-685, 1988.
84. Donald, L. J., et al. (HGM6) Cytogenet. Cell Genet. 32:268-269, 1982.
85. Donis-Keller, H., et al. (HGM10) Cytogenet. Cell Genet. 51:991, 1989. (abs.)
86. Donis-Keller, H., et al. Cell 51:319-337, 1987.
87. Dozier, C., et al. Nucleic Acids Res. 14:1928, 1986.
88. Drabkin, H. A., et al. ICSU Short Reports 4:112-113, 1986.
89. Drabkin, H., et al. Proc. Natl. Acad. Sci. USA 85:9258-9262, 1988.
90. Dracopoli, N. C., et al. Am. J. Hum. Genet. 43:462-470, 1988.
91. Dracopoli, N. C., et al. (HGM10) Cytogenet. Cell Genet. 51:993, 1989. (abs.)
92. Ekstrand, A. J. and Zech, L. Exp. Cell Res. 169:262-266, 1987.
93. Elbein, S. C., et al. Proc. Natl. Acad. Sci. USA 83:5223-5227, 1986.
94. Enns, C. A., et al. Proc. Natl. Acad. Sci. USA 79:3241-3245, 1982.
95. Erikson, J., et al. Proc. Natl. Acad. Sci. USA 83:1807-1811, 1986.
96. Evans, R. M. Science 240:889-895, 1988.
97. Fan, Y.-S., et al. (HGM10) Cytogenet. Cell Genet. 51:996, 1989. (abs.)
98. Faraco, J. H., et al. Nucleic Acids Res. 17:2150, 1989.
99. Fargin, A., et al. Nature 335:358-360, 1988.
100. Farrall, M., et al. Am. J. Hum. Genet. 39:713-719, 1986.
101. Fellous, M., et al. (HGM8) Cytogenet. Cell Genet. 40:627-628, 1985. (abs.)
102. Fitzgerald, L. A., et al. Biochem. 26:8158-8165, 1987.
103. Font, M. P., et al. J. Exp. Med. 168:1383-1394, 1988.
104. Forster, A., et al. EMBO J. 6:1945-1950, 1987.
105. Francke, U. and Foellmer, B. E. Genomics 4:610-612, 1989.
106. Francke, U., et al. Proc. Natl. Acad. Sci. USA 81:2826-2830, 1984.
107. Francke, U., et al. Cold Spring Harbor Symp. Quant. Biol. 51:855-866, 1986.
108. Frielle, T., et al. Proc. Natl. Acad. Sci. USA 84:7920-7924, 1987.
109. Fujii, D., et al. Somatic Cell Mol. Genet. 12:281-288, 1986.
110. Fulton, T. R., et al. Nucleic Acids Res. 17:271-284, 1989.
111. Garson, J. A., et al. Nucleic Acids Res. 15:4761-4770, 1987.
112. Gaudray, P., et al. Cancer Genet. Cytogenet. 38:192, 1989. (abs. 105)
113. Gehring, U., et al. Proc. Natl. Acad. Sci. USA 82:3751-3755, 1985.
114. Gelernter, J., et al. (HGM10) Cytogenet. Cell Genet. 51:1002, 1989. (abs.)
115. Gemmill, R. M., et al. Genomics 4:28-35, 1989.
116. Gerber, M. J., et al. Am. J. Hum. Genet. 43:442-451, 1988.
117. Gerhard, D. S., et al. ICSU Short Reports 1:172-173, 1984.
118. Gey, W. Humangenetik 10:362-365, 1970.
119. Gilliam, T. C., et al. Cell 50:565-571, 1987.

120. Gilliam, T. C., et al. Genomics, in press, 1989.
121. Giuffra, L. A., et al. Cytogenet. Cell Genet. 49:313-314, 1988.
122. Glaser, T., et al. Nature 321:882-887, 1986.
123. Goodfellow, P. N., et al. Somatic Cell Genet. 8:197-206, 1982.
124. Goodfellow, P. J., et al. Genomics 3:224-229, 1988.
125. Gosden, J. R., et al. Cytogenet. Cell Genet. 43:150-153, 1986.
126. Gosden, J. R., et al. Cytogenet. Cell Genet. 43:218-220, 1986.
127. Graham, M. and Adams, J. M. EMBO J. 5:2845-2851, 1986.
128. Grandy, D. K., et al. (HGM10) Cytogenet. Cell Genet. 51:1007, 1989. (abs.)
129. Gray, P. W., et al. Proc. Natl. Acad. Sci. USA 83:7547-7551, 1986.
130. Grenningloh, G., et al. Nature 328:215-220, 1987.
131. Groffen, J., et al. Nucleic Acids Res. 11:6331-6339, 1983.
132. Groffen, J., et al. J. Cell. Physiol. (suppl.) 3:179-191, 1984.
133. Grundy, H. O., et al. Am. J. Hum. Genet. 43:A145, 1988. (abs. 578)
134. Grzeschik, K. H. and Kazazian, H. H. (HGM8) Cytogenet. Cell Genet. 40:179-203, 1985.
135. Haines, J., et al. (HGM10) Cytogenet. Cell Genet. 46:625, 1987. (abs.)
136. Haluska, F. G., et al. Proc. Natl. Acad. Sci. USA 85:2215-2218, 1988.
137. Harper, M. E., et al. Nature 304:169-171, 1983.
138. Haselbacher, G. K., et al. Proc. Natl. Acad. Sci. USA 84:1104-1106, 1987.
139. Hatada, I., et al. Oncogene 3:537-540, 1988.
140. Hattori, K., et al. Proc. Natl. Acad. Sci. USA 85:9148-9152, 1988.
141. Heisterkamp, N., et al. Nature 299:747-749, 1982.
142. Heisterkamp, N., et al. Nucleic Acids Res. 16:10069-10081, 1988.
143. Henry, I., et al. (HGM10) Cytogenet. Cell Genet. 40:648-649, 1985. (abs.)
144. Hockett, R. D., et al. Proc. Natl. Acad. Sci. USA 85:9694-9698, 1988.
145. Hoehe, M. R., et al. Nucleic Acids Res. 16:9070, 1988.
146. Hofker, M. H., et al. (HGM10) Cytogenet. Cell Genet. 51:1014, 1989. (abs.)
147. Hollenberg, S. M., et al. Nature 318:635-641, 1985.
148. Hoppener, J. W. M., et al. Hum. Genet. 69:157-160, 1985.
149. Horn, T. M., et al. J. Virol. 58:955-959, 1986.
150. Huck, S., et al. EMBO J. 7:719-726, 1988.
151. Huck, S. and Lefranc, M.-P. FEBS Lett. 224:291-296, 1987.
152. Huebner, K., et al. Proc. Natl. Acad. Sci. USA 83:3934-3938, 1986.
153. Huebner, K., et al. Proc. Natl. Acad. Sci. USA 83:1403-1407, 1986.
154. Huebner, K., et al. Oncogene Res. 3:263-270, 1988.
155. Hulsebos, T., et al. Cytogenet. Cell Genet. 43:47-56, 1986.
156. Isobe, M., et al. Proc. Natl. Acad. Sci. USA 85:3933-3937, 1988.
157. Isobe, M., et al. Science 228:580-582, 1985.
158. Jackson, T. R., et al. Nature 335:437-440, 1988.
159. Janssen, J. W. G., et al. Cytogenet. Cell Genet. 41:129-135, 1986.
160. Jaye, M., et al. Science 233:541-545, 1986.
161. Jhanwar, S. C., et al. Proc. Natl. Acad. Sci. USA 79:7842-7846, 1982.
162. Jhanwar, S. C., et al. Cytogenet. Cell Genet. 38:73-75, 1984.
163. Julier, C., et al. Am. J. Hum. Genet. 42:297-308, 1988.
164. Jung, V., et al. Proc. Natl. Acad. Sci. USA 84:4151-4155, 1987.
165. Kanda, N., et al. Proc. Natl. Acad. Sci. USA 80:4069-4073, 1983.
166. Kaneda, Y., et al. Chromosoma 95:8-12, 1987.
167. Kere, J. Nucleic Acids Res. 17:1511-1520, 1989.
168. Kiess, W., et al. J. Biol. Chem. 263:9339-9344, 1988.
169. Kinzler, K. W., et al. (HGM9) Cytogenet. Cell Genet. 46:639, 1987. (abs.)
170. Kinzler, K. W., et al. Science 236:70-73, 1987.
171. Klinger, K., et al. Nucleic Acids Res. 14:8681-8686, 1986.
172. Klinger, K. W., et al. Proc. Natl. Acad. Sci. USA 84:8548-8552, 1987.
173. Kobilka, B. K., et al. J. Biol. Chem. 262:7321-7327, 1987.
174. Kobilka, B. K., et al. Proc. Natl. Acad. Sci. USA 84:46-50, 1987.
175. Kobilka, B. K., et al. Science 238:650-656, 1987.
176. Kobilka, B. K., et al. Nature 329:75-79, 1987.
177. Kohl, N. E., et al. Cell 35:359-367, 1983.
178. Kolble, K., et al. (HGM10) Cytogenet. Cell Genet. 51:1024, 1989. (abs.)
179. Krangel, M. S., et al. Science 237:64-67, 1987.
180. Kruh, G. D., et al. Science 234:1545-1548, 1986.

181. Kudoh, J., et al. (HGM9) Cytogenet. Cell Genet. 46:642, 1987. (abs.)
182. Lafreniere, R. G., et al. (HGM10) Cytogenet. Cell Genet. 51:1028, 1989. (abs.)
183. Langdon, W. Y., et al. Proc. Natl. Acad. Sci. USA 86:1168-1172, 1989.
184. Laureys, G., et al. Genomics 3:224-229, 1988.
185. Law, M. L., et al. Proc. Natl. Acad. Sci. USA 84:2877-2881, 1987.
186. Le Beau, M. M., et al. Cell 41:335, 1985.
187. Le Beau, M. M., et al. Nature 316:826-828, 1985.
188. Le Beau, M. M., et al. Cancer Genet. Cytogenet. 23:269-274, 1986.
189. Le Beau, M. M., et al. Proc. Natl. Acad. Sci. USA 83:9744-9748, 1986.
190. Le Beau, M. M., et al. Science 231:984-987, 1986.
191. Le Bouc, Y., et al. FEBS Lett. 196:108-112, 1986.
192. Lebo, R. V., et al. Amer. Soc. Hum. Genet., 1989. (abs.)
193. Lefranc, M.-P. (HGM10) Cytogenet. Cell Genet. 51:1031, 1989. (abs.)
194. Lefranc, M.-P. and Rabbitts, T. H. Nature 316:464-466, 1985.
195. Lefranc, M.-P., et al. Cell 45:237-246, 1986.
196. Lefranc, M.-P., et al. Nature 319:420-422, 1986.
197. Lefranc, M.-P., et al. Proc. Natl. Acad. Sci. USA 83:9596-9600, 1986.
198. Lemons, R. S., et al. Cell 12:251-262, 1977.
199. Lemons, R. S., et al. Cell 14:995-1005, 1978.
200. Leonard, W. J., et al. Science 228:1547-1549, 1985.
201. Levitan, E. S., et al. Nature 335:76-79, 1988.
202. Lewis, M., et al. (HGM9) Cytogenet. Cell Genet. 46:650, 1987. (abs.)
203. Lewis, W. H., et al. Nature 317:544-546, 1985.
204. Lichter, P., et al. (HGM10) Cytogenet. Cell Genet. 51:1033, 1989. (abs.)
205. Lingren, V., et al. Proc. Natl. Acad. Sci. USA 82:8567-8571, 1985.
206. Lobos, E. A., et al. Am. J. Hum. Genet. 44:522-533, 1989.
207. Lowe, D. G., et al. Cell 48:137-146, 1987.
208. Lubahn, D. B., et al. Science 240:327-330, 1988.
209. Lusis, A. J., et al. Proc. Natl. Acad. Sci. USA 83:3929-3933, 1986.
210. MacDonald, M. E., et al. Genomics 1:29-34, 1987.
211. Mannens, M., et al. Hum. Genet. 75:180-187, 1987.
212. Matsui, T., et al. Science 243:800-804, 1989.
213. Mattei, M. G., et al. (HGM9) Cytogenet. Cell Genet. 46:658, 1987. (abs.)
214. Mattei, M.-G., et al. Hum. Genet. 80:189-190, 1988.
215. Mayo, K. E., et al. Proc. Natl. Acad. Sci. USA 82:63-67, 1985.
216. McBride, O. W., et al. Nature 300:773-774, 1982.
217. McBride, O. W., et al. Nucleic Acids Res. 11:8221-8236, 1983.
218. McBride, O. W., et al. Am. J. Hum. Genet. 41:A177, 1987. (abs. 524)
219. McKeithan, T. W., et al. Proc. Natl. Acad. Sci. USA 83:6636-6640, 1986.
220. Mergia, A., et al. Biochem. Biophys. Res. Comm. 138:644-651, 1986.
221. Meyers, D. A., et al. Am. J. Hum. Genet. 39:539-541, 1986.
222. Meyers, D. A., et al. Am. J. Hum. Hered. 37:94-101, 1987.
223. Middleton-Price, H., et al. Ann. Hum. Genet. 52:189-195, 1988.
224. Middleton-Price, H., et al. (HGM9) Cytogenet. Cell Genet. 46:662, 1987.
225. Migeon, B. R., et al. Proc. Natl. Acad. Sci. USA 78:6339-6343, 1981.
226. Miller, C., et al. Nature 319:783-784, 1986.
227. Miller, Y. E., et al. Cancer Genet. Cytogenet. 35:179-197, 1988.
228. Mitchell, A. L., et al. (HGM10) Cytogenet. Cell Genet. 51:1045, 1989. (abs.)
229. Miyajima, N., et al. Nucleic Acids Res. 16:11057-11074, 1988.
230. Miyoshi, J., et al. Nucleic Acids Res. 12:1821-1828, 1984.
231. Modi, W. S., et al. (HGM10) Cytogenet. Cell Genet. 51:1046, 1989. (abs.)
232. Mogensen, K. E., et al. FEBS Lett. 140:285-287, 1982.
233. Morris, C. M., et al. Nature 320:281-283, 1986.
234. Morris, C. M., et al. Cancer Genet. Cytogenet. 40:703, 1985. (abs.)
235. Morrison, N., et al. (HGM10) Cytogenet. Cell Genet. 51:1048, 1989. (abs.)
236. Morton, C. C., et al. Science 223:173-175, 1984.
237. Morton, C. C., et al. Science 228:582-585, 1985.
238. Morton, C. C., et al. (HGM8) Cytogenet. Cell Genet. 40:703, 1985.
239. Morton, C. C., et al. Cytogenet. Cell Genet. 41:245-249, 1986.
240. Morton, C. C., et al. Genomics 4:367-375, 1989.
241. Moss, P. A. H., et al. Nucleic Acids Res. 14:9927-9932, 1986.
242. Motegi, T., et al. Hum. Genet. 60:193-195, 1982.

243. Munke, M., et al. (HGM8) Cytogenet. Cell Genet. 40:705-706, 1985. (abs.)
244. Murray, J. C., et al. Am. J. Hum. Genet. 42:490-497, 1988.
245. Murre, C., et al. Nature 316:549-552, 1985.
246. Myers, S., et al. (HGM10) Cytogenet. Cell Genet. 51:1050, 1989. (abs.)
247. Myklebost, O., et al. Genomics 5:65-69, 1989.
248. Nagarajan, L., et al. Proc. Natl. Acad. Sci. USA 83:2556-2560, 1986.
249. Nagarajan, L., et al. Proc. Natl. Acad. Sci. USA 83:6568-6572, 1986.
250. Naylor, S. L. and Shows, T. B. Somatic Cell Genet. 6:641-652, 1980.
251. Naylor, S. L., et al. Am. J. Hum. Genet. 39:707-712, 1986.
252. Naylor, S. L., et al. (HGM10) Cytogenet. Cell Genet. 51:1051, 1989. (abs.)
253. Neel, B. G., et al. Proc. Natl. Acad. Sci. USA 79:7842-7846, 1982.
254. Nguyen, C., et al. (HGM10) Oncogene 3:703-708, 1988.
255. Nimmo, E., et al. (HGM10) Cytogenet. Cell Genet. 51:1053, 1989. (abs.)
256. Nishimura, D., et al. Genomics 3:393-395, 1988.
257. Nishizawa, M., et al. Mol. Cell. Biol. 6:511-517, 1986.
258. Nordenskjold, M., et al. (HGM10) Cytogenet. Cell Genet. 51:1054, 1989. (abs.)
259. O'Brien, S. J., et al. Nature 302:839-842, 1983.
260. O'Brien, S. J., et al. Nature 303:74-77, 1983.
261. O'Connell, C., et al. J. Virology 138:225-235, 1984.
262. O'Connell, P., et al. (HGM9) Cytogenet. Cell Genet. 46:672, 1987. (abs.)
263. O'Connell, P., et al. Genomics 4:12-20, 1989.
264. O'Malley, K. L. and Rotwein, P. Nucleic Acids Res. 16:4437-4446, 1988.
265. Ochiya, T., et al. Proc. Natl. Acad. Sci. USA 83:4993-4997, 1986.
266. Ohno, H., et al. Int. J. Cancer 39:785-788, 1987.
267. Okada, A., et al. J. Exp. Med. 168:1481-1486, 1988.
268. Park, M., et al. Cell 45:895-904, 1986.
269. Park, M., et al. Proc. Natl. Acad. Sci. USA 85:2667-2671, 1988.
270. Parker, R. C., et al. Mol. Cell. Biol. 5:831-838, 1985.
271. Peltz, G. A., et al. Proc. Natl. Acad. Sci. USA 86:1013-1017, 1989.
272. Petkovich, M., et al. Nature 330:444-450, 1987.
273. Pettenati, M. J., et al. Cancer Genet. Cytogenet. 37:221-227, 1989.
274. Pfizenmaier, K., et al. J. Immunol. 141:856-860, 1988.
275. Pignatelli, M. and Bodmer, W. F. Proc. Natl. Acad. Sci. USA 85:5561-5565, 1988.
276. Pohlmann, R., et al. Proc. Natl. Acad. Sci. USA 84:5575-5579, 1987.
277. Popescu, N. C., et al. Somatic Cell Mol. Genet. 11:149-155, 1985.
278. Popescu, N. C., et al. (HGM10) Cytogenet. Cell Genet. 44:58-62, 1987.
279. Poustka, A. M., et al. Genomics 2:337-345, 1988.
280. Prakash, K., et al. Proc. Natl. Acad. Sci. USA 79:5210-5214, 1982.
281. Pritchett, D. B., et al. Science 242:1306-1308, 1988.
282. Rabbitts, T. H., et al. EMBO J. 4:1461-1465, 1985.
283. Rabin, M., et al. Oncogene Res. 1:169-178, 1987.
284. Rabin, M., et al. Am. J. Hum. Genet. 37:1112-1116, 1985.
285. Raeymaekers, P., et al. Am. J. Hum. Genet. 81:231-233, 1989.
286. Rao, V. N., et al. Science 244:66-70, 1989.
287. Rao, V. N., et al. Science 237:635-639, 1987.
288. Rashidbaigi, A., et al. Proc. Natl. Acad. Sci. USA 83:384-388, 1986.
289. Raziuddin, A., et al. Proc. Natl. Acad. Sci. USA 81:5504-5508, 1984.
290. Reeve, A. E., et al. Mol. Cell. Biol. 9:1799-1803, 1989.
291. Reeves, B. R., et al. Oncogene 4:373-378, 1989.
292. Rettig, W. J., et al. Somatic Cell Mol. Genet. 12:441-447, 1986.
293. Revan, M. J. and Reeves, B. R. Cytogenet. Cell Genet. 44:167-170, 1987.
294. Rey-Campos, J., et al. J. Exp. Med. 167:664-669, 1988.
295. Riddell, D. C., et al. Somatic Cell Mol. Genet. 11:189-195, 1985.
296. Rider, S. H., et al. (HGM10) Cytogenet. Cell Genet. 51:1066, 1989. (abs.)
297. Rider, S. H., et al. Ann. Hum. Genet. 51:153-160, 1987.
298. Roberts, W. M., et al. Cell 55:655-661, 1988.
299. Rodriguez de Cordoba, S. and Rubinstein, P. J. Exp. Med. 164:1274-1283, 1986.
300. Roebroek, A. J. M., et al. EMBO J. 5:2197-2202, 1986.
301. Roth, R. A. Science 239:1269-1271, 1988.
302. Rotwein, P., et al. J. Biol. Chem. 261:4828-4832, 1986.
303. Rouleau, G. A., et al. Genomics 4:1-6, 1989.
304. Rousseau-Merck, M. F., et al. (HGM9) Cytogenet. Cell Genet. 46:685, 1987. (abs.)
305. Rousseau-Merck, M. F., et al. (HGM10) Cytogenet. Cell Genet. 51:1070, 1989. (abs.)
306. Rousseau-Merck, M.-F., et al. Hum. Genet. 77:280-282, 1988.
307. Rousseau-Merck, M.-F., et al. Hum. Genet. 79:132-136, 1987.
308. Roussel, M. F., et al. J. Virol. 48:770-773, 1983.
309. Rovigatti, U., et al. Science 232:398-400, 1986.
310. Sacchi, N., et al. Science 231:379-382, 1986.
311. Sacchi, N., et al. Genomics 3:110-116, 1988.
312. Sakaguchi, A. Y., et al. Prog. Nucl. Acid Res. Mol. Biol. 29:279-283, 1983.
313. Sakaguchi, A. Y., et al. Science 219:1081-1082, 1983.
314. Sakaguchi, A. Y., et al. Mol. Cell. Biol. 4:989-993, 1984.
315. Sakaguchi, A. Y., et al. (HGM7) Cytogenet. Cell Genet. 37:572-573, 1984. (abs.)
316. Sammartino, L., et al. Immunogenetics 28:380-381, 1988.
317. Satyanarayana, K., et al. Proc. Natl. Acad. Sci. USA 85:8166-8170, 1988.
318. Scambler, P. J., et al. Nucleic Acids Res. 15:3639-3652, 1987.
319. Schmidtke, J., et al. Hum. Genet. 76:337-343, 1987.
320. Schneid, H., et al. Nucleic Acids Res. 16:9059, 1988.
321. Schnitzer, T. J., et al. J. Virol. 355:575-580, 1980.
322. Schofield, P. R., et al. Nucleic Acids Res. 15:3636, 1987.
323. Schonk, D., et al. Genomics 4:384-396, 1989.
324. Schulz, T. F., et al. Mol. Immunol. 23:1243-1248, 1986.
325. Schwab, M., et al. Nature 308:288-291, 1984.
326. Seldin, M. F., et al. (HGM10) Cytogenet. Cell Genet. 51:1077, 1989. (abs.)
327. Seligman, P. A., et al. Am. J. Hum. Genet. 38:540-548, 1986.
328. Semba, K., et al. Science 227:1038-1040, 1985.
329. Semba, K., et al. Proc. Natl. Acad. Sci. USA 83:5459-5463, 1986.
330. Semba, K., et al. Jpn. J. Cancer Res. 79:710-717, 1988.
331. Shaw, D. J., et al. J. Cancer Res. 74:267-269, 1986.
332. Sheer, D., et al. Ann. Hum. Genet. 49:167-171, 1985.
333. Sheppard, J. R., et al. Proc. Natl. Acad. Sci. USA 80:233-236, 1983.
334. Sherman, S. L., et al. Am. J. Hum. Genet. 42:830-838, 1988.
335. Shiang, R., et al. Genomics 4:82-86, 1989.
336. Shiang, R., et al. (HGM10) Cytogenet. Cell Genet. 46:691, 1987. (abs.)
337. Shiloh, Y., et al. Proc. Natl. Acad. Sci. USA 82:3761-3765, 1985.
338. Shimizu, N., et al. Proc. Natl. Acad. Sci. USA 77:3600-3604, 1980.
339. Shipley, J. M., et al. Ann. Hum. Genet. 51:153-160, 1987.
340. Shohat, M., et al. (HGM10) Cytogenet. Cell Genet. 51:1078, 1989. (abs.)
341. Shows, T. B., et al. (HGM10) Cytogenet. Cell Genet. 51:1079, 1989. (abs.)
342. Siddique, T., et al. Nucleic Acids Res. 17:1785, 1989.
343. Silver, J., et al. Mol. Cell. Biol. 7:1559-1562, 1987.
344. Simmers, R. N., et al. Hum. Genet. 78:144-147, 1987.
345. Sommerfelt, M. A., et al. Science 242:1557-1559, 1988.
346. Sosnoski, D. M., et al. J. Clin. Invest. 81:1993-1998, 1988.
347. Spence, J. E., et al. Am. J. Hum. Genet. 39:729-734, 1986.
348. Spurr, N. K., et al. Somatic Cell Mol. Genet. 12:637-640, 1986.
349. Spurr, N. K., et al. EMBO J. 3:159-163, 1984.
350. Staal, S. P., et al. Genomics 2:96-98, 1988.
351. Stallings, R. L., et al. Am. J. Hum. Genet. 43:144-151, 1988.
352. Steele, P. E., et al. J. Virol. 59:545-550, 1986.
353. Sternman, G., et al. Exp. Cell Res. 178:180-184, 1988.
354. Suter, U., et al. Nucleic Acids Res. 15:7295-7308, 1987.
355. Swan, D. C., et al. Proc. Natl. Acad. Sci. USA 79:4691-4695, 1982.
356. Szabo, P., et al. Mol. Biol. Med. 5:139-144, 1988.
357. Tan, Y. H., et al. J. Exp. Med. 137:317-330, 1973.
358. Taub, R., et al. Proc. Natl. Acad. Sci. USA 79:7837-7841, 1982.
359. Tempest, P. R., et al. Carcinogenesis 7:2051-2057, 1986.
360. ten Dijke, P., et al. Oncogene 3:721-724, 1988.
361. Tepler, I., et al. Am. J. Hum. Genet., in press, 1989.
362. Testa, J. R., et al. (HGM8) Cytogenet. 40:761, 1985. (abs.)
363. Teyssier, J. R., et al. J. Natl. Cancer Inst. 77:1187-1191, 1986.
364. Theriault, A., et al. (HGM10) Cytogenet. Cell Genet. 51:1089, 1989. (abs.)
365. Thompson, C. C., et al. Science 237:1610-1614, 1987.

366. Tokino, T., et al. Cytogenet. Cell Genet. 48:63-64, 1988.
367. Tricoli, J. V., et al. Cytogenet. Cell Genet. 42:94-98, 1986.
368. Tronick, S. R., et al. Proc. Natl. Acad. Sci. USA 82:6595-6599, 1985.
369. Tsui, L.-C., et al. Am. J. Hum. Genet. 39:720-728, 1986.
370. Turc-Carel, C., et al. Oncogene Res. 1:397-405, 1987.
371. Tycko, B., et al. J. Exp. Med. 169:369-377, 1989.
372. Ullrich, A., et al. EMBO J. 5:2503-2512, 1986.
373. Van Corg, N., et al. (HGM10) Cytogenet. Cell Genet. 51:1097, 1989. (abs.)
374. Van Corg, N., et al. Hum. Genet. 81:257-263, 1989.
375. van de Rijn, M., et al. Cytogenet. Cell Genet. 36:525-531, 1983.
376. van der Hout, A. H., et al. Hum. Genet. 80:161-164, 1988.
377. van Dongen, J. J. M., et al. J. Immunol. 137:1047-1053, 1986.
378. van Tuinen, P., et al. Genomics 1:374-381, 1987.
379. van't Veer, L. J., et al. Mol. Cell. Biol. 4:2532-2534, 1984.
380. Vitale, E., et al. Am. J. Hum. Genet. 39:832-836, 1986.
381. Vogelstein, B., et al. N. Engl. J. Med. 319:525-532, 1988.
382. Wada, A., et al. Biochem. Biophys. Res. Comm. 157:828-835, 1988.
383. Wainwright, B. J., et al. Cytogenet. Cell Genet. 44:101-102, 1987.
384. Ward, P., et al. J. Med. Genet.21:92-95, 1984.
385. Watkins, P. C., et al. Int. Cong. Hum. Genet. 7:614, 1986. (abs. M II.21)
386. Watkins, P. C., et al. Am. J. Hum. Genet. 39:735-743, 1986.
387a. Watson, D. K., et al. Proc. Natl. Acad. Sci. USA 82:7294-7298, 1985.
387b. Watson, D. K., et al. Proc. Natl. Acad. Sci. USA 83:1792-1796, 1986.
388. Weber, J. L. and May, P. E. (HGM10) Cytogenet. Cell Genet. 51:1103, 1989.
389. Wei, S., et al. Cancer Genet. Cytogenet., in press, 1989.
390. Weis, J. H., et al. J. Immunol. 138:312-315, 1987.
391. Westbrook, C. A., et al. Proc. Natl. Acad. Sci. USA 82:8742-8746, 1985.
392. Westbrook, C. A., et al. Proc. Natl. Acad. Sci. USA 84:251-255, 1987.
393. White, R. L., et al. ICSU Short Reports 1:86-89, 1984.

394. White, R., et al. Am. J. Hum. Genet. 39:694-698, 1986.
395. Wieacker, P., et al. Hum. Genet. 76:248-252, 1987.
396. Williams, B. P., et al. Oncogene 3:345-348, 1988.
397. Williams, B. P., et al. (HGM9) Cytogenet. Cell Genet. 46:717, 1987. (abs.)
398. Wong, W. W., et al. J. Exp. Med. 169:847-863, 1989.
399. Wu, J., et al. (HGM10) Cytogenet. Cell Genet. 51:1110, 1989. (abs.)
400. Xu, W., et al. Proc. Natl. Acad. Sci. USA 85:8563-8567, 1988.
401. Xue, F., et al. Genomics 2:288-293, 1988.
402. Yamanashi, Y., et al. Mol. Cell. Biol. 7:237-243, 1987.
403. Yang-Feng, T. L., et al. (HGM8) Cytogenet. Cell Genet. 40:782, 1985. (abs.)
404. Yang-Feng, T. L., et al. (HGM8) Cytogenet. Cell Genet. 40:784, 1985.
405. Yang-Feng, T. L., et al. (HGM9) Cytogenet. Cell Genet. 46:722, 1987. (abs.)
406. Yang-Feng, T. L., et al. (HGM9) Cytogenet. Cell Genet. 46:723, 1987. (abs.)
407. Yang-Feng, T. L., et al. Science 228:728-731, 1985.
408. Yarden, Y., et al. Nature 323:226-232, 1986.
409. Yoshida, M. C., et al. (HGM8) Cytogenet. Cell Genet. 40:786, 1985. (abs.)
410. Yoshida, M. C., et al. Proc. Natl. Acad. Sci. USA 85:4861-4864, 1988.
411. Yoshida, M. C., et al. Jpn. J. Cancer Res. (Gann) 77:1059-1061, 1986.
412. Zabel, B. U., et al. Proc. Natl. Acad. Sci. USA 81:4874-4878, 1984.
413. Zabel, B. U., et al. Somatic Cell Mol. Genet. 10:105-108, 1984.
414. Zabel, B. U., et al. Proc. Natl. Acad. Sci. USA 82:469-473, 1985.
415. Zelinski, T., et al. Genomics 2:154-156, 1988.
416. Zengerling, S., et al. Am. J. Hum. Genet. 40:228-236, 1987.
417. Zhang, Y., et al. Somatic Cell Mol. Genet. 14:99-104, 1988.
418. Zhang, Z.-X., et al. Proc. Soc. Exp. Biol. Med. 170:103-111, 1982.
419. Zuppan, P. J., et al. (HGM10) Cytogenet. Cell Genet. 51:1116, 1989. (abs.)

Human DNA Restriction Fragment Length Polymorphisms (RFLPs)

Kenneth K. Kidd[1,2], Rowena K. Track[1], Anne M. Bowcock[3], Florence Ricciuti[1], Gerard Hutchings[1], and Huey S. Chan[1]

1. Human Gene Mapping Library, 25 Science Park, New Haven, CT 06511. 2. Department of Human Genetics, Yale University School of Medicine, 333 Cedar Street, New Haven, CT 06510. 3. Stanford University School of Medicine, Stanford, CA 94305.

Introduction

The table accompanying this report is a highly abbreviated summary of information contained in the RFLP component of the Human Gene Mapping Library (HGML) database. This component (RFLP) of the HGML database represents the combined efforts of the staff of the Human Gene Mapping Library and the cumulative efforts of the successive DNA Committees from Human Gene Mapping Workshops 8, 9, and 10. This table was produced only 10 weeks after the most recent complete summary published as part of the Proceedings of the Tenth International Workshop on Human Gene Mapping (HGM10). Although it is based on a slightly larger data set than the (HGM10) report, this summary is intended to provide a succinct overview of the nearly nineteen hundred mapped RFLP loci. In contrast, extensive detail on band sizes and allele frequencies will be found in the HGM10 Report of the DNA Committee (Kidd et al., 1989) as well as the HGML database itself.

The summary of RFLPs prepared for <u>Genetic Maps 1987</u> contained information on 392 loci. With nearly five times the number of RFLP loci today, we have chosen to present an even more abbreviated description of each polymorphic system. This will simplify use of these tables as an overview of RFLPs. It is not our intent that this table provide documentation of each system sufficient for use in the laboratory; the references cited, the HGML database, and the HGM10 Report of the DNA Committee can all be consulted whenever more details are needed. For information on access to the HGML databases, contact the HGML by letter (25 Science Park, New Haven, CT 06511), by telephone ((203) 786-5515), by FAX ((203) 786-5534), or by email via BITNET (GENEMAP@YALEVM).

Criteria for Inclusion

The table contains information on only those RFLP loci that have been mapped to a chromosome or a subregion of a chromosome. The standard for mapping used by the chromosome committees at the Human Gene Mapping Workshops have been relaxed with respect to some of these loci; only very provisional mapping information, conveyed primarily by personal communication of the researcher, is sufficient to include an RFLP locus with markers on a specific chromosome. The chromosomal assignments presented here range in status from unreviewed to provisional to confirmed. The

Proceedings of the Tenth International Workshop on Human Gene Mapping (HGM10) (Kidd, Klinger, and Ruddle, Eds., 1989) and the on-line database of the HGML can provide both the most recent localization of each of these markers and the status of that chromosomal assignment. The regional localizations in this table are those existing in the HGM10 proceedings.

Explanation of Entries in the Table

Chromosome/Region

Loci are arranged in the table by chromosome (1-22,X,Y) and within each chromosome are arranged by location roughly from the tip of the short arm (pter) to the tip of the long arm (qter) with some exceptions. Since loci are assigned to regions or intervals, not points, based on the cytogenetically defined chromosome bands as in the figures in the accompanying article by Cohen et al. Genetic Maps, this volume, a complex algorithm orders loci based on both endpoints of each interval. Loci assigned only to a single chromosome arm are listed after those loci that may be mapped to a smaller segment of that arm. Loci with no regional localization, but assigned to the chromosome, are listed last. Loci with identical cytogenetically defined regional localizations are listed in alphabetical order. With increasing frequency other types of data (such as linkage mapping data) can order such loci, but that information has not been incorporated into this listing.

Because this table is arranged by position along the chromosome, other listings of loci must be consulted to determine where to find a specific locus in the table if the chromosome position is not known for that particular gene of interest. The Nomenclature Committee Report in HGM10 (McAlpine et al., 1989), the DNA Committee Report in HGM10 (Kidd et al.,1989), and the index accompanying the chromosomal diagrams in the accompanying article can all be used for this purpose. Alternatively, interactive searching of the HGML database can give the correct symbol for a gene known only by name or function as well as its location.

Locus Symbol

The locus symbol used is the most recently approved symbol from HGM10. The symbols for defined genes and pseudogenes are those approved by the Nomenclature Committee, and the D numbers for anonymous (arbitrary) loci defined by cloned DNA segments of unknown function are those assigned by the DNA Committee. Researchers are encouraged to contact the DNA Committee (through the HGML) to obtain an official locus symbol (the D number) for anonymous clones. Whenever possible, the DNA Committee hopes to assign these locus symbols in advance of publication so that a unique symbol will be associated with each RFLP from the beginning. Assignment of these D numbers is being handled on behalf of the DNA Committee by the staff of the HGML. In most cases for one to a few loci, a D number can be assigned over the telephone.

An asterisk (*) preceding the locus symbol in the table indicates that at least one probe for that locus is included in the NIH probe repository maintained by the American Type Culture Collection (ATCC). The PROBE component of the HGML database includes the ATCC accession number for easy cross reference. Additional information on these and on additional probes being added to the ATCC repository can be obtained by writing to Dr. Donna Maglott, American Type Culture Collection, 12301 Parklawn Drive, Rockville, MD 20852-1776, telephone (301) 231-5581, fax (301) 231-5826, bitnet: VYS@NIHCU.

Enzymes

Ten enzymes account for 82.9% of all probe/enzyme systems in the database. These enzymes, the number (%) of systems each identifies, and the abbreviation used in this table are

Enzyme	Abbreviation	Number of Systems (%)
TaqI	T	700 (20.2%)
MspI	M	650 (18.8%)
EcoRI	E	281 (8.1%)
HindIII	H	258 (7.5%)
PstI	Ps	224 (6.5%)
BglII	Bg	219 (6.3%)
BamHI, BstI	Ba	164 (4.8%)
PvuII	Pv	152 (4.4%)
RsaI	R	133 (3.8%)
SstI, SacI	S	87 (2.5%)
Other	O	593 (17.1%)

The table gives the enzymes known to reveal polymorphism at the locus, whether more than one probe is required or not. In the case of VNTR loci, multiple enzymes will often reveal the same polymorphism, but only the enzymes given as "primary" or best by the authors are listed.

Increasingly polymorphisms are being detected by PCR techniques in combination with direct size determination or detection by allele specific oligonucleotides. Strictly speaking, these are not RFLPs in this table. PCR-based typing is being developed for those loci detected only by PCR techniques, we have inserted the letters PCR in the enzyme column.

PIC or Heterozygosity

The polymorphism information content (PIC) is a measure of informativeness of a locus for linkage (Willard et al., 1985). It is similar to heterozygosity except that it is corrected for decreased linkage information in offspring from matings between identical heterozygotes. The value given is the largest value documented for the locus. For loci with multiple sites, PIC was calculated for haplotypes only if explicit haplotype frequencies were available. Consequently, haplotype loci often are tallied with the far lower PICs of a single system because haplotype studies have been done for relatively few loci. For VNTR loci in which it is usually impossible to identify specific alleles, the PIC cannot be calculated and the observed heterozygosity is given instead. If no value is given, the original authors provided neither a heterozygosity value nor allele frequencies from which a PIC could be calculated. If both values are given, they will usually be based on different studies and may be based on different (sets of) polymorphic systems at the locus. In a few cases the only information is that the heterozygosity is "high."

References

The references given are those describing the polymorphisms. Full titles for these and additional relevant references exist in the HGML database.

Special Annotations

Obsolete symbols have "see . . . " written after the locus symbol. DMD encompasses several D numbers that have been kept for convenience in dealing with the very large region. The D numbers that are known to be within the extent of the DMD locus have this indicated by the symbol DMD in parentheses. Other annotations should be self explanatory.

Acknowledgements

Production of this summary was supported in part by the Howard Hughes Medical Institute and in part by a USPHS grant from the National Institute of General Medical Sciences and the National Library of Medicine to support HGM10.

General References

Cohen et al., Genetic Maps, this volune.
Kidd, Klinger, and Ruddle, Cytogenet Cell Genet 51: 1-1147 (1989)
Kidd et al., Cytogenet Cell Genet 51: 622-947 (1989)
Willard et al., Cytogenet Cell Genet 40: 360-489 (1985)
McAlpine et al., Cytogenet Cell Genet 51: 13-66 (1989)

Chromosome/Region	Locus Symbol	Enzyme	PIC (HET)	Reference
1pter-p36	D1S32	many	high	457
1pter-p31	*D1S2	Bg	.28	668
1pter-p22	D1S15	Bg S	.65	329, 330
1pter-p22	D1S16	Bg	.30	330
1pter-p22	D1S17	T O	.62	329, 330
1pter-p22	D1S18	Bg	.23	330
1pter-p22	D1S19	S	.37	330
1pter-p22	D1S20	Bg	.11	330
1pter-p22	D1S21	T	.35	330
1pter-p22	D1S22	O	.37	330
1pter-p22	D1S23	E	.11	329
1p36.3	D1S98	O		1510
1p36.3	*D1Z2	M Ps Pv T	high	183
1p36.3-p36.13	PGD	Ba	.30	677
1p36	*PND	Bg T O	.37	403, 409, 415, 851, 1028
1p36	*D1F15S1	O	.36	205
1p36	D1S1	refers to single copy locus adjacent to D1F15S1		
1p36.2-p36.12	D1S95	E		1510
1p36.2-p36.12	D1S96	T O		1510
1p36.2-p36.12	D1S97	O		1510
1p36.2-p36.1	*FGR	E	.37(.51)	331
1p36.1-p35	HMG17	Ps O	.38	731
1p36.1-p34	*ALPL	S O	.34(.48)	1130, 1411
1p36.12-p36.11	D1S94	O		
1p35-p34	*FUCA1	Pv	.34(.47)	272
1p35-p33	D1S7	O	(.98)	859, 1439
1p35-p31.3	GLUT1	Bg O	.69	785, 1226, 1468
1p34	D1S72	M		231
1p32	MYCL1	E	.37(.46)	616, 654
1p31	ACADM	M Ps S T O	.68	667, 1375
1p31	GST1	Ba E	.38	1215
1p31-p21	D1S33	E H	.37	497
1p22-p21	F3	M	.33	1186
1p22-p13	D1S9	T	.36	330
1p22-p13	D1S10	T	.20	330
1p22-p13	D1S11	O	.37	330
1p22-p13	D1S12	O	.23	330
1p22-p13	D1S13	Pv	.37	330
1p22-p13	D1S14	T	.37	330
1p21	AMY1A, AMY1B, AMY2A, AMY2B	Ps O	.37	493, 499, 622
1p21-qter	D1S4	Bg	.18	617
1p13	*NGFB	Bg T O	.37	127, 148, 271, 330, 493
1p13	*NRAS	Bg E Ps T	.37	193, 506
1p13	TSHB	H T O	.61	332, 333
1p11-qter	EPHX	Bg E R O	.16	1239
1p	*D1S60	Ba	.37	555
1p	*D1S62	Pv	.37	577
1p	*D1S63	M	.31	1016
1p	*D1S64	M	.35	709
1p	*D1S73	M T	.37	421
1p	*D1S76	T	(.65)	1015, 1022
1p	*D1S77	Pv	(.67)	1014
1p	D1S79	Ps	(.59)	1013, 1022
1p	*D1S80	O	(.90)	1012
1p	D1S88	T	.37	1120
1p	D1Z4	E M O		953
1cen	D1Z3	Ba T	.37	623, 624, 625
1cen	D1Z5	Bg O	.60	1400, 1416
1q12-q23	*D1S26	O	.36	495
1q12-qter	D1S24	T	.07	329
1q21	*SPTA1	M Pv O	.38	566, 702, 1117
1q21-q23	*APCS	M	.37	620, 1444
1q21-q23	*APOA2	M	.65(.74)	381, 685, 884, 1212, 1406
1q21-q23	ATP1A2	Ps		1480
1q21-q23	MUC1	O	(.90)	444, 1295, 1296
1q22-q23	*D1S75	T	.50	1039
1q22-q25	ATP1B	H M	.37	849
1q23-q25.1	*AT3	Ba Ps	.38	110, 111, 493, 884, 1110, 1461
1q23-qter	D1F10S1	S	.32	617
1q25-q43	SNRPE	M	.16	94
1q31	LAMB2	M Ps O	.37	646, 1036
1q31-q32.1	F13B	Bg E O	.37	1405
1q32	*CR1	Ba E H Pv S O	.37	1431, 1440, 1441, 1442
1q32	DAF	Ba H	.54	1140, 1273
1q32 or 1q42	*REN	Bg H R T O	.37	407, 410, 493, 494, 592, 850, 852, 854, 903
1q41-q42	PPOL	Ps S	.37	867
1q42-q43	D1S8	O	(.97)	634, 859, 1439
1q	ANP	obsolete symbol; see PND at 1p36		
1q	D1F24S3	see DNF24 at end of table		
1q	D1S59	M	.16	560, 1046
1q	*D1S74	M	(.96)	940, 943
1q	*D1S81	R	(.85)	1011, 1022
1q	D1S82	M	.37	584
1q	D1S90	T O	.38	18
1	ERPL1	Ps	.34	507
1	D1S34	Ba	.22(.29)	323
1	D1S35	T	.35(.43)	323

Chromosome/Region	Locus Symbol	Enzyme	PIC (HET)	Reference
1	D1S36	M	.39	323
1	D1S37	T	.33(.52)	323
1	D1S38	M	.46(.55)	323
1	D1S39	M	.33(.40)	323
1	D1S40	T	.33(.31)	323
1	D1S41	T	.37(.36)	323
1	D1S42	M	.29(.36)	323
1	D1S43	E	.28(.26)	323
1	D1S44	H	.37(.50)	323
1	D1S45	M	.26(.33)	323
1	D1S46	M	.22(.19)	323
1	D1S47	R	.71(.79)	323, 1203
1	D1S48	R	.52(.52)	323, 687
1	D1S49	M	.74(.74)	323, 1203
1	D1S50	T	.74(.74)	323, 687
1	D1S51	M	.34	323
1	D1S52	Ba	.38(.45)	323
1	D1S53	M	.33	323
1	D1S54	M	.27(.29)	323
1	D1S55	E	.35(.52)	323
1	D1S56	E	.57	323
1	*D1S57	M R T	.58(.65)	330, 985, 1022
1	*D1S58	M	.37(.40)	955, 1022
1	D1S61	O	(.64)	1046
1	*D1S65	T O	.37	719
1	*D1S66	M	.32	956
1	*D1S67	M	.37	986
1	D1S68	M	.88(.90)	323
1	D1S69	T	.48	323
1	D1S70	T	.33(.50)	323
1	D1S71	M	.31(.40)	323
1	D1S78	M	.14	565
1	D1S84	M	.31	760
1	D1S85	T	.37	1348
1	D1S86	T	.37	1350
1	D1S87	T	.35	1501
1	D1S89	obsolete symbol; see D1S74		
1	D1S91	H	.30(.40)	135
2pter-p23	D2S12	Pv	.34	73
2pter-p23	*D2S49	M	.36	733
2pter-p22	D2S70	H	.34(.40)	1507
2pter-p12	TPO	Bg Ps	.37	857
2pter-q32	*D2S48	M	.37	416
2p25	ODC1	E	.32	1114
2p25	*D2S1	Bg M	.35	439, 817
2p25-p24	D2S10	Pv	.32	1077
2p24	*MYCN	Pv O	.59	724
2p24-p23	*APOB	E H M Pv O	.66	69, 95,96, 114, 115, 216, 273, 342, 343, 382, 406, 527, 593, 595, 640, 686, 740, 1107, 1108, 1260

Chromosome/Region	Locus Symbol	Enzyme	PIC (HET)	Reference
2p23	*POMC	R S	.36	127, 373, 374, 907
2p23-p15	*D2S6	H T	.38	278, 282
2p21	SPTBN1	O	.27	1391
2p16-p15	*D2S5	E M	.34	733, 1225
2p13	TGFA	Ba R T	.33	525, 927
2p13-q14	D2S62	T	.31	1385
2p12	CD8A	O	.37	127, 129
2p12	*IGKC	S	.15	383
2p12	IGKV	Bg	.22	681
2p	CPS1	O		77
2p	D2F24S2	see DNF24 at end of table		
2p	*D2S43	T	.37	733, 990
2p	*D2S44	M Ps	.79(.97)	48,948
2p	*D2S45	M	.33	552
2p	D2S46	T	.37	262
2p	*D2S47	Pv	(.60)	138
2p	*D2S51	M	.16	1009
2p	*D2S54	M	.37	553
2p	*D2S60	R	.32	710
2p	D2S61	Pv	(.32)	1008
2q11-q14	D2S16	T	.45	316
2q12-q14	D2S63	Ba M	.36	1318
2q12-q21	IL1A	T O	.32(.50)	894
2q13-q21	PROC	Ba M Pv O	.34	534, 1154
2q14-q21	GYPC	Ps T O	.37	29
2q14-q32	COL5A2	M	.22	1210
2q21-q33	*SCN2A	O	.31	800
2q24-q33	D2S17	E	.30	317
2q24-q34	D2S13	H	.25	312
2q31-q32.3	COL3A1	E O	.36	268, 269, 1322, 1325
2q31-qter	ELN	E H	.27(.40)	1078
2q33-q35	*CRYGP1	M		323
2q33-q35	*CRYG1, CRYG2, CRYG3, CRYG4, CRYG5	H T	.61	733, 819, 1425
2q34-q36	FN1	H M T O	.36	240, 241, 431, 432, 433
2q35-q37	*D2S3	H M Ps T	.81(.55)	733, 795, 797, 798, 806
2q37	*ALPP	Ps R T	.36	843, 1320, 1321
2q37	COL6A3	O	.37	166, 1404
2q	D2F24S1	see DNF24 at end of table		
2q	D2S50	M	(.80)	943, 1010
2q	D2S53	T	(.74)	943, 999
2q	*D2S55	R	.37	717
2	EN1	O	.37	815
2	LCT	M	.37	714
2	SFTP3	Ba	.16	355
2	D2S14	M	.35	153
2	D2S19	T	.36(.48)	323

Chromosome/ Region	Locus Symbol	Enzyme	PIC (HET)	Reference
2	D2S20	M	.40(.50)	323
2	D2S21	T	.37	323
2	D2S22	E H	.43(.52)	135, 323
2	D2S23	E	.21(.21)	135, 323
2	D2S24	E	.31(.36)	135, 323
2	D2S25	Ps	.30(.30)	135, 323
2	D2S26	Bg T	.30(.55)	323
2	D2S27	T	.30	323
2	D2S28	Bg M	.38	323
2	D2S29	T	.34(.48)	323
2	D2S30	M	.37	323
2	D2S31	E	.38(.50)	323
2	D2S32	Bg	.26(.24)	323
2	D2S33	M	.21(.21)	323
2	D2S34	M	.40(.55)	323
2	D2S35	T	.37	323
2	D2S36	T	.19(.26)	323
2	D2S37	T	.29(.29)	323
2	D2S38	M T	.57(.62)	323
2	D2S39	M	.36(.40)	323
2	D2S40	R	.26(.38)	323
2	D2S41	Bg	.45(.55)	323
2	D2S42	M	.31(.21)	323
2	D2S52	obsolete symbol; see D2S55		
2	D2S65	T	.37	1352
2	D2S66	obsolete symbol; see D2S50		
2	D2S67	obsolete symbol; see D2S53		
2	D2S68	R O	.32	1264
3pter-p25	D3S51	M	.34	446
3pter-p25	D3S95	M	.34	1255
3pter-p21	ERBA2	obsolete symbol; see THRB		
3pter-p21	D3S12	M	.38	323
3pter-p21	D3S17	T	.69(.76)	323, 1203
3pter-p21	D3S18	T	.23(.19)	323
3pter-p21	D3S22	M	.34	323
3pter-p21	D3S154	Bg	.30	471
3pter-p21	D3S157	E	.22	456
3pter-p14.2	D3S55	M	.24	446
3pter-p14.2	D3S94	H	(.47)	672
3pter-p14	D3S155	T	.36	471
3pter-q21	D3S4	M T	.23	763, 900
3pter-q21	D3S6	M	.28	617
3p25	*RAF1	E T O	.64	487, 1217
3p24.1-p22	THRB	Ba E H	.37	434, 881, 1464
3p21	D3F15S2E	H	.37	147, 693, 1499
3p21	D3S2	M	.24	63, 763, 1027, 1499
3p21	D3S39	M	.32	33
3p21-p14	D3S11	T	.40	314
3p21-p14	D3S34	Bg M	.07	33
3p21.2-p21.1	ITIH1	S O	.37	771, 774, 775
3p14.2	D3S3	M	.07	63
3cen	D3Z1	Ba Ps	.33	1399, 1416, 1494, 1495
3q11-q12	GPX1	O	.30	929
3q12	*D3S1	H	.29	1025, 1497

Chromosome/ Region	Locus Symbol	Enzyme	PIC (HET)	Reference
3q13.3-q22	*PCCB	E Ps	.37	730, 1402
3q21	*TF	H Ps	.07	597, 1337
3q21	D3S36	M	.32	33
3q21-q22	RBP1	T	.37	1097
3q21-q23	FTHL4	Ba	.34	436
3q21-q24	RHO	---PCR---	.31(.34)	1406
3q21-qter	ACPP	T	.36	1432
3q21-qter	D3S5	Ba E H M	.36	338
3q23-q25	CP	Ps	.36	763
3q26.2-qter	APOD	M T	.60	335, 336
3q26.2-qter	*TFRC	Pv S T	.15	666
3q27-q29	AHSG	Pv S T	.37	751
3q28	SST	Ba E	.46(.51)	820, 1406, 1498
3q	D3S43	M Ps	.26	1023
3	HPRTP1	H	.07	322
3	MOX2	E	.36	327
3	D3S13	M	.60(.64)	323
3	D3S14	Bg M	.53(.76)	323, 1203
3	D3S15	M	.37(.40)	323
3	D3S16	T	.36(.40)	323
3	D3S19	M	.31(.36)	323
3	D3S20	M		323, 1203
3	D3S21	M	.30(.43)	323
3	D3S23	M	.34(.40)	323
3	D3S26	T	.34	323
3	D3S29	Pv T	.37	763
3	*D3S30	M	.37	949
3	*D3S31	Pv	.50	203
3	*D3S32	R	.37	425
3	*D3S42	O	(.85)	962
3	*D3S44	Pv	(.85)	1007
3	*D3S45	M	(.80)	1006
3	*D3S46	M	(.80)	1005
3	D3S47	M	.33	135, 323
3	*D3S86	R	.36	805
3	*D3S91	obsolete symbol; see D3S44		
3	D3S169	T	.22	471
3	D3S170	T	.09	471
3	D3S171	Ps	.27	471
3	D3S172	Bg	.33	471
4pter-p15.1	*D4S52	Pv O	.37	464
4pter-p15.1	*D4S57	Bg	.30	464
4pter-p15.1	D4S61	E	.22	464
4pter-p15.1	D4S93	O	.34	1487
4pter-q21	*D4S123	T	.28	92
4pter-q26	D4S12	O	.37	1180
4pter-q26	D4S13	M Pv	.36	463
4p16.3	*D4S10	Bg E H M Ps T O	.46	45, 191, 263, 500, 501, 502, 829, 831, 1241, 1276, 1488
4p16.3	D4S43	Ba M T O	.58	464, 465
4p16.3	D4S81	Bg S	.28	1142
4p16.3	D4S82	obsolete symbol; see D4S81		
4p16.3	D4S90	Pv	.22	1489
4p16.3	D4S95	Ps T O	.63	1253, 1390
4p16.3	D4S96	M	.38	1253

Chromosome/Region	Locus Symbol	Enzyme	PIC (HET)	Reference
4p16.3	D4S97	R O	.30	1253
4p16.3	D4S98	S	.49	1253
4p16.3	*D4S125	M O	.37(.60)	174, 980, 1022
4p16.3	D4S126	S	.37	13
4p16.3	D4S127	Pv	.39	13
4p16.3	D4S141	H	.37	1256
4p16-q23	ATP1BL1	E O	.28	174, 445
4p16.2-p16.1	D4S62	O	.35	462, 523, 524
4p16.2-p15.1	*D4S18	Ba O	.32	464
4p16.2-p15.1	*D4S19	O	.36	464
4p16.2-p15.1	*D4S20	Bg O	.37	461, 464
4p16.2-p15.1	*D4S23	Bg O	.28	461, 464
4p16.2-p15.1	*D4S46	M	.35	464
4p16.2-p15.1	*D4S49	T	.37	464
4p16.2-p15.1	*D4S51	T	.28	464
4p16.2-p15.1	*D4S53	O	.28	464
4p16.2-p15.1	*D4S54	O	.30	464
4p16.2-p15.1	D4S59	H	.16	464
4p16.2-p15.1	D4S60	O	.16	464
4p16.2-p15.1	D4S84	H	.14	1487
4p16.2-p15.1	D4S86	Ps	.33	1487
4p16.1	*RAF1P1	Ba M O	.36	174, 462
4p16.1 or 4p15.1	*D4S15	O	.30	461, 464
4p16.1 or 4p15.1	D4S21	Ba O	.35	461
4p16.1 or 4p15.1	D4S41	S	.16	464
4p16.1 or 4p15.1	D4S42	Pv	.14	464
4p16.1 or 4p15.1	*D4S48	O	.30	464
4p16.1 or 4p15.1	*D4S50	M O	.30	464
4p15.3	QDPR	M O	.58	264, 831, 1367
4p15.1-q11	*D4S16	M	.37	464
4p15.1-q11	D4S17	Ba M R	.36	461, 464
4p15.1-q11	*D4S22	O	.34	464
4p15.1-q11	D4S25	E	.31	464
4p15.1-q11	*D4S26	Ps	.11	464
4p15.1-q11	*D4S39	O	.36	464
4p15.1-q11	*D4S47	O	.36	464
4p15.1-q11	D4S55	T	.36	464
4p15.1-q11	*D4S56	O	.29	464
4p13-q13	D4S67	H T	.36	1447
4p11-q11	*D4S35	M	.37	309, 1425
4p11-q21	*MT2P1	E	.38	1196
4p11-q22	*KIT	H	.19	93
4p	D4S104	Bg	.32(.31)	323
4p	D4S106	Bg	.33(.29)	323
4p	D4S107	M	.30	323
4p	D4S110	H	.37(.29)	323
4p	D4S117	Ba	.35(.33)	135, 323
4p	*D4S124	Ps	(.67)	943, 996
4cen	D4Z1	M S	.38	10,174
4q11-q13	*AFP	M Ps O	.37	925, 1425
4q11-q13	ALB	M Ps S O	.59	924, 930, 931
4q11-q21	D4S1	Bg H Ps	.36	463
4q11-q28	D4S138	M O	.67	904
4q11-qter	*D4S24	Pv O	.43	464
4q11-qter	*D4S27	M T	.37	464
4q11-qter	D4S28	Ps O	.38	461
4q11-qter	D4S29	T	.30	461
4q11-qter	D4S30	O	.29	461, 464
4q11-qter	*D4S31	Ps O	.38	464
4q11-qter	*D4S32	O	.27	461
4q11-qter	*D4S33	O	.36	461, 464
4q11-qter	D4S36	O	.16	461
4q11-qter	D4S37	Ps	.16	461
4q11-qter	*D4S38	O	.11	461, 464
4q11-qter	D4S44	Ps	.34	464
4q11-qter	D4S45	O	.28	464
4q12-q13	GC	Ba M Pv O	.37	221, 246, 1026, 1128, 1478
4q12-q21	PF4	E	.37	504
4q13-q21	AREG	M T	.37	321
4q13-q21	IL8	H	.37	895
4q21	IGJ	O	.23	866
4q21	INP10	E Ps O	.37	823, 922
4q21	MGSA	S	.31	83
4q21-q23	*ADH2	R O	.28	619, 1252, 1475
4q21-q23	*ADH3	M Pv S O	.59	924, 1082, 1252, 1475
4q21-q24	D4S143	T	.37	1139
4q21-q25	*D4S34	O	.25	461, 464
4q21-q31	LPC2A	S	.09	174, 780
4q21-qter	*D4S40	Ps O	.63	464
4q22-q31.2	D4S66	T	.39	318
4q24-q25	IF	O	.36	1224
4q25	EGF	S O	.56	926
4q25	FGFB	H	.27	878
4q26-q27	*IL2	O	.07	922
4q26-qter	D4S14	M		1180
4q26-qter	D4S112	M T	.15	1146
4q28	*FGA	T	.34	602
4q28	*FGB	O	.28	174, 244, 919
4q28	*FGG	O	.25	919
4q28-q31	GYPA	M O	.37(.50)	594, 923
4q28-q31	GYPB	O	.37	94
4q31	MLR	O	.36	174
4q32	HVBS6	R T	.37	174, 1089
4q	D4S9	M	.30	652
4q	D4S100	T	.34(.48)	323
4q	D4S101	E T	.37	323, 1203
4q	D4S102	T	.50(.67)	323
4q	D4S108	M	.37(.50)	323
4q	D4S109	H	.32(.40)	323
4q	D4S118	M	.25	135, 323
4q	D4S119	M	.24(.17)	323
4q	D4S120	T	.37(.45)	323
4q	D4S121	T	.21(.21)	323
4q	D4S139	T	(.89)	891
4	D4S2	M	.27	1240
4	D4S58	O	.27	464

Chromosome/Region	Locus Symbol	Enzyme	PIC (HET)	Reference
4	D4S64	Bg	.36	154
4	D4S103	M T	.36	323
4	D4S105	Bg	.57(.57)	323
4	D4S129	T	.34	1351
4	D4S130	T	.79(.83)	323
4	D4S140	obsolete symbol; see D4S124		
4	D4S144	H Ps	.35	174, 1145
5pter-p15.3	*D5S10	S O	.70	1069, 1070
5pter-p15.3	D5S11	M	.37	1070
5pter-p15	*D5S4	E M R	.37(.31)	40, 1240
5pter-p15	D5S47	E Ps	.34(.40)	323, 1280, 1408
5pter-p15	D5S48	M R	.71(.71)	323, 687, 1203, 1408
5pter-p15	D5S56	M	.69(.74)	135, 1280, 1408
5pter-p15	D5S59	M Ps	.60(.64)	323, 1408
5pter-p15	D5S73	T	.72	323, 1408
5pter-p15	D5S88	H T	.54	1119
5pter-p15	D5S90	M	.55(.60)	1408
5pter-p15	D5S92	M	.31(.45)	1408
5pter-p15	D5S95	M	.36(.50)	1408
5pter-p15	D5S96	M	.24(.33)	1408
5p15.3-p15.2	*D5S13	R O	.38	1068, 1070
5p15.3-p15.2	D5S14	O	.14	1068, 1070
5p15.3-p15.2	D5S15	E	.26	1070
5p15-p14	D5S57	H	.17(.19)	323, 1408
5p15-p14	D5S60	M	.43(.45)	323, 1408
5p15.2	D5S18	M	.33	1070
5p15.2	D5S23	M	.24	1070
5p15.2	D5S24	O	.07	1070
5p15.2-p15.1	*D5S12	M	.37	1068
5p15.1	*D5S16	R	.37	1070
5p15.1	D5S17	O	.47	1070
5p14	D5S7	Ps T O	.24	1068, 1070
5p14	*D5S19	E M	.37	1070
5p14	D5S25	S	.37	1070
5p14	*D5S26	E	.37	1070
5p14-p13	*HPRTP2	M	.27(.65)	1091
5p14-p13	*D5S27	O	.30	1070
5p14-p13	D5S69	T	.31(.36)	323, 1408
5p13	*D5S20	E O	.59	1070
5p13	D5S28	R	.19	1070
5p13	D5S29	H	.18	1070
5p13	D5S30	O	.14	1070
5p13-p11	*D5S21	M	.43(.38)	1070
5p13-p11	D5S31	S	.28	1070
5p13-p11	D5S32	E M	.24	1070
5p13-p11	*D5S33	M	.38	1070
5p13-q12	D5S93	M	.17(.17)	1408
5cen-q11.2	*D5S76	T	.55(.67)	767, 1271
5q11-q13	D5S99	E	.37	1139
5q11.2-q13.2	*DHFR	R O	.57	305, 376
5q11.2-q13.3	*D5S6	Ba	.52(.60)	311
5q11.2-q13.3	D5S51	T O	.36(.48)	323, 1280, 1408
5q11.2-q13.3	D5S63	T	.33(.50)	323, 1408
5q11.2-q13.3	*D5S78	M	.37(.58)	767
5q12-q14	*D5S39	M O	.34(.53)	767
5q13	*HEXB	Ps T O	.36	100, 1052, 1449
5q14-q21	D5S71	T	.40(.32)	8, 113, 767
5q14-q21	D5S82	Ps T	.36	942
5q14-q21	D5S83	M	.36	942
5q14-q21	D5S85	Ps T	.37	942
5q14-q22	D5S53	T	.57(.57)	323, 1280, 1408
5q15-q21	D5S98	Bg	.36	914
5q21	D5S37	Ps O	.75	874, 1286
5q21-q22	D5S81	T	.38	942
5q21-q22	D5S84	M T	.37(.55)	942
5q21-q22	D5S86	M T	.33	942
5q21-q33	D5S49	M	.35(.64)	323, 1280, 1408
5q21-q33	D5S50	Ps	.50(.67)	323, 1280, 1408
5q21-q33	D5S52	T	.57(.64)	323, 1280, 1408
5q21-q33	D5S58	M	.55(.55)	1280, 1408
5q21-q33	D5S64	T	.33(.52)	323, 1408
5q21-q33	D5S65	T	.33(.38)	323, 1408
5q21-q33	D5S89	S	.33	938
5q23-q31	*IL3	Bg Ps	.30	631
5q31-q32	*ADRB2R	O	.30	761, 762
5q31-q32	GRL	O	.37	928
5q31-q33	SPARC	M T	.64	1024
5q32-qter	*D5S36	M	.37	1070
5q33-q35	*CSF1R	E O	.27(.33)	19,296
5q33-q35	D5S54	M	.48(.55)	323, 1280, 1408
5q33-q35	D5S67	T	.34(.45)	323, 1408
5q33-q35	D5S68	M	.29(.33)	323, 1408
5q33-q35	*D5S70	T	.29	369, 767
5q33-q35	D5S91	E	.24(.36)	1408
5q33-q35	D5S100	T		135
5q34-qter	D5S2	M	.14(.23)	63, 1025, 1027, 1240
5q34-qter	*D5S22	M	.76	767, 1070
5q34-qter	D5S55	T	.57(.64)	135, 1280, 1408
5q34-qter	D5S61	M	.74(.76)	323, 1203, 1408
5q34-qter	D5S66	Bg	.29(.38)	323, 1408
5q34-qter	D5S72	T	.57(.57)	323, 1408
5q34-qter	D5S94	M	.19(.21)	1408
5q35-qter	D5S43	O	(.85)	1258, 1439

Chromosome/Region	Locus Symbol	Enzyme	PIC (HET)	Reference
5q35-qter	D5S62	T	.36(.40)	323, 1408
5q	D5S34	H R	.34	1070
5q	D5S35	O	.14	1070
5q	D5S77	obsolete symbol; see D5S39		
5	D5S1	Bg Ps R	.28	814, 1240
5	D5S3	T	.30	1392
5	D5S5	O	.36	1176
5	*D5S9	H	.09	1381
5	D5S80	obsolete symbol; see D5S22		
5	*D5S106	M	.34	375, 1264
5	D5S108		.45	
6pter-p25	D6F21S1	obsolete symbol; see D16F21S2		
6pter-p21.3	*D6S7	O	.41	103
6pter-p21.1	PGC	Ba	.24	36, 1079
6pter-q12	D6S61	Ps		46
6pter-q14	*D6F14S1	E H	.30	630
6p25-p24	F13A1	Ba E H M T O	.64	103, 109, 1504
6p23-q12	D6S16	Bg	.28	1356
6p23-q12	D6S17	H	.24	1356
6p23-q12	D6S18	E	.22	1356
6p23-q12	D6S35	E	.35	1356
6p23-q12	D6S36	T	.35	1356
6p21.3	*BF	T	.09	813
6p21.3	*CYP21P, CYP21	E H Pv O	.38	325, 902
6p21.3	*C2	Ba S T	.60	91, 261, 1450
6p21.3	*C4A,C4B	Bg Ba E H T O	.37	476, 1080, 1081, 1088, 1201, 1291, 1336, 1420, 1491, 1492, 1493
6p21.3	HLA-B	Bg H M T O	.66	250, 670, 1291
6p21.3	HLA-DPB1, HLA-DPB2	---PCR---		178, 359
6p21.3	HLA-DQA1	---PCR---		359
6p21.3	HLA-DQB1	Ba H		671
6p21.3	HLA-DRA	Bg E Ps T O	.44	1262, 1282
6p21.3	HLA-DRB1	T		1291
6p21.3	*HSPA1	Ps	.36	472
6p21.3	TNFB	E	.11	1087
6p21.3	*D6S8	M	.25	103, 743
6p21	MUT	H	.37(.41)	744, 1505
6p21	*PIM	M T O	.32	1503
6p21	D6S4	Bg O	.37	105, 678, 765, 1240
6p21.3-p12	FTHL5	Ba E H T	.36	259
6p21-q12	PGK1P2	O	.33	104
6p21.3-q12	D6S5	M O	.33	743, 765, 1503

Chromosome/Region	Locus Symbol	Enzyme	PIC (HET)	Reference
6p21-qter	*D6S2	Pv	.65	127, 375, 765, 1265
6p21-qter	D6S37	H	(.74)	984
6p21-qter	*D6S39	H Pv	(.55)	1022, 1047
6p21-qter	D6S43	E Pv O	.36	56
6p21-qter	*D6S44	T	(.70)	995, 1022
6p12-p11	*KRAS1P	T	.39	106, 1062
6p12.2	GST2	O	.38	220
6p11	D6S42	Ba T	.28	1384
6p	D6S9	obsolete symbol; see D11S288		
6p	*D6S10	T	.45	743, 765
6p	D6S19	M T	.57	323, 324, 687
6p	D6S20	M T	.55	323, 324
6p	D6S28	H	.37	323
6p	*D6S29	Ba	.38	562, 765
6p	D6S30	M	.37	765
6p	D6S38	M	.31	561, 1047
6p	*D6S41	Ps	.28	1021, 1022
6cen	D6Z1	Ba T	.37	624, 625
6q12-q21 or 6p23-p21.1	CGA	E H	.35	117, 591
6q21	SOD2	T	.36	1466
6q22-q23	*MYB	E	.37	328, 1482
6q23	ARG1	Pv	.21	669
6q24-q27	*ESR	Pv S O	.37	209, 238, 540
6q25-q27	*TCP1	Ps	.29	102, 1430
6q26-q27	PLG	M R S	.37	920, 921
6q	TCP10	Pv	.53	102
6q	D6S21	R	.74(.74)	323, 687, 1203
6q	D6S22	T	.43	323
6q	D6S23	M	.37	323
6q	D6S24	R	.07	323
6q	D6S25	H	.21	323
6q	D6S26	M	.48	323
6q	D6S27	Ps	.32	323
6q	D6S33	T	.30	323
6	*D6S3	H	.37	1240
6	*D6S6	E	.30	1240
6	D6S14	E		1356
6	D6S15	H	.36	1356
6	D6S32	M	.49	231
6	D6S47	M	.35	300
6	D6S48	T	(.78)	943
7pter-p14	D7S10	M	.38	70,71, 248
7pter-p14	D7S128	Ps	.22	1155
7pter-p14	D7S132	T	.28	1155
7pter-p14	D7S134	T	.36	1155
7pter-p14	*D7S135	T	.37	1045, 1155
7pter-p14	D7S142	H	.16	1155
7pter-p14	D7S149	Ps	.34	1155
7pter-p13	CALML1	T	.38	1179
7pter-q22	NPY	T	.22	875

Chromosome/Region	Locus Symbol	Enzyme	PIC (HET)	Reference
7pter-q22	*D7S11	H	.33	375, 1265
7pter-q22	D7S21	O	(.99)	859, 1439
7pter-q22	D7S55	Bg	.18(.17)	323
7pter-q22	D7S57	M	.49(.65)	323
7pter-q22	D7S58	H	.37(.40)	323
7pter-q22	D7S59	M	.49(.55)	323
7pter-q22	D7S62	H T	.62(.79)	323, 687, 1203
7pter-q22	D7S65	T	.34(.45)	323
7pter-q22	D7S66	M	.42(.28)	323
7pter-q22	D7S69	M	.30	323
7pter-q22	D7S74	M	.49	67,323
7pter-q22	D7S75	H	.37	67,323
7pter-q22	D7S76	Ps	.32	67,323
7pter-q22	D7S77	M	.24(.41)	67,323
7pter-q22	D7S78	T	.37	67,323
7pter-q22	D7S79	M	.35	67
7pter-q22	D7S80	M T	.21	67
7pter-q22	D7S81	M T	.30(.30)	67, 323, 1203
7pter-q22	D7S82	M	.73	67,323
7pter-q22	D7S83	M	.36(.39)	67,323
7pter-q22	D7S84	H	.19	67,323
7pter-q22	D7S85	M	.12	67,323
7pter-q22	D7S86	M	.34(.52)	67,323
7pter-q22	D7S88	T	.34(.26)	67,323
7pter-q22	D7S89	H M T	.58	67
7pter-q22	D7S90	M	.38(.39)	67,323
7pter-q22	D7S92	T	.37	67,323
7pter-q22	D7S94	E	.33(.30)	67,323
7pter-q22	D7S100	T	.37	67
7pter-q22	D7S102	M	.28(.41)	67,323
7pter-q22	D7S103	T	.36(.48)	67,323
7pter-q22	D7S105	T	.36(.17)	67,323
7pter-q22	D7S108	Ps	.23(.24)	67,323
7pter-q22	D7S109	H	.21	67
7pter-q22	D7S110	T	.30	67
7pter-q22	D7S112	T	.35	67
7p21-p14	IL6	M O	.79	128
7p21-p14	D7S410	T O	.51	58
7p15	*TCRG	E H Pv S T O	.57	180, 453, 454, 455, 573, 753, 756, 782, 1045, 1096
7p15-p13	D7S373	T	.29	1121
7p15-q22.1	ERV3	M	.37	1045
7p13-p12	*EGFR	Bg H Ps S O	.60	101, 748, 792, 906, 1254
7p13-p12	IBP1	Bg	.32	11
7p12	D7S96	T	.24	67
7p	D7S17	O	.36	123
7p	D7S147	Ps Pv S T	.34	1155
7p	*D7S150	T	.32	1155
7p	D7S369	M	.27	1045
7p	*D7S370	M	.36	933
7p	*D7S371	M	.33	583, 585
7p	D7S372	R	.33	1045
7cen	D7Z1	large fragment variation		1416
7cen	D7Z2	H	> .9	1398, 1416
7cen-q22.1	D7S397	M	.37	289
7cen-q22	*D7S129	E O	.33	1155
7cen-q32	*D7S6	T	.06	71
7q21.1	MDR2	incorrect nomenclature; see PGY3		
7q21	EPO	H O	.31	741, 1219, 1394
7q21	*PGY1	H	.37	1484
7q21	*PGY3	E M Pv	.25	225
7q21-q22	NKNA	R O	.31	295
7q21-q22	D7S15	H O	.37	323, 1329
7q21-q22	D7S64	E	.35	323
7q21.3-q22.1	*COL1A2	Bg E M R O	.60	7, 120, 121, 122, 150, 267, 268, 270, 367, 496, 1323, 1324, 1372
7q21.3-q22.1	CYP3	O	.16	165
7q21.3-q22	PLANH1	H S	.37	680, 695
7q22 or 7q31	LAMB1	O	.37	26, 618, 1103
7q22-q32	G7P1	T	.26	292
7q22-q32	D7S27	T O	.36	636
7q22-q32	D7S28	S	.36	636
7q22-q32	D7S29	E	.28	636
7q22-q32	D7S33	T	.37	636
7q22-q32	D7S391	Bg	.04	1155
7q22-q32	D7S418	T	.30	1205
7q22.3-q31.2	*D7S13	H M	.43	362, 734, 1364
7q31.1-q31.2	D7S16	obsolete symbol; see D7S18		
7q31.1-q31.2	*D7S18	E	.30	127, 370, 679, 1183
7q31	*MET	M T O	.81	78, 239, 298, 641, 862, 1261, 1328, 1419
7q31	*D7S8	E M Ps Pv S T O	.74	71, 127, 297, 298, 1042, 1328
7q31	D7S399	M		251
7q31	D7S402	M	.15	1126
7q31	D7S411	Ps O	.31	1127
7q31	D7S424	S		614
7q31	D7S426	Pv O	.36	614
7q31-q32	D7S23	Ps T O	.83	251, 361, 363, 379, 435, 519, 1275, 1407

Chromosome/Region	Locus Symbol	Enzyme	PIC (HET')	Reference
7q31-q32	D7S25	M	.24	1185
7q31-q32	D7S113	O	.36	1155
7q31-q32	D7S115	Ps	.36	1155
7q31-q32	D7S116	T	.16	1155
7q31-q32	D7S117	S	.36	1155
7q31-q32	D7S118	T O	.37	1155
7q31-q32	D7S119	M T O	.38	1155, 1327
7q31-q32	D7S120	T	.11	1155
7q31-q32	D7S121	M		1327
7q31-q32	D7S122	O	.35	1155
7q31-q32	D7S123	O		1327
7q31-q32	*D7S124	O	.36	1155
7q31-q32	D7S125	Ba Ps	.29	1155, 1327
7q31-q32	D7S126	Bg E H	.45	1045, 1155
7q31-q32	D7S127	M O	.35	1155
7q31-q32	D7S148	O		1327
7q31-q32	D7S152	M	.25	1155
7q31-q32	D7S201	O	.29	1155
7q31-q32	D7S202	O	.04	1155
7q31-q32	D7S206	O	.33	1155
7q31-q32	D7S252	Bg	.15	1155
7q31-q32	D7S283	O	.16	1155
7q31-q32	D7S316	O	.36	1155
7q31-q32	*D7S347	T	.25	1155
7q31-q32	D7S431	E	.18	297
7q31-q32	D7S432	T	.37	1506
7q31-q32	D7S433	Pv O	.28	297
7q31-q32	D7S434	M	.14	1506
7q31-q35	D7S63	M	.37(.36)	323
7q31-q35	D7S71	M	.34(.39)	67
7q31-q35	D7S72	H	.34	67,323
7q31-q35	D7S73	M	.26	67,323
7q31-q35	D7S87	T	.37	67
7q31-q35	D7S91	O	.36	67
7q31-q35	D7S93	M	.37	67
7q31-q35	D7S95	M	.35	67
7q31-q35	D7S97	T	.28	67
7q31-q35	D7S99	M	.34	67
7q31-q35	D7S101	M	.21	67
7q31-q35	D7S106	M	.33	67
7q31-q35	D7S107	T	.37	67
7q31-q35	D7S111	E	.57	67
7q31-qter	PRKAR2	O	.28	1178
7q32-q34	D7S130	Bg E	.35	1155
7q32-q34	D7S144	T	.30	1155
7q32-qter	PIP	R T	.30	932
7q32-qter	D7S114	Ps	.19	1155
7q32-qter	*D7S228	T	.33	1155
7q32-qter	*D7S258	Ps	.36	1155
7q34-qter	D7S392	Ps		1155
7q35	TCRB	Bg Ba H O	.75	97, 242, 586, 661, 842, 1045, 1149, 1150
7q35-qter	D7S54	M T	.72(.78)	323
7q35-qter	D7S56	M	.36(.52)	323
7q35-qter	D7S61	R	.69(.74)	323, 687, 1203
7q35-qter	D7S67	M	.37	323
7q35-qter	D7S68	Bg	.33(.40)	323
7q35-qter	D7S70	M	.51(.40)	323
7q35-qter	D7S98	R T	.37	67
7q36	EN2	S	.21	816
7q36	D7S104	T	.78	67
7q36-qter	D7S22	O	(.97)	634, 1439
7q	*D7S396	R	(.80)	982

Chromosome/Region	Locus Symbol	Enzyme	PIC (HET')	Reference
7q	D7S398	M	.59	943, 1020
7	*ASSP11	O	.24	77,266
7	D7F23S1	Ba E H M	.36	248
7	D7S1	H O	.19	206
7	D7S19	Ba Pv	.30	1278
7	D7S26	E Pv	.33	1185
7	D7S131	M R		1327
7	D7S133	R		1327
7	*D7S136	T	.38	1155
7	D7S137	Ps	.38	1155
7	*D7S139	T	.14	1155
7	D7S140	O		1327
7	*D7S141	Bg	.34	1155
7	*D7S143	Bg T	.33	1155
7	*D7S145	Ps	.34	1155
7	D7S146	T		1155
7	*D7S368	R	.35(.60)	983, 1022
7	D7S374	H	.25	1175
7	D7S395	Pv	.33	1023
7	D7S404	O	.40	1349
7	D7S408	obsolete symbol; see D7S398		
7	D7S423	M	.07(.07)	1409
7	D7S427	Bg R T	.43(.85)	310
8pter-q22	D8S74	Ps	.70	308
8p23	D8S7	E H Ps T O	.23	1448
8p23	D8S11	R O	.16	1447
8p22	LPL	Ba H Pv O	.57	387, 427, 428, 532, 783, 784
8p21	NEFL	T	.36	80,728
8p12-p11	POLB	Ba E H S		870
8p12-q11.2	PLAT	E	.37	88,89, 127, 1479
8p	SFTP2	E	.30	386
8p	D8S62	T	.30(.17)	
8q11	*MOS	E O	.11	307, 575, 1240
8q13	CRH	T	.11	657
8q13-q22	CA3	T		812
8q21-q22	CYP11B1, CYP11B2	M	.36	901
8q22	CA2	T	.38	747
8q22-q23	*D8S5	H	.30	313, 1047
8q22-qter	D8S75	Ba T	.74	308
8q23-q24	PENK	O	.43(.60)	794
8q23.2-q24.11	D8S48	H M	.25	822
8q24	MYC	Bg S T	.49	508, 509, 882, 1047
8q24	*TG	H Pv T O	.38	37, 38, 161, 1047, 1231
8	CALB	H	.32	1086
8	D8S2	T	.30	1182
8	*D8S3	M	.23	1184
8	D8S8	T	.45	151
8	D8S9	T	.30	1436
8	D8S10	H	.03	1447
8	D8S12	O	.07	1447
8	D8S13	M	.20	1470
8	D8S15	T	.37	51

Chromosome/Region	Locus Symbol	Enzyme	PIC (HET)	Reference
8	*D8S17	Ps	(.33)	950
8	D8S20			231
8	D8S21	R	.34	1047
8	D8S22	E	.69(.69)	323
8	D8S23	E	.34(.57)	323
8	D8S24	M	.29	323
8	D8S25	M	.37(.50)	323
8	D8S26	M T	.44(.74)	323, 1203
8	D8S27	H	.37(.50)	323
8	D8S28	T	.38	323
8	D8S29	M	.30(.62)	323
8	D8S30	M	.21(.19)	323
8	D8S31	T	.37(.55)	323
8	D8S32	Ps	.41(.64)	323
8	D8S33	T	.35(.40)	323
8	D8S34	T	.55(.57)	323
8	D8S35	Bg	.35(.40)	323
8	D8S36	Bg	.62(.70)	135, 323
8	*D8S39	Ps	(.65)	968
8	D8S72	Ba	.37	1233
8	D8S73	H	.25(.40)	135
8	D8S84		.58(.58)	1406
9pter-q11	D9S1	T	.37(.36)	63, 1027
9p22	IFNB1	M T O	.32	1061, 1240
9p22-p13	IFNA	M	.37	1061
9p	NRASL1	T	.37	883
9p	D9S18	T	.29	732
9q11-q22	ASSP3	Ba H	.37(.43)	266
9q12-q13	D9S5	T	.14	1067
9q21.3-q22.2	ALDOB	Pv	.18	1084, 1085
9q31	D9S29	Pv	.26	1083
9q32-q33	ITIL	O	.16	772, 773
9q32-q34	GSN	O	.36	727
9q34.1-q34.2	AK1	T	.28	81
9q34	*ABL	Bg Ps S T O	.26	133, 243, 1240, 1473
9q34	*D9S7	T O	(.75)	736, 951
9q34	D9S17	T	.34(.40)	736
9q34	D9S31	Pv	.47(.44)	736, 943, 960
9q34-qter	ASS	O	.36	635
9q	D9S6	T	.36	952
9q	*D9S10	Ps	.37(.50)	192, 736
9q	*D9S11	M Ps O	(.72)	736, 1022, 1023
9q	*D9S13	Ps	(.56)	991, 1022
9q	*D9S14	R O	.34(.48)	736, 1017
9q	*D9S15	M O	.34	732, 1369
9q	D9S16	M T	.51(.49)	200, 732, 736
9q	*D9S30	Pv	.36	961
9q	D9S39	M	.30	389
9	D9S3	H S	.32	617
9	*D9S4	T	.36	375, 1265
9	*D9S8	M	.33	732
9	*D9S9	M R	.32(.78)	718, 1022

Chromosome/Region	Locus Symbol	Enzyme	PIC (HET)	Reference
9	*D9S12	T	.15	581, 736
9	D9S19	T	.60(.64)	323
9	D9S20	E	.37	323
9	D9S21	T	.37	323
9	D9S22	E	.36	323
9	D9S23	H	.32	323
9	D9S25	Ps		323
9	D9S26	T	.35	323
9	D9S27	T	.29	554
9	D9S33	T	.37	323, 1280
9	D9S34	T	(.49)	943
9	D9S35	obsolete symbol; see D9S31		
9	D9S41	T	.37	1139
10pter-p13	*D10S17	M	.36	558
10pter-p13	D10S28	T	(.92)	145
10p15	ITIH2	M O	.33	776
10p15	D10S63	Bg	.45	134
10p15-p14	*IL2R	O	.12	24
10p14-p11.2	D10S49	Ps	.49	134
10p14-q11.2	D10S41	E T	.38	134
10p13-p12.2	*D10S24	M T	.55	1456
10p11.2	FNRB	Bg M S O	.71	1455
10p11-q23	*D10S13	Bg	.30	735
10p11-q23	*D10S14	M	.47	579, 735
10p11-q23	D10S16	M Pv	.35	735, 844
10p	D10S31	M	.36(.45)	143, 947
10p	D10S33	T	(.52)	141, 947
10p	D10S34	T	.37(.49)	947
10p	D10S35	M	.37(.55)	140, 947
10p	D10S39	Bg	.36	858
10cen	D10Z1	M		1416
10q11.2	*RBP3	Bg M T O	.44(.41)	224, 371, 793, 860, 947, 1232
10q11.2	TST1	T O	.33	1115
10q11.2	D10S11	T	.48	323
10q11.2	*D10S15	Pv	(.45)	735, 974
10q11.2	D10S59	Ps	.40	134
10q11.2-q22	D10S30	M	.34(.47)	139, 947
10q21.1	CDC2	T	.36	1267
10q21.1	EGR2	H	.16	1459
10q21.1	*D10S5	T O	.38	371, 873, 1232
10q21.1	*D10S22	M	.49	146
10q21.1-q22	*D10S19	Pv	.35	735, 981
10q21-q24	SFTP1	M	.36	385
10q21-q26	*D10S20	H R T O	.52	735, 887, 1039
10q22-q23	*D10S1	Bg T	.56	371
10q22-q23	*D10S3	Ps	.36	375, 1265
10q22-q23	*D10S4	S T	.59	371, 735, 802
10q23-q24	GLUD	T	.38	512
10q24.1-q24.3	CYP2C	O	.36	485
10q24-q26	*ADRA2R	O	.26	549
10q24-qter	*PLAU	Ba	.33	1213
10q24-qter	D10S18	M	.28	975

Chromosome/Region	Locus Symbol	Enzyme	PIC (HET)	Reference
10q24.3-qter	D10S21	Ba	.43	1118
10q25.1	D10S37	M	.37	1386
10q25-q26	G10P1	E	.07	725
10q26	OAT	M R O	.34	1122, 1125
10q	D10S7	M	.45	323
10q	D10S8	H	.21	323
10q	D10S9	M	.31	323
10q	D10S10	E	.31	323
10q	D10S12	H	.74	323
10q	*D10S23	Pv	.26	846
10q	*D10S25	T	(.60)	947
10q	*D10S26	Pv	(.52)	947, 959
10q	D10S27	M	.37(.50)	947
10q	D10S29	T	.35(.48)	144, 947
10q	D10S32	T	.52(.64)	142, 947
10q	D10S36	T	.35(.27)	947, 958
10q	D10S40	M	.33	134
10q	D10S42	Ba	.37	134
10q	D10S43	E	.25	134
10q	D10S44	H	.38	134
10q	D10S45	H	.31	134
10q	D10S46	T	.22	134
10q	D10S47	E	.59	134
10q	D10S48	H	.31	134
10q	D10S50	M O	.29	134
10q	D10S51	E Ps	.36	134
10q	D10S52	Bg	.18	134
10q	D10S53	Ps	.68	134
10q	D10S54	M	.55	134
10q	D10S55	M	.35	134
10q	D10S56	Bg	.34	134
10q	D10S57	T	.35	134
10q	D10S58	M	.37	134
10q	D10S62	H	.45	134
10	CYP2E	T	.16	869
10	D10S60	R	.36	134
10	D10S64	T O	.35	134
10	D10S65	T	.37	134
10	D10S66	M	.30	134
10	D10S68	T	.27	134
10	D10S69	H	.33	134
10	D10S70	Ba	.73	134
10	D10S88		.31(.34)	1406
11pter-p15.5	D11S19	O	.38	447
11pter-p15.4	*PTH	Ps T	.37	22, 1197, 1198
11pter-p15.1	D11S20	Ps	.15	447, 470
11pter-p15	D11S96	O	.29	448
11pter-p13	D11S17	Bg Ba	.33	447, 470
11pter-p13	D11S18	S	.37	447, 470, 1376
11pter-p11	SAA	H Ps T	.36	682, 683
11p15.5	*HBB	Ba H R O	.42	20, 34,43, 339, 347, 378, 651, 1043, 1331
11p15.5	*HBB@	several dozen haplotypes have been identified for the 60 kb long beta hemoglobin cluster		17, 655, 1063, 1363
11p15.5	HBBP1	O	.37	1043, 1362
11p15.5	*HBD	Ps R O	.38	339
11p15.5	*HBE1	T O	.37	20, 1043, 1125
11p15.5	HBG1	H Pv	.36	20, 1043, 1063
11p15.5	HBG2	H T O	.37	466, 1043, 1124, 1361
11p15.5	*HRAS	Ba T O	.67(.84)	217, 473, 712, 788, 1101, 1102, 1116, 1136
11p15.5	*IGF2	Ba E M S O	.37	214, 229, 348, 470, 675, 1200, 1467, 1469
11p15.5	*INS	Bg E Ps Pv R S T O	.57(.80)	86, 87, 214, 256, 303, 346, 348, 349, 546, 1053, 1071, 1159, 1161, 1427
11p15.5	*TH	Bg E Ps R S T O	.41	214, 348, 659, 701, 905, 1053, 1105, 1317
11p15.5	*D11S12	M T O	.37	62, 1027
11p15.5-p15.4	RRM1	S	.37	75
11p15	MUC2	O	(.50)	492
11p15-p13	D11S150	O	(.90)	164
11p-q12	*D11S149	Pv	.21	580
11p15.4	CALCA	T	.35	470, 588, 589
11p15.1-p14	LDHA	T	.37	448
11p14-p13	D11S21	M T	.37	469
11p14.2-p12	D11S152	Pv	.24	170
11p14.2-p12	D11S153	E Ps	.34	170
11p14.2-p12	D11S155			170
11p13	*CAT	M T O	.38	136, 171, 899, 1112, 1113, 1357
11p13	FSHB	H	.37	1395
11p13	*D11S16	M O	.54	375, 1265
11p13	D11S151	E	.24	598

Chromosome/ Region	Locus Symbol	Enzyme	PIC (HET)	Reference
11p13	*D11S347	S	.36	127, 375, 1264
11p13	D11S406	O	.14	468
11p13	D11S407	M	.34	468
11p13	D11S408	M	.35	468
11p13	D11S409	M	.35	468
11p13	D11S410	O	.23	468
11p13	D11S411	O	.20	468
11p13	D11S412	T O	.35	468
11p13	D11S413	M	.14	468
11p13	D11S414	H M T O	.50	468
11p12	D11S28	M	.30	354
11p12.08-p11	D11S9	T	.27	447
11p11-q11	D11S33	E	.42	52
11cen	D11Z1	O	.41	1399, 1401, 1416
11q12-q13.1	C1NH	O	.34	112
11q12-q13.2	*PYGM	M	.46	194
11q12-q13.2	*D11S146	M T	.37	944, 976
11q13	INT2	Ba Ps T	.36	52, 208, 508
11q13	*PGA3, *PGA4, *PGA5	Bg E		1299, 1300
11q13	SEA	O		1429
11q13	D11S97	O	(.77)	634, 1438
11q13-q14	D11S37	H	.37	315
11q13-q23	D11S23	O	.18	469
11q13-q23	D11S24	Ba	.24	469, 470
11q13.2-q22	D11S141	T	.37	323
11q13.2-q23.3	D11S351	M Ps	.37(.50)	39
11q14-q21	*TYR	T	.37	1263
11q21-q22	*D11S35	M	.80(.88)	855
11q22	STMY	T	.37	1266
11q22	*D11S84	M T	.32	855
11q22-q23	DRD2	T	.30	486
11q22-q23	PGR	H	.28	742
11q22-q23	*D11S85	M	.37	855
11q22-q23	*D11S98	M T	.37	855
11q22-q23	D11S132	E H	.55(.63)	323
11q22-q23	D11S133	E M	.37(.59)	323
11q22.3-q23	THY1	M	.33(.45)	437
11q22.3-q23.3	*D11S144	M	.37	198
11q23	*CD3D	M	.11	835
11q23	*CD3E	T	.34	219
11q23	CD3G	E M	.27	218, 711
11q23-q24	*APOA1	E M Ps S T O	.73	179, 230, 233, 235, 405, 414, 662, 908, 1066, 1094, 1135, 1152, 1216, 1288, 1315, 1347
11q23-q24	NCAM	Ba	.29	885
11q23-qter	APOA4	Bg E H Ps S T O	.55	228, 234, 236, 411, 653, 885, 1059
11q23-qter	*APOC3	Pv S O	.36	233, 237, 535, 663, 1058, 1094, 1134, 1315
11q23-qter	*D11S29	T	.30(.30)	1383
11q23-qter	*D11S83	M	.34	855
11q23-qter	*D11S286	Ba	.34	826
11q23.2-qter	PBGD	M Ps O	.37	745, 750, 810, 811
11q23.3	*ETS1	S O	.36	491, 660, 673, 1170
11q23.3-qter	*D11S147	Ps	.30	977
11q	*D11S34	M	.35	855, 871
11q	*D11S36	Pv R	.46	855
11	D11S15	O	.36	617
11	D11S127	M	.48(.71)	323, 1203
11	D11S128	M	.62(.71)	323, 687, 1203
11	D11S129	Bg	.48	323
11	D11S130	R	.54	1280
11	D11S131	T	.35	323
11	D11S134	M	.36	323
11	D11S135	obsolete symbol; see D9S33		
11	D11S136	M T	.38	323
11	D11S137	H M T	.37	323
11	D11S138	Ps	.30(.21)	323
11	D11S139	Ps	.34	135, 1280
11	D11S140	O	.26	323
11	D11S142	M	.34	323
11	*D11S145	M	.50	201
11	D11S148	obsolete symbol; see D11S144		
11	D11S284	T	.42	231
11	D11S288	M	.33	743
11	D11S378	O	.20	1358
11	D11S379	Ps	.36(.31)	135
12pter-p12	*F8VWF	Bg Ba E R S T O	.36	99, 364, 615, 698, 738, 739, 791, 839, 1037, 1111, 1171, 1353
12pter-p12	*D12S2	E	.27	49, 1050
12p13	C1R	Ps Pv		232, 1342
12p13	C1S	Ps Pv		1342
12p13.3-p12.3	*A2M	Bg Pv T O	.37	127, 284, 848, 1181

Chromosome/ Region	Locus Symbol	Enzyme	PIC (HET)	Reference
12p13.2	*PRB1, PRB2, PRB3, PRB4	E	.60	41, 827, 1050
12p12.1	*KRAS2	Ps T O	.38	64, 531, 1050, 1062, 1240
12p	*D12S16	T	.32	578
12cen-q14	*D12S4	M T	.35	1050, 1240
12q12-q13.1	D12S32	O	.32	1056
12q12-q13	HOX3	T	.34	650
12q14	*D12S6	E M T	.37	182, 184, 1050
12q14-q24.1	*D12S7	T	.53	1050, 1326
12q14-qter	*D12S8	M T	.36	1048, 1050
12q14.3	COL2A1	Ba E H Pv O	.69	356, 396, 1289, 1292, 1297, 1338, 1403, 1462
12q22-q23	D12S33	Ps	.21	1144
12q22-q24.2	PAH	Bg Ba E H M Pv O	.82	35, 212, 319, 503, 537, 787, 789, 1445, 1446
12q23	IGF1	H Pv O	.53(.54)	590, 1160, 1199, 1406
12q24.2	*ALDH2	E	.35	1251
12q24.3-qter	D12S11	O	(.94)	1439
12qter	D12S37	M	.20	1165
12q	D12S12	M	.48	47
12q	*D12S14	M	.59	1000
12q	*D12S15	T	.22	845, 1050
12q	*D12S17	M	.48	1001, 1050
12q	D12S36	M	.37(.57)	1409
12	ELA1	T	.25	1050
12	TUBAL1	M	.36	438
12	VDR	O	.37	368
12	*D12S9	Bg	.29	1379
12	D12S10	E	.37	1380
12	D12S18	R	.22	582
12	D12S19	Ps	.38	135, 1038
12	D12S20	E Ps	.46(.67)	135, 323
12	D12S21	M	.34(.50)	323
12	D12S22	T	.36	323
12	D12S23	M	.37	323
12	D12S24	M	.19	323
12	D12S25	M	.45	323
12	D12S26	E	.31	323
12	D12S27	E O	.36	323
12	D12S28	T	.37	323
12	D12S29	M		323
12	*D12S31	O	.29	1050
12	*D12S34	T	.37	375, 1264

Chromosome/ Region	Locus Symbol	Enzyme	PIC (HET)	Reference
12	D12S39	H	.37(.67)	1509
13pter-q14.1	D13S36	E	.37	51
13cen	D13Z1	Ba T	.22	837
13q12-q13	*D13S11	M Ps O	.35	1189
13q12-q14	*D13S1	M T O	.56	127, 210, 505, 764
13q12-q14	*D13S6	E O	.36	127, 340, 764
13q12-q14	D13S33	E T	.49	51
13q12-q14	*D13S62	O	.30	375, 1264
13q12-q21	*D13S37	Bg	(.69)	764
13q13-q14	D13S64	T	.28	1139
13q14.1	D13S21	M O	.35	1187, 1188
13q14.1	*D13S22	H O	.47	127, 130
13q14.1-q14.2	*ESD	O	.56	505, 749, 1269
13q14.1-q14.2	D13S10	T O	.56	125, 127, 505, 764, 766
13q14.1-q14.2	D13S30	obsolete symbol; see RB1		
13q14.1-q14.2	D13S56	H	.22	323
13q14.1-q14.2	D13S59	O	.37	124
13q14.2	*RB1	Ba R O	.80(.94)	118, 1423
13q14.2-q21	D13S31	Pv T	.65	51
13q14.2-q21	D13S55	M T	.36	323
13q21.1-q21.2	*D13S26	E O	.34	131
13q21	D13S12	M	.67	1190
13q21-q22	D13S39	M	.37	764
13q21-q22	D13S41	T	.37	764
13q21-q34	*D13S3	H M	.37	127, 210, 764
13q21-q34	D13S61	O	.24	124, 133
13q22	*D13S2	M Ps T	.39	127, 210, 746, 764
13q22	D13S7	Bg	.25	340, 764
13q22-q31	*D13S4	M	.37	127, 210, 764
13q22-q34	*PCCA	H O	.36	124, 730
13q22-q34	*D13S5	E H	.37	127, 340, 764
13q22-q34	D13S24	O	.26	124
13q22-q34	D13S32	Bg T O	.37	51
13q22-q34	*D13S34	T	.24	51
13q31	D13S63	Bg	.20	1214
13q34	COL4A1	H O	.43	126, 127, 132, 1316
13q34	COL4A2	T O	.38	132
13q34	F10	E Ps T O	.29	522, 633, 1177
13q34	PPOLP1	H Ps	.19	223, 867
13q34	*D13S35	E	.35	51
13q	D13F13S1	E	.11	630

Chromosome/Region	Locus Symbol	Enzyme	PIC (HET)	Reference
13q	*D13S49	T	(.50)	987, 1022
13q	D13S50	H	.38	957
13q	D13S51	T	.19	557
13q	*D13S52	M	(.80)	965
13q	D13S53	M	(.63)	943, 972
13q	D13S54	Ps	(.57)	943, 954
13	D13S15	E	.37	1064
13	D13S38	Bg	.36	764
13	D13S40	E	.24	764
13	D13S42	E	.30	764
13	D13S57	obsolete symbol; see D13S53		
13	D13S58	obsolete symbol; see D13S54		
14q11-q13	ANG	O	.33	1413
14q11-q24.3	COLL2	obsolete symbol; see D14S11		
14q11-q24.3	D14S11	Pv	.37	254
14q11.2	TCRA	Bg E H M Pv S T O	.41	55, 586, 587, 778, 842, 865, 893, 1151, 1257
14q11.2	TCRD	Ba O	.46	226, 227
14q11.2-q13	MYH6	Ba	.34	253
14q11.2-q24.3	PYGL	Ba M Ps T O	.37	1029
14q11.2-q24.3	D14S3	Ba E	.14	1240
14q22-q24	*HSPA2	Pv	.36	472
14q24.3	*FOS	T	.12	253
14q24.3-q32.1	D14S12	Bg	.36	152
14q32.1	AACT	T	.33	658
14q32.1	*PI	Ba E M Ps S T O	.81	2,255, 257, 547, 648, 864, 1030, 1044
14q32.1	*PIL	Bg E M Ps T	.54	1, 252, 255, 548, 647, 1104
14q32.1-q32.32	*D14S18	E	.15	978
14q32.1-q32.32	*D14S21	H	.30	847
14q32	*D14S13	Ps R	.80(.95)	48,979
14q32	*D14S16	Ps	.39	970
14q32.3	CKBB	E	.36	254, 840
14q32.3	IGHJ	M	.68	1228
14q32.32-q32.33	AKT1	Ba	.37	253
14q32.32-q32.33	*D14S1	E H Ps	.79(.64)	42, 48, 1270, 1465
14q32.32-q32.33	*D14S17	T	.34	422
14q32.32-q32.33	*D14S19	Ba	.49	556
14q32.32-q32.33	*D14S23	M	(.83)	943, 966
14q32.33	IGHA1	Pv S	.46	536, 664, 665
14q32.33	IGHA2	Pv S O	.79	536, 664, 665, 754
14q32.33	IGHD	T O	.37	253, 254
14q32.33	IGHE	Ba		755
14q32.33	*IGHGP	Ba O	.39	79, 82, 211, 449, 452, 638
14q32.33	*IGHG1	Ba O	.53	449, 452, 638
14q32.33	*IGHG2	Ba O	.49	79, 82, 211, 449, 452, 638
14q32.33	*IGHG3	Ba O	.53	449, 452, 596, 638
14q32.33	*IGHG4	Ba E S O	.56	79, 82, 449, 452
14q32.33	*IGHM	Pv S	.48	886
14q32.33	*IGHV	Bg E H R S O	.52	639, 1172, 1332, 1373, 1374
14q32.33	*D14S20	M	(.70)	943, 998
14q	*D14S22	Ps	(.85)	997
14	*SPTB	H Ps Pv	.36	1106
14	*D14S4	Ps S	.62	375, 1265
14	D14S14	Ps	.34	1470, 1472
14	D14S24	M	.37	135, 323
14	D14S25	T	.37	323
14	D14S26	T	.37	323
14	D14S27	Bg M	.40	323
14	D14S28	Ps	.33	323
14	D14S29	obsolete symbol; see D14S20		
14	D14S30	obsolete symbol; see D14S23		
14	D14S31	Bg	.37	127, 375, 1264
14	D14S32	T	.36(.50)	1509
14	D14S33	E	.69(.26)	135
15pter-q13	*D15S24	E T	.75(.85)	946, 1141
15p11-q11	D15S58	E	.33	1306
15q11-q12	*D15S9	O	.33	737
15q11-q12	*D15S10	T	.25	737
15q11-q12	*D15S11	R	.38	737
15q11-q12	*D15S12	O	.44	737
15q11-q12	*D15S13	T	.33	737
15q11-q12	*D15S18	Bg	.21	737
15q11-q12	D15S53	O	.37	1306
15q11-qter	*ACTC	---PCR---	.84	796
15q11-qter	D15S44	M	.31(.32)	417
15q11-qter	*D15S45	H	.33	420
15q12-q24	*D15S4	Ba O	.20	375, 1265
15q13-qter	PEPN	O	.30	713
15q14-q21	*D15S1	M	.37	63

Chromosome/ Region	Locus Symbol	Enzyme	PIC (HET)	Reference
15q15-q22	*D15S2	E	.33	160, 946
15q21-q23	LIPC	Bg H M O	.38	276, 533, 786
15q21-q23	D15S56	Ba H	.22	1306
15q22-q24	*CYP1	Bg E M	.29	626, 1268
15q22-q24	D15S8	obsolete symbol; see CYP1		
15q24-qter	D15S57	Ba H M Pv O	.35	1306
15q25-qter	*FES	H	.09	1312
15	CSPG1	Bg O	.36(.50)	384
15	*D15S3	E	.37	248
15	*D15S25	R	.36	946
15	D15S26	Pv	.20	197
15	*D15S27	M	.34	576
15	*D15S28	Ba	.30	1002
15	*D15S29	M	.38	419, 946
15	D15S30	M	.37	720
15	D15S31	obsolete symbol; see D15S33		
15	D15S32	Ps T	.36	231
15	D15S33	Pv	.29	196
15	D15S34	M T	.38(.24)	195, 946
15	D15S35	M T	.39	946, 969
15	D15S36	T	.30	971
15	*D15S37	E	.37	424
15	D15S38	M T	.37	418, 946
15	D15S46	H M	.55	323, 324
15	D15S47	Bg	.29	323, 324
15	D15S48	M	.37	323
15	D15S49	M	.34	323
15	D15S50	Ba	.31	323
15	D15S51	M	.24	323
15	D15S52	M	.63	323
15	D15S75	T	.37	375, 1264
16pter-p13.13	D16S126	T	.32	156
16pter-p13.13	D16S127	T	.37	156, 158
16pter-p13.11	D16S24	Ba Pv	.37	808, 809
16pter-p13	*D16S32	T O	.37	518
16pter-p13	*D16S33	M	.24	518
16pter-p13	*D16S34	Ps	.36	518
16pter-p13	*D16S35	T O	.44	518
16pter-p13	*D16S36	T	.34	518
16pter-p13	*D16S37	M O	.36	518
16pter-p13	D16S45	E O	.37	323, 1132, 1153
16pter-p13	*D16S92	O	.26	518
16pter-p13.1	D16S143	O	.35	516
16pter-p13.1	D16S144	H O	.36	516, 768
16p13.3	*HBA1, HBA2, HBAP1	Ps R O	.45	32,538
16p13.3	*HBZ,HBZP	Bg Pv S O	.77	380, 477, 538, 541, 1032, 1319, 1360
16p13.3	D16S21	Ba H M R	.41	1470
16p13.3	*D16S85	E Pv S O	.82	538, 632, 1133
16p13.3	D16S94	M	.37	608, 609
16p13.3	D16S145	Pv	.26	515
16p13.3	D16S146	O	.29	515
16p13.3-p13.13	D16S51	Ba	.51	323
16p13.3-p13.13	D16S80	T	.31	157
16p13.3-p13.13	D16S81	S	.11	188
16p13.3-p13.13	D16S82	S	.27	188
16p13	D16S22	S	.28	1132
16p13	D16S49	E	.36	323
16p13	D16S53	Ps	.37	323
16p13	D16S55	O	.25	323, 1132
16p13	D16S56	Bg E	.30	323
16p13	D16S58	H	.27	323, 1132, 1153
16p13	D16S60	O	.57	323, 656
16p13	D16S63	H	.32	323, 1153
16p13	D16S66	Ps	.36	323
16p13	D16S72	M	.35	323
16p13	D16S74	Bg	.38	323, 1132
16p13	D16S138	O	.25	543
16p13.2-q13	D16S147	T	.47	642
16p13.2-q13	D16S148	M	.36(.35)	642
16p13.2-q13	D16S149	M	.19(.23)	642
16p13.2-q13	D16S150	T	.31(.42)	642
16p13.2-q21	D16S158	T	.14(.16)	642
16p13.13-p13.11	D16S8	Pv	.37	399
16p13.13-p13.11	D16S79	T O	.37	188, 441
16p13.13-p13.11	D16S96	M T	.72	608
16p13.11	D16S64	H	.35	323
16p13.11	D16S75	H M	.41	323
16p13.11	D16S131	T	.35	608, 612
16p12-qter	D16S68	Bg Ps	.36	323
16p11.2-q22.1	D16S91	R	.11	291
16p11-cen	D16F21S2	O		1438
16p	D16S70	M	.45	656
16p	D16S73	T	.38	323
16p	D16S76	T	.35	323
16p	*D16S83	R O	(.89)	1437
16p	D16S84	M	.36	967
16cen	D16Z2			1416
16q12-q13	D16S39	Ps	.37	323
16q13-q21	D16S23	Pv	.35	808
16q13-q22.1	MT2	T	.22	611
16q13-q22.1	D16S10	M R T	.36	400, 440
16q13-q22.1	D16S27	Pv	.29	808
16q21	CETP	M T O	.53	334, 402
16q21-q22	D16S151	M	.29(.33)	642
16q21-q22	D16S152	T	.37(.51)	642
16q21-q22	D16S160	M	.37(.43)	642
16q22.1	*HP	Bg Ba E H Ps T O	.53	539, 613, 833, 1065
16q22.1	HPR	Ps	.25	90, 1065, 1414
16q22.1	TAT	Ba M O	.44	1414, 1415, 1463
16q22.1	*D16S4	M T	.62	610

Chromosome/Region	Locus Symbol	Enzyme	PIC (HET)	Reference
16q22.1	*D16S14	O	.22	610
16q22-q24	*D16S20	Bg Pv S O	.64	856
16q22.1-q24	D16S26	O	.06	808
16q22.1-q24	D16S28	O	.30	808
16q22.1-q24	D16S30	Ps O	.33	808
16q22.1-q24	D16S69	Ba	.31	323
16q22-qter	D16S153	T	.35(.50)	642
16q22-qter	D16S154	T	.65(.55)	642
16q22-qter	D16S155	M	.34(.51)	642
16q22-qter	D16S156	T	.23(.29)	642
16q22-qter	D16S157	T	.22(.22)	642
16q22.3-q23.2	CTRB	Pv T	.59(.38)	1414
16q23.1-q24	*D16S5	R	.34	610
16q24	APRT	Bg T O	.32	27, 1274
16q24	*D16S7	T	.78	177
16q	D16S38	Ba	.29	323
16q	D16S40	T	.37	323
16q	D16S41	T	.33	323
16q	D16S42	Ps	.30	323
16q	D16S43	H Ps	.37	323, 656
16q	D16S44	Bg	.38	323
16q	D16S46	M T	.30	323
16q	D16S47	E	.37	323
16q	D16S48	H	.36	323
16q	D16S50	T	.35	323
16q	D16S52	H	.33	323
16q	D16S54	E O	.33	323, 656
16q	D16S57	Ba	.32	323, 656
16q	D16S59	M	.28	323, 656
16q	D16S61	E	.35	323
16q	D16S62	T	.40	323
16q	D16S65	Bg T O	.37	323
16q	D16S67	T O	.39	323
16q	D16S71	Ba	.36	323
16q	D16S77	H O	.37	323
16	D16S1	Bg	.09	1240
16	D16S136	T	(.46)	943
16	D16S137	Ps T	.37(.60)	642, 943
16	D16S260		.43(.57)	1406
16	D16S261		.66	1508
17pter-p12	D17S65	H	.30	61
17pter-p12	D17S66	Bg	.32	61
17pter-p12	D17S67	M	.33	61
17pter-p12	D17S68	M	.31	61
17pter-p12	D17S70	T	.14	61
17pter-p12	D17S126	H	.32	59
17pter-p12	D17S127	M	.33	59
17pter-p12	D17S128	E	.35	59
17pter-p12	D17S129	R	.31	59
17pter-p11	MYH3	E	.37	1234
17p13.3	*D17S5	M R	(.86)	629, 945, 994, 1022
17p13.3	*D17S28	T	(.64)	1003, 1345
17p13	GLUT4	O	.36	779
17p13	*D17S1	M S	.35(.40)	63, 1208
17p13	*D17S34	R	.86	696
17p13.1	*MYH2	E H M	.43	1208, 1344
17p13.1	*POLR2	E	.28	1344
17p13.1	*TP53	Bg O	.20	172, 853
17p13.1-p11.2	D17S31	M	.30(.43)	202
17p13.1-q21	*D17S86	T	.26	804
17p12-p11.2	D17S61	M	.34	61

Chromosome/Region	Locus Symbol	Enzyme	PIC (HET)	Reference
17p12-p11.2	D17S62	M	.30	61
17p12-p11.2	D17S63	M	.33	61
17p12-p11.2	D17S64	M	.16	61
17p12-p11.2	*D17S71	M Pv	.37(.56)	68,945
17p12-p11.2	D17S121	M	.23	59
17p12-p11.2	D17S122	M	.45	59
17p12-p11.2	D17S123	Bg T	.27	59
17p12-p11.2	D17S124	Bg M	.33	59
17p12-p11.2	D17S125	Bg M	.33	59
17p11.2	*D17S29	R T	.33(.31)	945
17p11.2-cen	D17S58	Bg T	.36(.54)	61
17p11.2-cen	D17S59	M	.07	61
17p11.1-qter	PPY	M	.37	388
17cen	D17Z1	E H Pv	.35	366, 1397, 1416, 1426, 1494
17cen-q12	D17S37	T	.50(.56)	475, 1281, 1334
17cen-q12	D17S38	M	.17	488
17cen-q12	D17S54	Bg T	.35(.48)	61,366
17cen-q12	D17S55	Bg M	.25(.28)	61
17cen-q12	D17S56	R	.16	61
17cen-q12	D17S57	M	.33(.40)	61
17cen-q12	D17S72	R	.34	366
17cen-q12	D17S73	Bg H	.30(.33)	366
17cen-q12	D17S82	Bg	(.52)	1049
17cen-q12	D17S83	Ba T	(.45)	1049
17cen-q12	D17S84	M	(.30)	1049
17cen-q12	D17S85	Bg H Pv T	(.50)	1049
17cen-q12	*D17S115	M	.33	59
17cen-q12	*D17S116	M	.14	59
17cen-q12	*D17S117	Ps	.30	59
17cen-q12	D17S118	M Pv	.37	59
17cen-q12	D17S119	M	.30	59
17cen-q12	*D17S120	T	.36	59
17q11.1-q11.2	D17S137	M	.36	1474
17q11-q12	D17S226	Pv	.36	1474
17q11.2	D17S81	M	(.40)	1049
17q11.2-q12	*CRYB1	M Pv	.37	65,66
17q11.2-q12	ERBA1	obsolete symbol; see THRA1		
17q11.2-q12	*THRA1	E M Pv	.18(.37)	861, 1049, 1143
17q11.2-q12	D17S33	R	.37(.49)	563, 945, 1049, 1418
17q12-q24	D17S40	M	.33(.43)	61
17q12-q24	D17S41	Ps T	.30(.47)	61,945
17q12-q24	D17S42	T	.28	61
17q12-q24	D17S43	M	.19	61
17q12-q24	D17S44	H	.33	61
17q12-q24	D17S45	Bg	.05	61
17q12-q24	D17S46	Bg	.52	61
17q12-q24	D17S47	Bg	.37	61
17q12-q24	D17S48	Bg	.16	61
17q12-q24	D17S49	Bg	.35	61
17q12-q24	D17S50	T	.35	61
17q12-q24	D17S51	R	.37	61
17q12-q24	D17S2	Ps	.16	61
17q12-q24	D17S60	T	.28	61
17q12-q24	D17S92	Ps	.30	59
17q12-q24	D17S93	M	.35	59
17q12-q24	D17S94	T	.23	59
17q12-q24	D17S95	R	.07	59
17q12-q24	D17S96	M	.28	59
17q12-q24	D17S97	T	.25	59
17q12-q24	D17S98	M	.29	59
17q12-q24	D17S99	M T	.25	59
17q12-q24	D17S100	M	.35	59
17q12-q24	D17S101	M	.31	59

Chromosome/Region	Locus Symbol	Enzyme	PIC (HET)	Reference
17q12-q24	D17S102	M	.21	59
17q12-q24	D17S103	T	.14	59
17q12-q24	D17S104	R	.32	59
17q12-q24	D17S105	M	.37	59
17q12-q24	D17S106	T	.35	59
17q12-q24	D17S107	Ps	.19	59
17q12-q24	D17S108	Bg	.30	59
17q12-q24	D17S109	O	.19	59
17q12-q24	D17S110	M	.30	59
17q12-q24	D17S111	Pv	.24	59
17q12-q24	D17S112	Pv	.32	59
17q12-q24	D17S113	T	.38	59
17q12-q24	D17S114	M	.27	59
17q21.1	*RARA	Ps	.33	31,53
17q21	MTBT1	H	.16	1305
17q21-q22	HOX2	S O	.22(.28)	917
17q21-q22	NGFR	H T O	.37(.31)	149, 366, 1451, 1452
17q21-q23	D17S76	M	.32	807
17q21-q23	*D17S78	M	.37	825
17q21-qter	EPB3	Ps	.37(.49)	1283
17q21.3-q22	COL1A1	M R	.43	1060, 1298
17q21.3-q23	*MPO	Ps	.25	888
17q21.32	GP3A	T	.37	181
17q22-q24	*GH1,GH2, CSH1, CSH2, CSHP1	Bg Ba M O	.38	213
17q23-q25.3	D17S4	Pv T	(.76)	945, 973, 1022
17q23-qter	*D17S20	Bg E T	.36(.41)	186
17q23-qter	*D17S21	M	.35	697
17q23-qter	*D17S77	M T	.37	803
17q23.2-q25.3	*TK1	T O	.38(.60)	916, 918
17q24-qter	D17S53	M	.16	61
17q24-qter	D17S130	T	.34	59
17q24-qter	D17S131	Ps	.42	59
17q	GAS	Ba	.09	1195
17q	*D17S24	Pv T	(.65)	934, 945
17q	*D17S26	Pv	(.83)	1004
17q	D17S27	Ps	.30(.38)	935
17q	D17S32	Ba	.36(.45)	564, 945
17q	*D17S74	O	(.91)	945, 989
17q	D17S75	M	.39	943, 945, 993
17q	D17S79	Ps Pv	.60(.73)	48,945
17q	*D17S80	Pv	.30	1019
17q	D17S132	obsolete symbol; see D17S75		
17q	D17S231	E	.33(.45)	1409
17q	D17S232	Bg	.26(.29)	1409
17	ACTG	Ba O	.38	357, 358
17	MYL4	Ba	.19	1346
17	PENT	O	.34	550, 551
17	*D17S2	Bg Ba	.70	1206
17	D17S3	obsolete symbol; see D18S6		
17	D17S35	Ba M	.37(.60)	488
17	D17S36	Bg M	.41(.40)	323, 1281, 1334
17	D17S230	Bg	.40(.50)	1409
18p11.3	D18S3	M T	.37(.40)	863, 900, 1051
18p11	*D18S6	Ps T	.34(.42)	443, 1051
18q11.1-q11.2	D18S7	M	.37	299, 1051
18q11.2-q12.1	*PALB	M O	.38	1485, 1486
18q21	GRP	Pv	.37	304
18q21	SSAV1	Bg E	.37	757
18q21.3	BCL2	E	.36	1330
18q21.3	D18S8	M	.57	841
18q21.3-qter	*D18S5	Ps T	.37(.33)	1051, 1307
18q22-qter	*MBP	Ba Ps Pv	.11	137, 649
18q23	D18S11	H Ps O	.95(.96)	909, 1051
18	*DHFRP1	Ps	.38	14,15, 16
18	D18S1	T	.37(.57)	63, 1051
18	D18S10	Ps	.16	1470
18	D18S12	Ps	.55	323
18	D18S13	O	.43	488
18	D18S14	M	(.57)	323
18	D18S15	obsolete symbol; see D21S114		
18	D18S16	M		323
18	D18S17	Ps	.74(.74)	323, 687, 1203
18	D18S18	M	.31	323
18	*D18S19	Ps	.38(.48)	1051, 1496
18	*D18S20	obsolete symbol; see D18S22		
18	*D18S21	Pv	.37(.37)	559
18	D18S22	Pv	(.70)	707
18	*D18S23	Ps	.28	423
18	*D18S24	M	.19	199, 1051
18	D18S25	Ba	.36(.56)	1051
18	D18S27	Ps	(.86)	621
18	D18S29	O	.34	1496
19p13.2cen	*MEL	O	.32	1035
19p13.3-p13.2	*C3	Bg S T	.35	309, 450
19p13.3-p13.2	*INSR	Bg E H Ps Pv R S T O	.80	3,258, 350, 351, 781, 1090, 1173, 1220, 1223, 1279, 1301
19p13.2-p13.1	*LDLR	M Ps Pv R T O	.60	159, 426, 442, 528, 529, 530, 544, 545, 604, 703, 704, 705, 706, 758, 759, 941, 1095, 1287, 1302, 1476, 1477

Chromosome/ Region	Locus Symbol	Enzyme	PIC (HET')	Reference
19p13.2-q12	D19S27	O	.28	162
19p13.2-cen	*D19S11	Ba H M T	.37	175, 176, 185
19p13.1-q12	D19S30	M	.36	1245
19p12	D19S5	M Ps	.35	215
19p	D19S14	Bg T	.36	601
19cen-q12	*D19S7	M	.34	1222
19cen-q12	D19S29	M	.37	1246
19cen-q13.1	D19S13	Bg T	.34	600
19cen-q13.1	D19S18	E M	.36	1244
19q12-q13.1	D19S31	M	.12	1248
19q12-q13.2	*D19S9	E	.22	163, 1222
19q13.1	TGFB1	O	.36	25
19q13.1	*D19F11S1	E	.34	630
19q13.1	D19S28	T	.37	1247
19q13.1	D19S32	E	.34	1249
19q13.1-q13.2	ATP1A3	Ps S	.36	514
19q13.1-q13.2	BCL3	O	.36	513, 832
19q13.1-q13.2	CYP2A	S	.37	290, 599, 1365
19q13.1-q13.2	CYP2B	Ba M	.37	889
19q13.1-q13.2	D19S58	R	.37	752
19q13.2	*APOC1	O	.28	412, 413
19q13.2	*APOC2	T O	.79(.80)	54, 404, 408, 605, 699, 1191, 1250, 1371, 1406, 1427
19q13.2	*APOE	E O	.36	353, 676, 1243, 1410
19q13.2	*D19S8	M S T	.51	1221, 1222
19q13.2	D19S19	Bg M Ps	.34	72, 1157, 1368
19q13.2-q13.3	*CKMM	T O	.34	1034, 1098
19q13.2-q13.3	D19S15	T	.19	1193
19q13.2-q13.3	D19S16	M T	.35	832, 1192
19q13.3-qter	D19S6	T O	.37	1365, 1366
19q13.4	*PRKCG	M	.55	637
19q	D19S36	O	.35	700
19	TRSP	S	.23	868
19	D19S3	T	.11	1240
19	D19S17	E Pv T	.37	1235
19	D19S20	Ps O	(.83)	943, 964
19	D19S21	Ps	(.70)	943
19	*D19S22	Pv	.40	832, 963
19	*D19S24	Ba	.35(.53)	941
19	D19S25	O		836
19	D19S34	obsolete symbol; see D19S21		
19	D19S47	---PCR---	.69(.72)	1406
19	D19S48	---PCR---	.42(.39)	1406
19	D19S49	---PCR---	.71(.78)	1406
19	D19S50	T	.46	832
20pter-p12	PDYN	T	.33	799
20pter-p12	*PRIP	Pv	.16	1458

Chromosome/ Region	Locus Symbol	Enzyme	PIC (HET')	Reference
20p12	*D20S5	M Pv	.34	478, 479
20p12	*D20S6	T	.37	479
20p	D20S14	Ba	.37	482
20q13.11 or 20q13.2-qter	ADA	O	.27	490
20q13	*D20S8	R T	.37	801
20q13	D20S24	T	.49	1165
20q13.2	*D20S4	H M	.48	63, 479, 1040
20q13.3	D20S21	T	.36	1290
20q	*D20S19	T	(.87)	988
20q	*D20S20	T	.34	936
20	D20S15	Bg	.71(.71)	323, 687, 1203
20	D20S16	Bg	.98(.98)	687, 1203
20	D20S17	M	.37	323
20	D20S18	M	.35	323
20	D20S22	M Ps	.51	1290
20	D20S23	M	.24	1290
21pter-q21.1	D21S6	Ba	.07	1240, 1396
21pter-q21.1	*D21S26	Bg Ps	.56	890
21pter-q21.2	D21S149	M T	.41	1340
21p11.2	D21S5	M T O	.43	1396, 1425
21p	*D21S24	Ps	.34	890
21cen	D21Z1	Ba T	.22	837
21cen-q11	D21S111	S O	.40	1389, 1443
21q11.1	D21S52	Bg H	.56	1393
21q11.1	D21S59	T	.30	1393
21q11.2	D21S72	obsolete symbol; see D21S11		
21q11.2 or 21q21.2	D21S110	M	.30	723
21q11.2-q21	D21S1	Ba M	.71	674, 1396
21q11.2-q21	D21S11	E Pv T O	.65	480, 1272, 1396
21q11.2-q22.2	D21S4	Ps	.20	1396
21q11.2-q22.2	D21S12	T	.15	1100, 1240
21q21	D21S8	H	.37	1396
21q21	D21S54	M	.34	913, 1393
21q21	D21S58	Ps	.37	1393
21q21-q22.1	D21S22	M R O	.35	248, 249
21q21-q22.1	D21S95	O	.33	430
21q21-q22.1	D21S97	Ps	.23	430
21q21-q22	D21S154	R	.26	1165
21q21.2	APP	Bg E H Pv T O	.45	285, 429, 474, 1174, 1303, 1304, 1309, 1310, 1339, 1387, 1460
21q21.2	D21S16	O	.30	1272, 1284
21q21.2	*D21S13E	M Ps T	.49	287, 1284
21q21.2-qter	*D21S17	Bg	.37	1284
21q21.2-qter	D21S147	M	.37	1340
21q21.2-qter	D21S148	M	.30	1340

Chromosome/Region	Locus Symbol	Enzyme	PIC (HET)	Reference
21q22.1	SOD1	Bg M	.37	23, 127, 283, 674
21q22.1-q22.2	D21S25	H	.36	890
21q22.1-q22.2	D21S93	M	.34	430
21q22.1-q22.2	D21S144	M	.25	1340, 1341
21q22	*D21S113	M	(.50)	1018
21q22.1-q22.3	D21S143	T	.37	301
21q22.1-qter	D21S94	Pv O	.24	430
21q22.3	*BCEI	Ba	.30	627, 896, 897
21q22.3	ETS2	M		260
21q22.3	PFKL	O	.35	1388
21q22.3	D21S3	T O	.20	674, 1396
21q22.3	D21S15	M	.37	1284
21q22.3	D21S23	E M	.35	1396
21q22.3	D21S53	S O	.56	1272, 1393
21q22.3	D21S55	O	.64	1393
21q22.3	D21S56	E O	.61	1393
21q22.3	D21S57	T	.37	1393
21q22.3	D21S73	T	.35	306
21q22.3	D21S112	R	.93	323, 687, 1203
21q22.3-qter	D21S19	Ps	.27	1285
21	*HSPA3	O	.35	472
21	D21F24S4	see DNF24 at end of table		
21	D21S7	R	.36	1396
21	D21S82	Ba	.53	1470, 1471
21	D21S109	E	.39	1382
21	D21S114	M	.26(.26)	323
21	D21S150	O	.35	1158
22pter-q11	D22S3	Ba	.16	248
22pter-q11	D22S57	M O	.37	173
22q11	*BCR	Bg Ba H T	.34	1156, 1169
22q11.1-q11.2	*GGT1, GGT2	Pv	.37	1162
22q11	IGKVP3, IGKVP4, IGKVP5	O	.37	76
22q11.1-q11.2	*IGLC1, IGLC2, IGLC3, IGLC4, IGLC5, IGLC6	E H	.46	451, 645, 1218
22q11.1-q11.2	IGLV	H T O	.65	644, 645
22q11.1-q11.2	D22S9	M T	.33	872, 1218
22q11.1-q11.2	*D22S10	Ps T	.35	572
22q11	D22S43	M T	.57	173
22q11-q12	D22S33	M	.20	173
22q11-q12	D22S41	T	.36	173
22q11-q12	D22S46	Ba	.12	173
22q11-q12	D22S56	R	.37	173
22q11-qter	D22S5	Ba E	.14	248
22q11.2	D22S20	M	.33	1500
22q11.2-q12.1	CRYB2	Ps	.38	390
22q11.2-qter	MB	T	.36	643, 1412
22q11.2-qter	D22S1	Bg T	.37	63, 645, 1218
22q11.2-qter	D22S15	S O	.28	1163, 1164
22q11.2-qter	*D22S32	M	.36	708
22q11.2-qter	D22S35	Ps T	.36	173
22q11.2-qter	D22S38	E	.04	173
22q11.2-qter	D22S39	M	.22	173
22q11.2-qter	D22S40	H Pv	.37	173
22q11.2-qter	D22S44	M	.32	173
22q11.2-qter	D22S45	E Pv	.48	173
22q11.2-qter	D22S49	T	.22	173
22q11.2-qter	D22S52	Ps R	.22	173
22q11.2-qter	D22S53	Bg T	.36	173
22q11.2-qter	D22S54	M	.06	173
22q11.2-qter	D22S55	T	.46	173
22q11.2-qter	D22S58	Ba	.06	173
22q11.2-qter	D22S59	Ps Pv	.37	173
22q11.2-qter	D22S62	R	.19	173
22q11.2-qter	D22S64	Pv	.37	173
22q11.22-q11.23	F8VWFL	T	.21	839
22q12-q13	D22S85	T	.36	341
22q12-q13	D22S86	M	.28	341
22q12-q13	D22S87	M	.18	341
22q12-q13	D22S88	Bg	.32	341
22q12-q13	D22S89	Ps	.36	341
22q12-q13	D22S90	T	.38	341
22q12-q13	D22S91	M	.37	341
22q12-q13	D22S92	M	.37	341
22q12-q13	D22S93	M	.32	341
22q12-q13	D22S98	M O	.24	341
22q12-q13	D22S99	E Pv	.24	341
22q12-q13	D22S100	M	.24	341
22q12-q13	D22S101	M	.37	341
22q12-q13	D22S102	T O	.38	341
22q12-q13	D22S103	Ps O	.28(.20)	341
22q12-q13	D22S104	O	.31	341
22q12-q13	D22S105	O	.16	341
22q12-q13	D22S106	O	.18	341
22q12.3-q13.1	*PDGFB	H	.37	393, 644, 645
22q13	D22S80	M	.37	1502
22q13-qter	D22S82	T	.36	341
22q13-qter	D22S83	Ps	.18	341
22q13-qter	D22S84	Pv	.32	341
22q13-qter	D22S94	T	.28	341
22q13-qter	D22S95	O	.34	341
22q13-qter	D22S96	O	.38	341
22q13-qter	D22S97	Bg T	.37	341
22	HCF2	Ba	.37	108
22	D22S14	Ps	.37(.49)	398
22	D22S17	T	.33(.40)	323
22	D22S18	Bg	.36(.40)	323
22	D22S19	obsolete symbol; see D4S130		
22	D22S68	M	.38	173
22	D22Z1	Pv O		204
Xq11.2-q12	*DXS1	H T	.34	365, 1422
Xq26-Xqter	DXS311	T	.38	1259
Xq11.2-q13	DXS339	Ps	.49(.63)	60
Xq11.2-q13	DXS348	Bg	.44(.25)	60
Xpter-p21.1	DXS257	T	.48(.40)	308
Xpter-p21.1	DXS264	Bg	.21	308
Xpter-p21.1	DXS314	Ba	.27(.32)	60
Xpter-p21.1	DXS315	M	.48(.58)	60
Xpter-p21.1	DXS316	M	.15(.16)	60
Xpter-p21.1	DXS317	Bg Ba	.47(.52)	60
Xpter-p21.1	DXS318	T	.42(.44)	60
Xpter-p21.1	DXS319	Bg M	.49(.40)	60
Xpter-p21.1	DXS336	T	.11(.16)	60
Xpter-p21.1	DXS342	M	.57(.50)	60
Xpter-p21.1	DXS359	Bg	.11	60
Xpter-p11	DXS198	O	.18	1277
Xpter-p11.1	DXS361	M	.35(.30)	60
Xpter-p11.1	DXS365	M		60
Xpter-q26	HRASP	Ba		1482
Xp22.3	*DXS143	O	.49	880
Xp22.3	DXS237	H	.50	460, 1481

Chromosome/Region	Locus Symbol	Enzyme	PIC (HET)	Reference
Xp22.3	DXS277	Pv T		777
Xp22.3	DXYZ2X	E T		1099, 1166, 1230
Xp22.3-p22.2	*DXS85	E	.48	569
Xp22.3-p22.2	DXS89	H M T	.31	5,12, 715
Xp22	DXS2	O	.24	294
Xp22.3-p21	DXS104	Bg M	.06	569
Xp22-q26	DXS428	Ps	.37(.40)	135
Xp22.32; Yp11.3	*MIC2	M Pv T		483, 1075
Xp22.32	PABX	M	.50	352
Xp22.32	STS	O	.48	1435
Xp22.32	DXS31	O	.31	691, 1453
Xp22.32	DXS278	E	.90	688
Xp22.32; Yp11.3	DXYS14	E Ps		48, 247, 1075, 1167, 1417
Xp22.32; Yp11.3	DXYS15	Ps T		483, 1166, 1167, 1229, 1230
Xp22.32; Yp11.3	DXYS17	E T	.42	483, 1166, 1167, 1229
Xp22.32; Yp11.3	*DXYS20	T		1073, 1166
Xp22.32; Yp11.3	*DXYS28	E T	.56	1073
Xp22.32; Yp11.3	DXYS75	T	(.95)	1377
Xp22.2	*DXS9	T	.28	119, 337
Xp22.2	*DXS16	Bg M O	.50	44, 167, 286, 293, 1355
Xp22.2	*DXS43	Pv	.50	9, 1424
Xp22.2	DXS197	Bg	.38	1277
Xp22.2	DXS207	O	.50	4
Xp22.2-p22.1	DXS274	M	.50	323
Xp22.1	DXS41	Ps	.49	9,1424
Xp22.1	*DXS92	H T	.58	280
Xp22.1	DXS103	Ps	.50	391
Xp22.1-p21.3	ZFX	M	.50	320
Xp22.1-p21.2	GLR	O	.49	1227
Xp22.1-q26	DXS275	T	.42	323
Xp21.3	DXS28	O	.28	294, 395
Xp21.3	*DXS67	M O	.26	9, 1041, 1211
Xp21.3-p21.2	DXS235	M	.18	360
Xp21.3-p21.2	DXS268 (DMD)	Ps	.54	1343
Xp21.3-p21.1	*DMD	Bg H M Ps T O	.50	274, 275, 392, 690, 729, 876, 1109, 1131, 1359
Xp21.3-p21.1	DXS208	Ps	.49	517, 715
Xp21.2	*DXS164 (DMD)	Bg Ba M Ps T O	.51	721, 722, 898, 1147, 1204
Xp21.2	DXS269 (DMD)	M O	.48	1148, 1378
Xp21.2	*DXS270 (DMD)	Ba	.41	898
Xp21.1	CYBB	O	.40	74
Xp21.1	*OTC	Ba M T	.68	326, 394, 1168
Xp21.1	*DXS84	Bg E H Ps	.66	569, 570
Xp21.1	DXS141	M T	.38	222
Xp21.1	DXS142 (DMD)	T	.28	790
Xp21.1	*DXS148	M	.46	571
Xp21.1	DXS206 (DMD)	T O	.48	1129
Xp21.1-p11.2	DXS260	Ba	.27(.20)	308
Xp21.1-p11.2	DXS320	Bg Ba	.52	60
Xp21.1-p11.2	DXS321	Bg	.45(.54)	60
Xp21.1-p11.2	DXS322	M	.11(.12)	60
Xp21.1-p11.21	DXS337	T	.33(.20)	60
Xp21.1-p11.21	DXS338	M	.24(.27)	60
Xp21.1-p11.21	DXS352	Bg T	.46(.50)	60
Xp21.1-p11.21	DXS353	Bg H	.46(.43)	60
Xp21.1-p11.21	DXS357	Ps	.49	60
Xp21.1-p11.21	DXS426	---PCR---	(.52)	824
Xp11.4	DXS209	T	.10	481
Xp11.4-p11.3	*DXS7	E H T	.50	44, 1335
Xp11.4-p11.3	DXS77	O		107
Xp11-q13	DXS168	E	.38	838
Xp11.3-p11.23	MAOA	O	.46	1072
Xp11.3-p11.23	*TIMP	Bg	.45	345
Xp11.3-p11.21	OATL1	O	.48	1123
Xp11.22	DXS146	O	.46	694, 716
Xp11.22	DXS255	E	(.90)	397
Xp11.21	*DXS14	M	.46	1424
Xp11.21-p11.1	DXS323	Ba	.15(.16)	60
Xp11.21-p11.1	DXS324	M Ps	.40(.24)	60
Xp11.21-p11.1	DXS343	T	.13(.15)	60
Xcen	DXZ1	H O	.50	344, 1416, 1426
Xcen-q12	DXS18	M Pv	.04	574
Xq11-q13	DXS91	M	.23	279
Xq11.2-q12	DXS62	H	.18	1421
Xq12	*DXS106	Bg	.46	568
Xq12	DXS159	Ps	.44	30, 511
Xq12	AR	H	.18	168
Xq12	DXS132	O	.50	834
Xq12	DXS153	O	.34	834
Xq13.1	DXS135	T		567
Xq13.1	DXS162	M		567
Xq13	*PGK1	Ps O	.50	372, 607, 1242, 1355
Xq13	DXS325	Bg R	.15(.16)	60
Xq13	DXS347	Bg	.30(.18)	60
Xq13	DXS356	R	.24(.18)	60
Xq13-q21.1	DXS326	M	.47(.60)	60
Xq13-q21.1	DXS355	H	.28	60
Xq13-q21.2	DXS349	M	.51(.46)	60
Xq13-q22.1	DXS364	T	.49(.56)	60
Xq13-q24	DXYS30X	O	.16	892
Xq13-q24	DXYS31X	R O	.49	892
Xq13-q24	DXYS33X	O	.08	892
Xq13-q24	DXYS47X	obsolete symbol; see D8S75		

Chromosome/Region	Locus Symbol	Enzyme	PIC (HET)	Reference
Xq21.1	*DXS72	H	.50	1194
Xq21.1	DXS346	Bg	.47(.59)	60
Xq21.1-q21.2	DXS367	E	.50	879
Xq21	DXYS5X	T O	.50	939, 1209
Xq21.1-q22.1	DXS345	Bg	.38	60
Xq21.2-q21.3	DXS95	T	.18	277
Xq21.2-q22.1	DXS265	M	.49(.48)	308
Xq21.2-q22.1	DXS327	M	.42(.44)	60
Xq21.2-q22.1	DXS328	H Ps	.48(.24)	60
Xq21.2-q22.1	DXYS46X	T	.08	308
Xq21.2-q22.1	DXYS67X	Ps	.44(.44)	60
Xq21.2-q22.1	DXYS68X	M	.42	60
Xq21.2-q22	DXS262	Bg	.46(.24)	308
Xq21.2-q24	DXS366	T	.49(.47)	60
Xq21.2-qter	DXS362	T	.50(.56)	60
Xq21.31	*DXYS1X	T	.48	1074, 1076, 1370
Xq21.3	DXS3	M T	.47	9,726, 830, 877
Xq21.3	*DXYS2X	Ps	.44	689, 1209
Xq21.31-q21.33	DXYS12X	T	.35	689, 692, 1425
Xq21.3	DXYS13X	M O	.16	5,1425
Xq21.3-q22.1	DXS96	Ps	.42	277
Xq21.3-q22	GLA	S O	.23	302
Xq21.3-q22	PLP	M	.15	1457
Xq21.3-qter	*DXF3S2	S	.46	50
Xq21.33	DXS88	Bg	.34	828
Xq21.33-q22	*DXS87	Bg	.49	828, 1335
Xq21.33-q22	*DXS178	T	.42	245, 1425
Xq21.33-q27.3	DXS19	Pv T	.20	910
Xq22	DXS17	M T	.46	169, 337, 520, 726
Xq22	*DXS94	Ps	.50	281
Xq22	*DXS101	M	.46	568
Xq22-q24	DXS97	Ps	.18	277, 1425
Xq22-q24	*DXS287	R	.47	992
Xq22.1-q24	DXS350	M	.49(.33)	60
Xq22-qter	DXS424	---PCR---	(.79)	824
Xq22-qter	DXS425	---PCR---	(.79)	824
Xq22-qter	DXYS44X	Bg Ba	.56	308
Xq22-qter	DXYS45X	Bg Ps	.21	308
Xq22-qter	DXYS48X	obsolete symbol; see D8S74		
Xq23-qter	DXS330	M	.30	60
Xq24-q25	*DXS11	T	.27	9,84
Xq24-q25	DXS42	Bg O	.31	9, 1137, 1238, 1335
Xq24-q26	DXS8	M	.33	542
Xq24-q26	DXS58	Pv	.15	84
Xq24-q26	DXS261	T	.28(.39)	308
Xq24-q26	DXS267	M Ps	.30	308
Xq24-q26	DXS286	M	.10	248
Xq24-q26	DXS329	H	.23	60
Xq24-q26	DXS427	E	.15(.21)	135
Xq24-q27	DXS12	E	.31	1237
Xq24-qter	DXS358	H T	.45	60
Xq24-qter	DXS363	T	.42(.56)	60
Xq25	DXS100	Ps T	.27	910, 1454
Xq25-q26	DXS37	Pv T	.36	910

Chromosome/Region	Locus Symbol	Enzyme	PIC (HET)	Reference
Xq26	*HPRT	Ba O	.38	372, 459, 489, 1355
Xq26	*DXS10	T	.44	116
Xq26	DXS86	Bg	.40	1054, 1425
Xq26	*DXS107	T	.46	568
Xq26	*DXS177	E	.32	245
Xq26	DXS144E	T	.50	510
Xq26-q27.1	DXS99	S	.50	910, 911
Xq26-q27	*DXS79	O	.31	915
Xq26-qter	*DXF1S2	M	.68	50
Xq26-qter	DXS256	M	.47(.48)	308
Xq26-qter	DXS258	M	.04(.04)	308
Xq26-qter	DXS259	T	.23(.22)	308
Xq26-qter	DXS263	T	.26(.17)	308
Xq26-qter	DXS266	T	.44(.46)	308
Xq26-qter	*DXS294	T	.34	1294
Xq26-qter	DXS331	T	.18	60
Xq26-qter	DXS332	M	.29(.20)	60
Xq26-qter	DXS333	M	.47(.44)	60
Xq26-qter	DXS334	H T	.50(.50)	60
Xq26-qter	DXS335	T	.34	60
Xq26-qter	DXS340	H T	.50	60
Xq26-qter	DXS341	M Ps	.47	60
Xq26-qter	DXS344	T	.30(.30)	60
Xq26-qter	DXS351	Bg	.66(.68)	60
Xq26-qter	DXS354	Ps	.64(.45)	60
Xq26-qter	DXS360	M	.46	60
Xq26.2-q26.3	*DXS51	T	.50	288, 337, 458, 517
Xq26.2-q27.1	*DXS102	M T O	.49	28, 568, 1425
Xq26.3-q27.1	F9	Ba M T O	.51	189, 190, 401, 458, 517, 521, 684, 818, 1308, 1433
Xq27.1-q27.2	DXS105	E H Ps T O	.48(.40)	28, 568, 684, 1092, 1138, 1425
Xq27	CDR	R O	.38	1236
Xq27	DXS119	O	.08	28
Xq27	DXS152	O	.46	28
Xq27	DXS312	O	.45	912
Xq27-q28	DXF22S3	T	.38	57
Xq27-q28	*DXS90	M	.38	277, 517, 1425
Xq27.2	*DXS98	Bg M O	.31	116, 1202
Xq27.2-q27.3	DXS369	O	.48	606
Xq27.3-q28	DXS296	M T	.27	1293
Xq28	*F8C	H M O	.56	6, 21,98, 467, 1207, 1311, 1314, 1434, 1490

Chromosome/ Region	Locus Symbol	Enzyme	PIC (HET)	Reference
Xq28	GCP,RCP	S	.57	666, 1314
Xq28	*G6PD	Pv O	.61	1483
Xq28	*DXS15	Bg	.50	337, 520
Xq28	DXS33	O	.15	1093
Xq28	DXS52	M T O	.82(.84)	187, 484, 937, 1054, 1055, 1057, 1335
Xq28	DXS115	Ps O	.35(.38)	28, 1092, 1311
Xq28	DXS134	M	.49	567, 568, 684
Xq28	DXS304	T	.31	265
Xq28	DXS305	Ps T	.58	1354
Xq28	DXS374	Ps	.49	85, 1313
Xq	DXS250	T O	.49	377
X	*DXS108	E	.48	568, 1425

Chromosome/ Region	Locus Symbol	Enzyme	PIC (HET)	Reference
X	DXS273	M	.34(.50)	135, 323
X	DXYS23X	Ba		1209
X	DXYS43X	obsolete symbol; RFLP for probe QST-18 is now unassigned		
Yp11.3	PABY	see symbol PABX at Xp22.32		
Yp11	*DXYS1Y	M	.36	1076
Yp11	DXYS5Y	O	.32	939
Yp	DYZ5	O		1333
Yp	DXYS8Y	T		1033
Ycen	DYZ3			1416
Yq11	DYS1	T	.73	155, 498, 526, 821, 1031
Yq11	DYS11	E T	.36	207, 769, 770
Yq	DYZ8	T	.09	628
Y	DXYS43Y	obsolete symbol; RFLP for probe QST-18 is now unassigned		
Multilocus	DNF24	Ps	high	621

1 Abbott et al., Lancet i:1425-1426,1987.
2 Abbott et al., Lancet i:763-764,1988.
3 Accili et al., Nucleic Acids Res. 17:821,1989.
4 Ahrens and Kruse, Nucleic Acids Res. 14:7819,1986.
5 Ahrens et al., (HGM8) Cytogenet. Cell Genet. 40:567,1985.
6 Ahrens et al., Hum. Genet. 76:127-128,1987.
7 Ahti et al., Hum. Genet. 75:79-80,1987.
8 Aldred et al., Lancet ii:565,1988.
9 Aldridge et al., Am. J. Hum. Genet. 36:546-564,1984.
10 Alexandrov et al., (HGM10) Cytogenet. Cell Genet.
    51:949,1989.
11 Alitalo et al., (HGM10) Cytogenet. Cell Genet. 51:950,1989.
12 Alitalo et al., Am. J. Hum. Genet. 43:476-483,1988.
13 Allitto et al., (HGM10) Cytogenet. Cell Genet. 51:950,1989.
14 Anagnou et al., Am. J. Hum. Genet. 42:345-352,1988.
15 Anagnou et al., Nucleic Acids Res. 15:5501,1987.
16 Anagnou et al., Proc. Natl. Acad. Sci. USA
    81:5170-5174,1984.
17 Anathanassiadou et al., Brit. J. Haematol. 66:379-383,1987.
18 Anderson et al., (HGM10) Cytogenet. Cell Genet.
    51:951,1989.
19 Angert et al., Nucleic Acids Res. 17:2153,1989.
20 Antonarakis et al., Hum. Genet. 69:1-14,1985.
21 Antonarakis et al., N. Engl. J. Med. 313:842-848,1985.
22 Antonarakis et al., Proc. Natl. Acad. Sci. USA
    80:6615-6619,1983.
23 Antonarakis et al., Proc. Natl. Acad. Sci. USA
    82:3360-3364,1985.
24 Ardinger and Murray, Nucleic Acids Res. 16:8201,1988.
25 Ardinger et al., Nucleic Acids Res. 16:8202,1988.
26 Ardinger et al., Nucleic Acids Res. 16:8742,1988.
27 Arrand et al., Nucleic Acids Res. 15:9615,1987.
28 Arveiler et al., Am. J. Hum. Genet. 42:380-389,1988.
29 Arveiler et al., Nucleic Acids Res. 15:1880,1987.
30 Arveiler et al., Nucleic Acids Res. 15:5903,1987.
31 Arveiler et al., Nucleic Acids Res. 16:6252,1988.
32 Assum et al., Hum. Genet. 69:144-146,1985.
33 Atchison et al., Cytogenet. Cell Genet. 48:156-159,1988.
34 Atweh and Forget, Am. J. Hum. Genet. 38:855-859,1986.
35 Aulehla-Scholz et al., Hum. Genet. 78:353-355,1988.
36 Azuma et al., Nucleic Acids Res. 16:9372,1988.
37 Baas et al., Hum. Genet. 67:301-305,1984.
38 Baas et al., Hum. Genet. 69:138-143,1985.
39 Bader et al., (HGM10) Cytogenet. Cell Genet. 51:995,1989.
40 Bailey et al., Nucleic Acids Res. 15:6762,1987.
41 Baird et al., (HGM8) Cytogenet. Cell Genet. 40:573-574,1985.
42 Baird et al., Am. J. Hum. Genet. 39:489-501,1986.
43 Baird et al., Proc. Natl. Acad. Sci. USA 78:4218-4221,1981.
44 Bakker et al., Lancet i:655-658,1985.
45 Bakker et al., Nucleic Acids Res. 15:9100,1987.
46 Balazs et al., (HGM10) Cytogenet. Cell Genet. 51:956,1989.
47 Balazs et al., (HGM9) Cytogenet. Cell Genet. 46:574,1987.
48 Balazs et al., Am. J. Hum. Genet. 44:182-190, 1989.
49 Balazs et al., Hum. Genet. 68:57-61,1984.
50 Balazs et al., Somatic Cell Mol. Genet. 10:385-397,1984.
51 Bale et al., (HGM9) Cytogenet. Cell Genet. 46:575,1987.
52 Bale et al., Genomics 4:320-322,1989.
53 Bale et al., Nucleic Acids Res. 16:7755,1988.
54 Ball et al., Ann. Hum. Genet. 49:129-134,1985.
55 Ball et al., Immunogenetics 26:48-55,1987.
56 Ballantyne et al., Nucleic Acids Res. 16:1650,1988.
57 Bardoni et al., (HGM9) Cytogenet. Cell Genet. 46:575,1987.
58 Bardoni et al., Hum. Genet. 81:23-25,1988.
59 Barker et al., (HGM10) Cytogenet. Cell Genet. 51:957,1989.
60 Barker et al., (HGM10) Cytogenet. Cell Genet. 51:958,1989.
61 Barker et al., (HGM9) Cytogenet. Cell Genet. 46:576,1987.
62 Barker et al., Am. J. Hum. Genet. 36:1159-1171,1984.
63 Barker et al., Cell 36:131-138,1984.
64 Barker et al., Mol. Biol. Med. 1:199-206,1983.
65 Barker et al., Nucleic Acids Res. 17:826,1989.
66 Barker et al., Nucleic Acids Res. 17:827,1989.
67 Barker et al., Proc. Natl. Acad. Sci. USA 84:8006-8010,1987.
68 Barker et al., Science 236:1100-1102,1987.
69 Barni et al., Hum. Genet. 73:313-319,1986.
70 Bartels et al., (HGM8) Cytogenet. Cell Genet. 40:577,1985.
71 Bartels et al., Am. J. Hum. Genet. 38:280-287,1986.
72 Bartlett et al., Science 235:1648-1650,1987.
73 Bates and Williamson, Nucleic Acids Res. 15:864,1987.
74 Battat and Francke, Nucleic Acids Res. 17:3619,1989.
75 Battat and Francke, Nucleic Acids Res. 17:4005,1989.

76 Bauer et al., Biol. Chem. Hoppe-Seyler 367:751-755,1986.
77 Beaudet et al., (HGM8) Cytogenet. Cell Genet. 40:579,1985.
78 Beaudet et al., Am. J. Hum. Genet. 39:681-693,1986.
79 Bech-Hansen and Cox, Am. J. Hum. Genet. 38:67-74,1986.
80 Bech-Hansen et al., Nucleic Acids Res. 16:4183,1988.
81 Bech-Hansen et al., Nucleic Acids Res. 17:4004,1989.
82 Bech-Hansen et al., Proc. Natl. Acad. Sci. USA
    80:6952-6956,1983.
83 Beck et al., personal communication,1989.
84 Beckett et al., J. Neurogenet. 3:225-231,1986.
85 Bell et al., (HGM10) Cytogenet. Cell Genet. 51:960,1989.
86 Bell et al., Diabetes 33:176-183,1984.
87 Bell et al., Nature 284:26-32,1980.
88 Benham et al., (HGM8) Cytogenet. Cell Genet. 40:581,1985.
89 Benham et al., Mol. Biol. Med. 2:251-259,1984.
90 Bensi et al., EMBO J. 4:119-126,1985.
91 Bentley et al., Immunogenetics 22:377-390,1985.
92 Berdahl et al., Nucleic Acids Res. 16:2743,1988.
93 Berdahl et al., Nucleic Acids Res. 16:4740,1988.
94 Berdahl et al., personal communication,1989.
95 Berg et al., Proc. Natl. Acad. Sci. USA 83:7367-7370,1986.
96 Berg, Clin. Genet. 30:515-520,1986.
97 Berliner et al., J. Clin. Invest. 76:1283-1285,1985.
98 Bernardi et al., Hum. Genet. 78:359-362,1988.
99 Bernardi et al., Nucleic Acids Res. 15:1347,1987.
100 Bikker et al., Nucleic Acids Res. 16:8198,1988.
101 Biunno et al., Nucleic Acids Res. 16:7753,1988.
102 Blanche et al., (HGM10) Cytogenet. Cell Genet. 51:963,1989.
103 Blanche et al., (HGM10) Cytogenet. Cell Genet. 51:963,1989.
104 Blanche et al., Nucleic Acids Res. 15:3941,1987.
105 Blanche et al., Nucleic Acids Res. 15:5902,1987.
106 Blanche et al., Nucleic Acids Res. 16:1652,1988.
107 Bleeker-Wagemakers et al., (HGM10) Cytogenet. Cell Genet.
    51:964,1989.
108 Blinder et al., Biochem. 27:752-759,1988.
109 Board et al., Am. J. Hum. Genet. 42:712-717,1988.
110 Bock and Levitan, Nucleic Acids Res. 24:8569-8582,1983.
111 Bock and Prochownik, Blood 70:1273-1278,1987.
112 Bock et al., Biochem. 25:4292-4301,1986.
113 Bodmer et al., Nature 328:614-618,1987.
114 Boerwinkle and Chan, Nucleic Acids Res. 17:4003,1989.
115 Boerwinkle et al., Proc. Natl. Acad. Sci. USA
    86:212-216,1989.
116 Boggs and Nussbaum, Somatic Cell Mol. Genet.
    10:607-613,1984.
117 Boime et al., ICSU Short Reports 4:204-207,1986.
118 Bookstein et al., Proc. Natl. Acad. Sci. USA
    85:2210-2214,1988.
119 Borresen et al., Clin. Genet. 27:411-413,1985.
120 Borresen et al., Hum. Genet. 78:216-221,1988.
121 Borresen et al., Medical Genetics: Past, Present, Future,
    pp. 37-51, Alan R. Liss, Inc., 1985.
122 Borresen et al., Mut. Res. 202:77-83,1988.
123 Bowcock and Hebert, Nucleic Acids Res. 17:1787,1989.
124 Bowcock et al., (HGM10) Cytogenet. Cell Genet. 51:966,1989.
125 Bowcock et al., Cytogenet. Cell Genet. 44:236-237,1987.
126 Bowcock et al., Cytogenet. Cell Genet. 45:234-236,1987.
127 Bowcock et al., Gene Geography 1:47-64,1987.
128 Bowcock et al., Genomics 3:8-16,1988.
129 Bowcock et al., Nucleic Acids Res. 14:7817,1986.
130 Bowcock et al., Nucleic Acids Res. 15:382,1987.
131 Bowcock et al., Nucleic Acids Res. 16:2745,1988.
132 Bowcock et al., Proc. Natl. Acad. Sci. USA
    85:2701-2705,1988.
133 Bowcock, personal communication,1989.
134 Bowden et al., (HGM10) Cytogenet. Cell Genet. 51:967,1989.
135 Bowden et al., Am. J. Hum. Genet. 44:671-678,1989.
136 Boyd et al., Hum. Genet. 73:171-174,1986.
137 Boylan et al., Am. J. Hum. Genet. 40:387-400,1987.
138 Bragg et al., Nucleic Acids Res. 15:10072,1987.
139 Bragg et al., Nucleic Acids Res. 16:10406,1988.
140 Bragg et al., Nucleic Acids Res. 16:11390,1988.
141 Bragg et al., Nucleic Acids Res. 16:11391,1988.
142 Bragg et al., Nucleic Acids Res. 16:11392,1988.
143 Bragg et al., Nucleic Acids Res. 16:11393,1988.
144 Bragg et al., Nucleic Acids Res. 16:11394,1988.
145 Bragg et al., Nucleic Acids Res. 16:11395,1988.
146 Bragg et al., Nucleic Acids Res. 16:4185,1988.
147 Brauch et al., N. Engl. J. Med. 317:1109-1113,1987.

148 Breakefield et al., Biochemical and Clinical Aspects of Neuropeptides: Synthesis, Processing, and Gene Structure, pp. 113-128, Acad. Press, NY, 1983.
149 Breakefield et al., Mol. Biol. Med. 3:483-494,1986.
150 Brebner et al., Hum. Genet. 70:25-27,1985.
151 Brennan et al., Nucleic Acids Res. 14:9224,1986.
152 Brennan et al., Nucleic Acids Res. 14:9225,1986.
153 Brennan et al., Nucleic Acids Res. 15:1341,1987.
154 Brennan et al., Nucleic Acids Res. 15:1342,1987.
155 Breuil et al., Ann. Genet. 30:209-212,1987.
156 Breuning et al., (HGM10) Cytogenet. Cell Genet. 51:969,1989.
157 Breuning et al., Lancet ii:1359-1361,1987.
158 Breuning et al., Nucleic Acids Res. 17:5872,1989.
159 Brink et al., Hum. Genet. 77:32-35,1987.
160 Brissenden et al., Genet. Epidemiol. 3:231-239,1986.
161 Brocas et al., Hum. Genet. 74:178-180,1986.
162 Brook and Shaw, Nucleic Acids Res. 16:7751,1988.
163 Brook et al., Hum. Genet. 68:282-285,1984.
164 Brookes et al., Nucleic Acids Res. 17:1792,1989.
165 Brooks et al., Am. J. Hum. Genet. 43:280-284,1988.
166 Brotherton et al., Nucleic Acids Res. 17:1274,1989.
167 Brown and Willard, Nucleic Acids Res. 15:9614,1987.
168 Brown et al., Am. J. Hum. Genet. 44:264-269,1989.
169 Brunner et al., Hum. Genet. 80:337-340,1988.
170 Bruns et al., (HGM9) Cytogenet. Cell Genet. 46:588,1987.
171 Bruns et al., Am. J. Hum. Genet. 36:25S,1984.
172 Buchman et al., Gene 70:245-252,1988.
173 Budarf et al., Am. J. Hum. Genet. 43:A139,1988.
174 Buetow et al., (HGM10) Cytogenet. Cell Genet. 51:973,1989.
175 Bufton et al., (HGM8) Cytogenet. Cell Genet. 40:596,1985.
176 Bufton et al., Am. J. Hum. Genet. 38:447-460,1986.
177 Bufton et al., Hum. Genet. 74:425-431,1986.
178 Bugawan et al., J. Immunol. 141:4024-4030,1988.
179 Buraczynska et al., Hum. Genet. 74:165-167,1986.
180 Buresi et al., Immunogenetics 29:161-172,1989.
181 Burk et al., Nucleic Acids Res. 16:7216,1988.
182 Buroker and Litt, Genetics 110:s15,1985.
183 Buroker et al., Genetics (suppl.) s64,1980.
184 Buroker et al., Hum. Genet. 72:86-94,1986.
185 Buroker et al., Hum. Genet. 76:90-95,1987.
186 Buroker et al., Nucleic Acids Res. 15:9097,1987.
187 Caccia et al., J. Exp. Med. 161:1255-1260,1985.
188 Callen, Ann. Genet. 29:235-239,1986.
189 Camerino et al., Hum. Genet. 71:79-81,1985.
190 Camerino et al., Proc. Natl. Acad. Sci. USA 81:498-502,1984.
191 Carlock et al., Nucleic Acids Res. 15:377,1987.
192 Carlson et al., Nucleic Acids Res. 15:10613,1987.
193 Carlson et al., Nucleic Acids Res. 15:9623,1987.
194 Carlson et al., Nucleic Acids Res. 16:10403,1988.
195 Carlson et al., Nucleic Acids Res. 16:10941,1988.
196 Carlson et al., Nucleic Acids Res. 16:1225,1988.
197 Carlson et al., Nucleic Acids Res. 16:1226,1988.
198 Carlson et al., Nucleic Acids Res. 16:378,1988.
199 Carlson et al., Nucleic Acids Res. 16:4188,1988.
200 Carlson et al., Nucleic Acids Res. 16:4744,1988.
201 Carlson et al., Nucleic Acids Res. 16:4745,1988.
202 Carlson et al., Nucleic Acids Res. 16:783,1988.
203 Carlson et al., Nucleic Acids Res. 16:9358,1988.
204 Carritt and Welch, (HGM8) Cytogenet. Cell Genet. 40:599,1985.
205 Carritt et al., (HGM8) Cytogenet. Cell Genet. 40:599-600,1985.
206 Carritt, Hum. Genet. 72:83-85,1986.
207 Casanova et al., Science 230:1403-1406,1985.
208 Casey et al., Mol. Cell. Biol. 6:502-510,1986.
209 Castagnoli et al., Nucleic Acids Res. 15:866,1987.
210 Cavenee et al., Am. J. Hum. Genet. 36:10-24,1984.
211 Chaabani et al., Hum. Genet. 73:110-113,1986.
212 Chakraborty et al., Hum. Genet. 76:40-46,1987.
213 Chakravarti et al., Proc. Natl. Acad. Sci. USA 81:6085-6089,1984.
214 Chakravarti et al., Proc. Natl. Acad. Sci. USA 83:1045-1049,1986.
215 Chamberlain et al., Hum. Genet. 76:186-190,1987.
216 Chan et al., Biochem. Biophys. Res. Comm. 133:248-255,1985.
217 Chandler et al., Cell 50:711-717,1987.
218 Charmley et al., Nucleic Acids Res. 17:2372,1989.
219 Charmley et al., Nucleic Acids Res. 17:2374,1989.
220 Chen and Board, Nucleic Acids Res. 15:6306,1987.
221 Chen and Board, Nucleic Acids Res. 16:8199,1988.

222 Chen et al., (HGM10) Cytogenet. Cell Genet. 51:976,1989.
223 Cherney et al., Proc. Natl. Acad. Sci. USA 84:8370-8374,1987.
224 Chin et al., Nucleic Acids Res. 16:1645,1988.
225 Choi et al., Nucleic Acids Res. 15:6305,1987.
226 Chuchana et al., Nucleic Acids Res. 17:1275,1989.
227 Chuchana et al., Nucleic Acids Res. 17:3622,1989.
228 Civeira et al., Nucleic Acids Res. 17:6428,1989.
229 Cocozza et al., Nucleic Acids Res. 16:2737,1988.
230 Cohen et al., Nucleic Acids Res. 14:1924,1986.
231 Cohen-Haguenauer et al., (HGM9) Cytogenet. Cell Genet. 46:596,1987.
232 Cohen-Haguenauer et al., Int. Cong. Hum. Genet. 7:617,1986.
233 Coleman et al., Mol. Biol. Med. 3:213-228,1986.
234 Coleman et al., Nucleic Acids Res. 14:7818,1986.
235 Coleman et al., Nucleic Acids Res. 16:1221,1988.
236 Coleman et al., Nucleic Acids Res. 16:1222,1988.
237 Coleman et al., Nucleic Acids Res. 16:2364,1988.
238 Coleman et al., Nucleic Acids Res. 16:7208, 1988.
239 Collins et al., Science 235:1046-1049,1987.
240 Colombi et al., Nucleic Acids Res. 15:6761,1987.
241 Colombi et al., Nucleic Acids Res. 16:9074,1988.
242 Concannon et al., J. Exp. Med. 165:1130-1140,1987.
243 Connor et al., J. Med. Genet. 24:544-546,1987.
244 Cook et al., Biochem. Soc. Trans. 16:541-542,1988.
245 Cooke et al., (HGM8) Cytogenet. Cell Genet. 40:607,1985.
246 Cooke et al., Hum. Genet. 73:225-229,1986.
247 Cooke et al., Nature 317:687-692,1985.
248 Cooper et al., Hum. Genet. 69:201-205,1985.
249 Cooper et al., Hum. Genet. 75:129-135,1987.
250 Coppin et al., Proc. Natl. Acad. Sci. USA 82:8614-8618,1985.
251 Coutelle et al., (HGM10) Cytogenet. Cell Genet. 51:979,1989.
252 Cox and Coulson, Nucleic Acids Res. 15:4701,1987.
253 Cox et al., (HGM10) Cytogenet. Cell Genet. 51:980,1989.
254 Cox et al., (HGM9) Cytogenet. Cell Genet. 46:599,1987.
255 Cox et al., Am. J. Hum. Genet. 41:891-906,1987.
256 Cox et al., Am. J. Hum. Genet. 43:495-501,1988.
257 Cox et al., Nature 316:79-81,1985.
258 Cox et al., Nucleic Acids Res. 16:8204,1988.
259 Cragg et al., Hum. Genet. 80:63-68,1988.
260 Creau-Goldberg et al., Hum. Genet. 76:396-398,1987.
261 Cross et al., Immunogenetics 21:39-48,1985.
262 Culver et al., Nucleic Acids Res. 15:10074,1987.
263 Curtis et al., Hum. Genet. 81:188-190,1989.
264 Dahl et al., Nucleic Acids Res. 15:1921-1932,1987.
265 Dahl et al., Nucleic Acids Res. 17:2884,1989.
266 Daiger et al., Am. J. Hum. Genet. 36:736-749,1984.
267 Dalgleish et al., Hum. Genet. 73:91-92,1986.
268 Dalgleish et al., J. Med. Genet. 24:148-151,1987.
269 Dalgleish et al., Nucleic Acids Res. 13:4609,1985.
270 Dalgleish, Hum. Genet. 78:109,1988.
271 Darby et al., Am. J. Hum. Genet. 37:52-59,1985.
272 Darby et al., Nucleic Acids Res. 14:9543,1986.
273 Darnfors et al., Nucleic Acids Res. 14:7135,1986.
274 Darras and Francke, Am. J. Hum. Genet. 43:612-619,1988.
275 Darras et al., Am. J. Med. Genet. 29:713-726,1988.
276 Datta et al., J. Biol. Chem. 263:1107-1110,1988.
277 Davatelis et al., (HGM8) Cytogenet. Cell Genet. 40:611,1985.
278 Davatelis et al., (HGM8) Cytogenet. Cell Genet. 40:612,1985.
279 Davatelis et al., Nucleic Acids Res. 13:7539,1985.
280 Davatelis et al., Nucleic Acids Res. 13:7540,1985.
281 Davatelis et al., Nucleic Acids Res. 15:4694,1987.
282 Davatelis, Am. J. Hum. Genet. 37:1015-1021,1985.
283 David et al., C. R. Acad. Sci. Paris, Ser. III 306:1-4,1988.
284 David et al., Nucleic Acids Res. 15:374,1987.
285 David et al., Nucleic Acids Res. 15:9103,1987.
286 Davies et al., (HGM8) Cytogenet. Cell Genet. 40:613,1985.
287 Davies et al., Hum. Genet. 66:54-56,1984.
288 Davies et al., Hum. Genet. 70:249-255,1985.
289 Davies et al., Hum. Genet. 77:122-126,1987.
290 Davis, Ann. Hum. Genet. 51:9-12,1987.
291 Davison et al., (HGM9) Cytogenet. Cell Genet. 46:604-605,1987.
292 De Benedetti et al., (HGM10) Cytogenet. Cell Genet. 51:984,1989.
293 de Martinville and Uhrhammer, Nucleic Acids Res. 16:10949,1988.
294 de Martinville et al., Am. J. Hum. Genet. 37:235-249,1985.
295 de Miguel et al., Nucleic Acids Res. 16:1644,1988.

296 De Qi Xu et al., Proc. Natl. Acad. Sci. USA
    82:2862-2865,1985.
297 Dean et al., (HGM10) Cytogenet. Cell Genet. 51:985,1989.
298 Dean et al., J. Pediatr. 111:490-495,1987.
299 Delattre et al., Nucleic Acids Res. 15:1343,1987.
300 Delattre et al., Nucleic Acids Res. 17:1789,1989.
301 Delattre et al., Nucleic Acids Res. 17:1790,1989.
302 Desnick et al., Enzyme 38:54-64,1987.
303 Detera-Wadleigh et al., Nature 325:806-808,1987.
304 Detera-Wadleigh et al., Nucleic Acids Res. 15:375,1987.
305 Detera-Wadleigh et al., Nucleic Acids Res. 17:6432,1989.
306 Devin-Gage et al., Nucleic Acids Res. 13:7909,1985.
307 Dietrich et al., Nucleic Acids Res. 17:1273,1989.
308 Dietz Band et al., (HGM9) Cytogenet. Cell Genet.
    46:606,1987.
309 Dietz et al., Genet. Epidemiol. 3:313-321,1986.
310 Dietz-Band et al., (HGM10) Cytogenet. Cell Genet.
    51:989,1989.
311 Dietzsch et al., Nucleic Acids Res. 14:1923,1986.
312 Dietzsch et al., Nucleic Acids Res. 14:6780,1986.
313 Dietzsch et al., Nucleic Acids Res. 14:6781,1986.
314 Dietzsch et al., Nucleic Acids Res. 14:8698,1986.
315 Dietzsch et al., Nucleic Acids Res. 15:2400,1987.
316 Dietzsch et al., Nucleic Acids Res. 15:5907,1987.
317 Dietzsch et al., Nucleic Acids Res. 15:5908,1987.
318 Dietzsch et al., Nucleic Acids Res. 15:861,1987.
319 DiLella et al., Biochem. 25:743-749,1985.
320 Disteche et al., (HGM10) Cytogenet. Cell Genet. 51:989,1989.
321 Disteche et al., (HGM10) Cytogenet. Cell Genet. 51:990,1989.
322 Dobrovic et al., Nucleic Acids Res. 15:1346,1987.
323 Donis-Keller et al., Cell 51:319-337,1987.
324 Donis-Keller, personal communication,1988.
325 Donohue et al., Biochem. Biophys. Res. Comm.
    136:722-729,1986.
326 Dorkins et al., Hum. Genet. 71:103-107,1985.
327 Douglas et al., Nucleic Acids Res. 16:9067,1988.
328 Dozier et al., Nucleic Acids Res. 14:1928,1986.
329 Dracopoli et al., (HGM9) Cytogenet. Cell Genet. 46:608,1987.
330 Dracopoli et al., Am. J. Hum. Genet. 43:462-470,1988.
331 Dracopoli et al., Genomics 3:124-128,1988.
332 Dracopoli et al., Genomics 3:161-167,1988.
333 Dracopoli et al., Proc. Natl. Acad. Sci. USA
    83:1822-1826,1986.
334 Drayna and Lawn, Nucleic Acids Res. 15:4698,1987.
335 Drayna et al., J. Biol. Chem. 261:16535-16539,1986.
336 Drayna et al., Nucleic Acids Res. 15:9617,1987.
337 Drayna et al., Proc. Natl. Acad. Sci. USA 81:2836-2839,1984.
338 Driesel et al., (HGM8) Cytogenet. Cell Genet. 40:620,1985.
339 Driscoll et al., J. Clin. Invest. 68:915-919,1981.
340 Dryja et al., Hum. Genet. 65:320-324,1984.
341 Dumanski et al., (HGM10) Cytogenet. Cell Genet.
    51:993,1989.
342 Dunning et al., Clin. Genet. 33:181-188,1988.
343 Dunning et al., Hum. Genet. 78:325-329,1988.
344 Durfy and Willard, Am. J. Hum. Genet. 41:391-401,1987.
345 Durfy et al., Nucleic Acids Res. 14:9226,1986.
346 Egeland et al., Nature 325:783-787,1987.
347 El-Hazmi and Warsy, Hum. Hered. 37:237-240,1987.
348 Elbein et al., Diabetes 34:1139-1144,1985.
349 Elbein et al., Diabetes 34:433-439,1985.
350 Elbein et al., Proc. Natl. Acad. Sci. USA 83:5223-5227,1986.
351 Elbein, personal communication,1989.
352 Ellis et al., (HGM10) Cytogenet. Cell Genet. 51:994,1989.
353 Emi et al., Genomics 3:373-379,1988.
354 Emrie and Fisher, Nucleic Acids Res. 14:1919,1986.
355 Emrie et al., Somatic Cell Mol. Genet. 14:105-110,1988.
356 Eng and Strom, Am. J. Hum. Genet. 37:719-732,1985.
357 Erba et al., Mol. Cell. Biol. 8:1775-1789,1988.
358 Erba et al., Nucleic Acids Res. 14:5275-5294,1986.
359 Erlich and Bugawan, In: PCR Technology: Principles and
    Applications. H. Erlich (ed.), Stockton Publishing,
    (in press), 1989.
360 Estivill and Scambler, Nucleic Acids Res. 16:2740,1988.
361 Estivill et al., Genomics 1:257-263,1987.
362 Estivill et al., Hum. Genet. 74:320-322,1986.
363 Estivill et al., Nature 326:840-845,1987.
364 Ewerhardt et al., Nucleic Acids Res. 17:5416,1989.
365 Fadda et al., Nucleic Acids Res. 15:4695,1987.
366 Fain et al., Genomics 1:340-345,1987.
367 Falk et al., Am. J. Hum. Genet. 38:269-279,1986.

368 Faraco et al., Nucleic Acids Res. 17:2150,1989.
369 Farber et al., Nucleic Acids Res. 16:2360,1988.
370 Farrall et al., Am. J. Hum. Genet. 41:286-287,1987.
371 Farrer et al., Genomics 3:72-77,1988.
372 Fearon et al., N. Engl. J. Med. 316:427-431,1987.
373 Feder et al., Am. J. Hum. Genet. 35:1090-1096,1983.
374 Feder et al., Am. J. Hum. Genet. 37:286-294,1985.
375 Feder et al., Am. J. Hum. Genet. 37:635-649,1985.
376 Feder et al., Nucleic Acids Res. 15:5906,1987.
377 Feil et al., Nucleic Acids Res. 17:1279,1989.
378 Feldenzer et al., J. Clin. Invest. 64:751-755,1979.
379 Feldman et al., Lancet ii:102,1988.
380 Felice et al., Blood 63:1253-1257,1984.
381 Ferns et al., Hum. Genet. 74:302-306,1986.
382 Ferns et al., Hum. Genet. 81:76-80,1988.
383 Field et al., Nucleic Acids Res. 15:3942,1987.
384 Finkelstein et al., (HGM10) Cytogenet. Cell Genet.
    51:998,1989.
385 Fisher et al., Am. J. Hum. Genet. 40:503-511,1987.
386 Fisher et al., Am. J. Hum. Genet. 43:436-441,1988.
387 Fisher et al., Nucleic Acids Res. 15:7657,1988.
388 Fletcher et al., Nucleic Acids Res. 15:7650,1987.
389 Florian et al., (HGM10) Cytogenet. Cell Genet. 51:999,1989.
390 Fontaine et al., (HGM10) Cytogenet. Cell Genet. 51:999,1989.
391 Forrest et al., (HGM8) Cytogenet. Cell Genet. 40:631,1985.
392 Forrest et al., Lancet ii:1294-1297,1987.
393 Fourney et al., Nucleic Acids Res. 16:8197,1988.
394 Fox et al., Am. J. Hum. Genet. 38:841-847,1986.
395 Francke et al., Am. J. Hum. Genet. 37:250-267,1985.
396 Francomano et al., Genomics 1:293-296,1987.
397 Fraser et al., Nucleic Acids Res. 15:9616,1987.
398 Fratini et al., Nucleic Acids Res. 16:9064,1988.
399 Fratini et al., Nucleic Acids Res. 16:9065,1988.
400 Fratini et al., Nucleic Acids Res. 16:9066,1988.
401 Freedenberg et al., Hum. Genet. 76:262-264,1987.
402 Freeman et al., Nucleic Acids Res. 17:2880,1989.
403 Frossard and Coleman, Nucleic Acids Res. 14:9223,1986.
404 Frossard et al., Gene 51:103-106,1987.
405 Frossard et al., Nucleic Acids Res. 14:1922,1986.
406 Frossard et al., Nucleic Acids Res. 14:4373,1986.
407 Frossard et al., Nucleic Acids Res. 14:4380,1986.
408 Frossard et al., Nucleic Acids Res. 14:5120,1986.
409 Frossard et al., Nucleic Acids Res., 14:5121,1986.
410 Frossard et al., Nucleic Acids Res. 14:6778,1986.
411 Frossard et al., Nucleic Acids Res. 14:8699,1986.
412 Frossard et al., Nucleic Acids Res. 15:1344,1987.
413 Frossard et al., Nucleic Acids Res. 15:1884,1987.
414 Frossard et al., Nucleic Acids Res. 15:381,1987.
415 Frossard et al., Nucleic Acids Res. 15:7656,1987.
416 Fujimoto et al., Nucleic Acids Res. 15:10078,1987.
417 Fujimoto et al., Nucleic Acids Res. 16:10942,1988.
418 Fujimoto et al., Nucleic Acids Res. 16:10943,1988.
419 Fujimoto et al., Nucleic Acids Res. 16:10944,1988.
420 Fujimoto et al., Nucleic Acids Res. 16:10946,1988.
421 Fujimoto et al., Nucleic Acids Res. 16:3110,1988.
422 Fujimoto et al., Nucleic Acids Res. 16:380,1988.
423 Fujimoto et al., Nucleic Acids Res. 16:4748,1988.
424 Fujimoto et al., Nucleic Acids Res. 16:6256,1988.
425 Fujimoto et al., Nucleic Acids Res. 16:9357,1988.
426 Funke et al., Nucleic Acids Res. 14:7820,1986.
427 Funke et al., Nucleic Acids Res. 15:9102,1987.
428 Funke et al., Nucleic Acids Res. 16:2741,1988.
429 Furuya et al., Biochem. Biophys. Res. Comm. 150:75-81,1988.
430 Galt et al., Hum. Genet. 81:113-119,1989.
431 Gardella et al., Nucleic Acids Res. 16:11388,1988.
432 Gardella et al., Nucleic Acids Res. 16:11388,1988.
433 Gardella et al., Nucleic Acids Res. 16:1651,1988.
434 Gareau et al., Nucleic Acids Res. 16:1223,1988.
435 Gasparini et al., Prenat. Diag. 9:349-355,1989.
436 Gatti et al., Am. J. Hum. Genet. 41:654-667,1987.
437 Gatti et al., Hum. Immunol. 22:145-150,1988.
438 Gatti et al., Nucleic Acids Res. 15:8119,1987.
439 Gedde-Dahl et al., (HGM8) Cytogenet. Cell Genet.
    40:637,1985.
440 Gedeon et al., Nucleic Acids Res. 17:2146,1989.
441 Gedeon et al., Nucleic Acids Res., in press, 1989.
442 Geisel et al., Nucleic Acids Res. 15:3943,1987.
443 Geitvik et al., Int. Cong. Hum. Genet. 7:687,1986.
444 Gendler et al., Proc. Natl. Acad. Sci. USA
    84:6060-6064,1987.

445 Georgiou et al., personal communication,1989.
446 Gerber et al., Am. J. Hum. Genet. 43:442-451,1988.
447 Gerhard and Housman, (HGM8) Cytogenet. Cell Genet.
    40:640,1985.
448 Gerhard et al., (HGM9) Cytogenet. Cell Genet. 46:619,1987.
449 Ghanem et al., Eur. J. Immunol. 18:1059-1065,1988.
450 Ghanem et al., Exp. Clin. Immunogenet. 4:222-226,1987.
451 Ghanem et al., Exp. Clin. Immunogenet. 5:186-195,1988.
452 Ghanem et al., Exp. Clin. Immunogenet. 6:39-54,1988.
453 Ghanem et al., Nucleic Acids Res. 17:1270,1989.
454 Ghanem et al., Nucleic Acids Res. 17:1271,1989.
455 Ghanem et al., Nucleic Acids Res. 17:471,1989.
456 Ghosh-Choudhury et al., (HGM10) Cytogenet. Cell Genet.
    51:1004,1989.
457 Giannakudis et al., (HGM9) Cytogenet. Cell Genet.
    46:619-620,1987.
458 Giannelli et al., Ann. Hum. Genet. 51:107-124,1987.
459 Gibbs et al., J. Inher. Metab. Dis. 9:45-58,1986.
460 Gillard et al., Nucleic Acids Res. 15:3977-3985,1987.
461 Gilliam et al., (HGM8) Cytogenet. Cell Genet. 40:641,1985.
462 Gilliam et al., Cell 50:565-571,1987.
463 Gilliam et al., Hum. Genet. 68:154-158,1984.
464 Gilliam et al., Nucleic Acids Res. 15:1445-1458,1987.
465 Gilliam et al., Science 238:950-952,1987.
466 Gilman et al., Ann. N. Y. Acad. Sci. 445:235-247,1985.
467 Gitschier et al., Nature 314:738-740,1985.
468 Glaser et al., (HGM10) Cytogenet. Cell Genet. 51:1005,1989.
469 Glaser et al., (HGM8) Cytogenet. Cell Genet. 40:642,1985.
470 Glaser et al., Nature 321:882-887,1986.
471 Glenn et al., (HGM10) Cytogenet. Cell Genet. 51:1005,1989.
472 Goate et al., Hum. Genet. 75:123-128,1987.
473 Goldfarb et al., Nature 296:404-409,1982.
474 Goldgaber et al., Science 235:877-880,1987.
475 Goldgar et al., Am. J. Hum. Genet. 44:6-12,1989.
476 Goldstein et al., Nucleic Acids Res. 14:5570,1986.
477 Goodbourn et al., Mol. Biol. Med. 2:223-238,1984.
478 Goodfellow et al., Am. J. Hum. Genet. 37:890-897,1985.
479 Goodfellow et al., Cytogenet. Cell Genet. 44:112-117,1987.
480 Goodfellow et al., Nucleic Acids Res. 14:4375,1985.
481 Goodfellow et al., Nucleic Acids Res. 14:8693,1986.
482 Goodfellow et al., Nucleic Acids Res. 15:7213,1987.
483 Goodfellow et al., Science 234:740-743,1986.
484 Goonewardena et al., Clin. Genet. 30:249-254,1986.
485 Gough et al., Nucleic Acids Res. 17:4426,1989.
486 Grandy et al., (HGM10) Cytogenet. Cell Genet. 51:1007,1989.
487 Graziano et al., Oncogene Res. 3:99-103,1988.
488 Green et al., (HGM9) Cytogenet. Cell Genet. 46:623,1987.
489 Greer et al., Genomics 4:60-67,1989.
490 Gribbin et al., Nucleic Acids Res. 17:3626,1989.
491 Griffin et al., Proc. Natl. Acad. Sci. USA
    83:6122-6126,1986.
492 Griffiths et al., (HGM10) Cytogenet. Cell Genet.
    51:1008,1989.
493 Griffiths et al., Am. J. Hum. Genet. 42:756-771,1988.
494 Griffiths et al., Cytogenet. Cell Genet. 45:231-233,1987.
495 Griffiths et al., Nucleic Acids Res. 16:7752,1988.
496 Grobler-Rabie et al., EMBO J. 4:1745-1748,1985.
497 Grossman et al., Nucleic Acids Res. 15:5904,1987.
498 Guerin et al., Nucleic Acids Res. 16:7759,1988.
499 Gumucio et al., Am. J. Hum. Genet. 37:A155,1985.
500 Gusella et al., (HGM8) Cytogenet. Cell Genet. 40:646,1985.
501 Gusella et al., Cold Spring Harbor Symp. Quant. Biol.
    51:359-364,1986.
502 Gusella et al., Nature 306:234-238,1983.
503 Guttler and Woo, J. Inher. Metab. Dis. (suppl.)
    9:58-68,1986.
504 Guzzo et al., Nucleic Acids Res. 15:380,1987.
505 Haines et al., Genet. Epidemiol. 5:375-380,1988.
506 Hall and Brown, Nucleic Acids Res. 13:5255-5268,1985.
507 Hall et al., (HGM9) Cytogenet. Cell Genet. 46:626,1987.
508 Hall et al., Am. J. Hum. Genet. 44:577-584,1989.
509 Haluska et al., Nucleic Acids Res. 15:865,1987.
510 Hanauer et al., (HGM8) Cytogenet. Cell Genet.
    40:647-648,1985.
511 Hanauer et al., Hum. Genet. 80:177-180,1988.
512 Hanauer et al., Nucleic Acids Res. 15:6308,1987.
513 Harley et al., (HGM10) Cytogenet. Cell Genet. 51:1011,1989.
514 Harley et al., Genomics 3:380-384,1988.
515 Harris et al., (HGM10) Cytogenet. Cell Genet. 51:1011,1989.
516 Harris et al., (HGM10) Cytogenet. Cell Genet. 51:1011,1989.

517 Harris et al., Clin. Genet. 33:162-168,1988.
518 Harris et al., Hum. Genet. 77:95-103,1987.
519 Harris et al., Hum. Genet. 79:76-79,1988.
520 Hartley et al., Nucleic Acids Res. 12:5277-5285,1984.
521 Hay et al., Blood 67:1508-1511,1986.
522 Hay et al., Nucleic Acids Res. 14:5118,1986.
523 Hayden et al., Am. J. Hum. Genet. 42:125-131,1988.
524 Hayden et al., Nucleic Acids Res. 15:3938,1987.
525 Hayward et al., Nucleic Acids Res. 15:5503,1987.
526 Hazout and Lucotte, Ann. Genet. 29:246-252,1986.
527 Hegele et al., N. Engl. J. Med. 315:1509-1515,1986.
528 Hegele et al., Nucleic Acids Res. 16:7214,1988.
529 Hegele et al., Nucleic Acids Res. 17:1786,1989.
530 Hegele et al., Nucleic Acids Res. 17:470,1989.
531 Heighway and Geurts van Kessel, Nucleic Acids Res.
    14:8700,1986.
532 Heinzmann et al., Nucleic Acids Res. 15:6763, 1987.
533 Heinzmann et al., Nucleic Acids Res. 16:4739,1988.
534 Hekkert et al., Nucleic Acids Res. 16:11849,1988.
535 Henderson et al., Hum. Genet. 75:62-65,1987.
536 Hendriks et al., Nucleic Acids Res. 16:2365,1988.
537 Herrmann et al., Clin. Genet. 34:176-180,1988.
538 Higgs et al., Proc. Natl. Acad. Sci. USA 83:5165-5169,1986.
539 Hill et al., Am. J. Hum. Genet. 38:382-389,1986.
540 Hill et al., Cancer Res. 49:145-148,1989.
541 Hill et al., Cell 42:809-819,1985.
542 Hill et al., Hum. Genet. 60:222-226,1982.
543 Himmelbauer et al., (HGM10) Cytogenet. Cell Genet.
    51:1014,1989.
544 Hobbs et al., Nucleic Acids Res. 15:379,1987.
545 Hobbs et al., Proc. Natl. Acad. Sci. USA 82:7651-7655,1985.
546 Hodgkinson et al., Nature 325:805-806,1987.
547 Hodgson and Kalsheker, J. Med. Genet. 24:47-51,1987.
548 Hodgson and Kalsheker, Nucleic Acids Res. 14:6779,1986.
549 Hoehe et al., Nucleic Acids Res. 16:9070,1988.
550 Hoehe et al., Nucleic Acids Res. 17:2148,1989.
551 Hoehe et al., Nucleic Acids Res. 17:828,1989.
552 Hoff et al., Nucleic Acids Res. 15:10075,1987.
553 Hoff et al., Nucleic Acids Res. 15:10077,1987.
554 Hoff et al., Nucleic Acids Res. 15:10606,1987.
555 Hoff et al., Nucleic Acids Res. 15:9619,1987.
556 Hoff et al., Nucleic Acids Res. 16:10400,1988.
557 Hoff et al., Nucleic Acids Res. 16:10401,1988.
558 Hoff et al., Nucleic Acids Res. 16:373,1988.
559 Hoff et al., Nucleic Acids Res. 16:4189,1988.
560 Hoff et al., Nucleic Acids Res. 16:4741,1988.
561 Hoff et al., Nucleic Acids Res. 16:4742,1988.
562 Hoff et al., Nucleic Acids Res. 16:5217,1988.
563 Hoff et al., Nucleic Acids Res. 16:781,1988.
564 Hoff et al., Nucleic Acids Res. 16:787,1988.
565 Hoff et al., Nucleic Acids Res. 16:9366,1988.
566 Hoffman et al., Nucleic Acids Res. 15:4696,1987.
567 Hofker et al., Am. J. Hum. Genet. 39:438-451,1986.
568 Hofker et al., Am. J. Hum. Genet. 40:312-328,1987.
569 Hofker et al., Hum. Genet. 70:148-156,1985.
570 Hofker et al., Hum. Genet. 74:270-274,1986.
571 Hofker et al., Hum. Genet. 74:275-279,1986.
572 Hofker et al., Nucleic Acids Res. 13:7167,1985.
573 Holcombe et al., Genomics 1:287-291,1987.
574 Holden et al., Am. J. Hum. Genet. 35:174A,1983.
575 Hollstein et al., Nucleic Acids Res. 14:8695,1986.
576 Holm et al., Nucleic Acids Res. 16:10949,1988.
577 Holm et al., Nucleic Acids Res. 16:3115,1988.
578 Holm et al., Nucleic Acids Res. 16:3118,1988.
579 Holm et al., Nucleic Acids Res. 16:372,1988.
580 Holm et al., Nucleic Acids Res. 16:4748,1988.
581 Holm et al., Nucleic Acids Res. 16:5216,1988.
582 Holm et al., Nucleic Acids Res. 16:5701,1988.
583 Holm et al., Nucleic Acids Res. 16:5709,1988.
584 Holm et al., Nucleic Acids Res. 16:9887,1988.
585 Holm et al., Nucleic Acids Res. 16:9887,1988.
586 Hoover et al., Cold Spring Harbor Symp. Quant. Biol.
    51:803-809,1986.
587 Hoover et al., J. Exp. Med. 162:1087-1092,1985.
588 Hoppener et al., FEBS Lett. 233:57-63,1988.
589 Hoppener et al., Hum. Genet. 66:309-312,1984.
590 Hoppener et al., Hum. Genet. 69:157-160,1985.
591 Hoshina et al., Proc. Natl. Acad. Sci. USA
    81:2504-2507,1984.
592 Houot et al., Nucleic Acids Res., in press, 1987.

593 Huang and Breslow, J. Biol. Chem. 262:8952-8955,1987.
594 Huang et al., Blood 70:1830-1835,1987.
595 Huang et al., Proc. Natl. Acad. Sci. USA 83:644-648,1986.
596 Huck et al., FEBS Lett. 208:221-230,1986.
597 Huerre et al., Ann. Genet. (Paris) 27:5-10,1984.
598 Huff et al., Nucleic Acids Res. 15:7651,1987.
599 Hulsebos et al., Cytogenet. Cell Genet. 43:47-56,1986.
600 Hulsebos et al., Nucleic Acids Res. 14:7137,1986.
601 Hulsebos et al., Nucleic Acids Res. 15:378,1987.
602 Humphries et al., Hum. Genet. 68:148-153,1984.
603 Humphries et al., Lancet i:1003-1005,1985.
604 Humphries et al., Lancet ii:794-795,1988.
605 Humphries et al., Mol. Biol. Med. 1:463-471,1983.
606 Hupkes et al., (HGM10) Cytogenet. Cell Genet. 51:1016,1989.
607 Hutz et al., Hum. Genet. 66:217-219,1984.
608 Hyland et al., (HGM10) Cytogenet. Cell Genet. 51:1017,1989.
609 Hyland et al., (HGM10) Cytogenet. Cell Genet. 51:1017,1989.
610 Hyland et al., Hum. Genet. 79:277-279,1988.
611 Hyland et al., Nucleic Acids Res. 15:1350,1987.
612 Hyland et al., Nucleic Acids Res. 17:6430,1989.
613 Hyland, Nucleic Acids Res. 16:8203,1988.
614 Iannuzzi et al., Am. J. Hum. Genet. 44:695-703,1989.
615 Iannuzzi et al., Nucleic Acids Res. 15:5909,1987.
616 Ibson et al., J. Cell. Biochem. 33:267-288,1987.
617 Icking et al., (HGM8) Cytogenet. Cell Genet. 40:659,1985.
618 Ikonen et al., Nucleic Acids Res. 17:473,1989.
619 Ikuta et al., Biochem. Genet. 26:519-525,1988.
620 Ionasescu et al., Cytogenet. Cell Genet. 45:240-241,1987.
621 Ip et al., (HGM10) Cytogenet. Cell Genet. 51:1018,1989.
622 Ishizaki et al., Hum. Genet. 71:261-262,1985.
623 Jabs et al., Am. J. Hum. Genet. 36:141S,1984.
624 Jabs et al., Am. J. Hum. Genet. 38:297-308,1986.
625 Jabs et al., Proc. Natl. Acad. Sci. USA 81:4884-4888,1984.
626 Jaiswal and Nebert, Nucleic Acids Res. 14:4376,1986.
627 Jakowlew et al., Nucleic Acids Res. 12:2861-2878,1984.
628 Jakubiczka et al., (HGM10) Cytogenet. Cell Genet.
         51:1018,1989.
629 James et al., Cancer Res. 48:5546-5551,1988.
630 Jandel et al., (HGM8) Cytogenet. Cell Genet.
         40:660-661,1985.
631 Jaquet et al., Nucleic Acids Res. 17:3620,1989.
632 Jarman et al., EMBO J. 5:1857-1863,1986.
633 Jaye et al., Nucleic Acids Res. 13:8286,1985.
634 Jeffreys et al., Nucl. Acids Res. 16:10953-10971,1988.
635 Jinno et al., J. Inher. Metab. Dis. 9:317-320,1987.
636 Jobs et al., (HGM9) Cytogenet. Cell Genet. 46:634,1987.
637 Johnson et al., (HGM9) Cytogenet. Cell Genet.
         46:634-635,1987.
638 Johnson et al., Am. J. Hum. Genet. 38:617-640,1986.
639 Johnson et al., Proc. Natl. Acad. Sci. USA
         81:7840-7844,1984.
640 Jones et al., Nucleic Acids Res. 17:472,1989.
641 Julier and White, Am. J. Hum. Genet. 42:45-48,1988.
642 Julier et al., (HGM10) Cytogenet. Cell Genet. 51:1020,1989.
643 Julier et al., (HGM8) Cytogenet. Cell Genet.
         40:663-664,1985.
644 Julier et al., (HGM8) Cytogenet. Cell Genet. 40:664,1985.
645 Julier et al., Am. J. Hum. Genet. 42:297-308,1988.
646 Kallunki et al., Nucleic Acids Res. 17:4423,1989.
647 Kalsheker and Watkins, Hum. Genet. 80:108-109,1988.
648 Kalsheker, personal communication,1987.
649 Kamholz et al., Am. J. Hum. Genet. 40:365-373,1987.
650 Kamino et al., Nucleic Acids Res. 16:11387,1988.
651 Kan et al., N. Engl. J. Med. 302:185-188,1980.
652 Kao et al., (HGM7) Cytogenet. Cell Genet. 37:506,1984.
653 Karathanasis et al., Proc. Natl. Acad. Sci. USA
         83:8457-8461,1986.
654 Kawashima et al., Proc. Natl. Acad. Sci. USA
         85:23533-2356,1988.
655 Kazazian et al., EMBO J. 3:593-596,1984.
656 Keith et al., (HGM9) Cytogenet. Cell Genet. 46:638,1987.
657 Kellogg et al., (HGM10) Cytogenet. Cell Genet. 51:1022,1989.
658 Kelsey et al., J. Med. Genet. 25:361-368,1988.
659 Kelsoe et al., Nucleic Acids Res. 16:7760,1988.
660 Kerckaert et al., Nucleic Acids Res. 15:5905,1987.
661 Kere, Nucleic Acids Res. 17:1511-1520,1989.
662 Kessling et al., Clin. Genet. 28:296-306,1985.
663 Kessling et al., Hum. Genet. 78:237-239,1988.
664 Keyeux et al., Nucleic Acids Res. 17:3623,1989.
665 Keyeux et al., Nucleic Acids Res. 17:3624,1989.

666 Kidd et al., (HGM9) Cytogenet. Cell Genet. 46:639,1987.
667 Kidd et al., Genomics, in press, 1989.
668 Kidd et al., Hum. Hered. 38:22-26,1988.
669 Kidd et al., Nucleic Acids Res. 14:9544,1986.
670 Killeen et al., Am. J. Med. Genet. 29:703-712,1988.
671 Kim et al., Proc. Natl. Acad. Sci. USA 82:8139-8143,1985.
672 Kiousis et al., Nucleic Acids Res. 17:5876,1989.
673 Kittur et al., (HGM8) Cytogenet. Cell Genet. 40:669,1985.
674 Kittur et al., EMBO J. 4:2257-2260,1985.
675 Kittur et al., Proc. Natl. Acad. Sci. USA 82:5064-5067,1985.
676 Klasen et al., Hum. Genet. 75:244-247,1987.
677 Kleyn et al., (HGM10) Cytogenet. Cell Genet. 51:1023,1989.
678 Klinger et al., ICSU Short Reports 1:190-191,1984.
679 Klinger et al., Nucleic Acids Res. 14:8681-8686,1986.
680 Klinger et al., Proc. Natl. Acad. Sci. USA
         84:8548-8552,1987.
681 Klobeck et al., Nucleic Acids Res. 15:9655-9665,1987.
682 Kluve-Beckerman et al., Biochem. Biophys. Res. Comm.
         137:1196-1204,1986.
683 Kluve-Beckerman et al., Biochem. Genet. 24:795-803,1986.
684 Knoers et al., Hum. Genet. 80:31-38,1988.
685 Knott et al., Biochem. Biophys. Res. Comm. 125:299-306,1984.
686 Knott et al., Nucleic Acids Res. 14:9215-9216,1986.
687 Knowlton et al., Blood 68:378-385,1986.
688 Knowlton et al., Nucleic Acids Res. 17:423-437,1989.
689 Koenig et al., (HGM8) Cytogenet. Cell Genet.
         40:670-671,1985.
690 Koenig et al., Cell 50:509-517,1987.
691 Koenig et al., Nucleic Acids Res. 12:4097-4109,1984.
692 Koenig et al., Nucleic Acids Res. 13:5485-5501,1985.
693 Kok et al., Nature 330:578-581,1987.
694 Kolvraa et al., Hum. Genet. 74:284-287,1986.
695 Kondo and Berg, Nucleic Acids Res. 17:2375,1989.
696 Kondoleon et al., Nucleic Acids Res. 15:10605,1987.
697 Kondoleon et al., Nucleic Acids Res. 15:9096,1987.
698 Konkle et al., Nucleic Acids Res. 15:6766,1987.
699 Korneluk et al., Nucleic Acids Res. 15:6769,1987.
700 Korneluk et al., Nucleic Acids Res. 17:5876,1989.
701 Korner et al., Nucleic Acids Res. 16:9078,1988.
702 Kotula and Curtis, Nucleic Acids Res. 16:10950,1988.
703 Kotze et al., J. Med. Genet. 24:750-755,1987.
704 Kotze et al., Nucleic Acids Res. 15:10067,1987.
705 Kotze et al., Nucleic Acids Res. 15:376,1987.
706 Kotze et al., S. Afr. Med. J. 70:77-79,1986.
707 Krapcho et al., Nucleic Acids Res. 16:1227,1988.
708 Krapcho et al., Nucleic Acids Res. 16:5221,1988.
709 Krapcho et al., Nucleic Acids Res. 16:5704,1988.
710 Krapcho et al., Nucleic Acids Res. 16:9360,1988.
711 Krissansen et al., Immunogenetics 26:258-266,1987.
712 Krontiris et al., Nature 313:369-374,1985.
713 Kruse et al., (HGM10) Cytogenet. Cell Genet. 51:1026,1989.
714 Kruse et al., (HGM10) Cytogenet. Cell Genet. 51:1026,1989.
715 Kruse et al., Int. Cong. Hum. Genet. 7:681-682,1986.
716 Kruse et al., Nucleic Acids Res. 14:1921,1986.
717 Kumlin-Wolff et al., Nucleic Acids Res. 15:10076,1987.
718 Kumlin-Wolff et al., Nucleic Acids Res. 15:10611,1987.
719 Kumlin-Wolff et al., Nucleic Acids Res. 15:9621,1987.
720 Kumlin-Wolff et al., Nucleic Acids Res. 16:1224,1988.
721 Kunkel and L. M., Nature 322:73-77,1986.
722 Kunkel et al., Proc. Natl. Acad. Sci. USA 82:4778-4782,1985.
723 Kurnit et al., Cytogenet. Cell Genet. 43:109-116,1986.
724 Kurosawa et al., Oncogene 2:85-90,1987.
725 Kusari et al., J. Interferon Res. 7:53-59,1987.
726 Kwan et al., J. Clin. Invest. 77:649-652,1986.
727 Kwiatkowski et al., Nucleic Acids Res. 17:4425,1989.
728 Lacoste-Royal et al., Nucleic Acids Res. 16:4184,1988.
729 Laing et al., Nucleic Acids Res. 16:7209,1988.
730 Lamhonwah et al., Proc. Natl. Acad. Sci. USA
         83:4864-4868,1986.
731 Landsman et al., Nucleic Acids Res. 17:2301-2314,1989.
732 Lathrop et al., (HGM9) Cytogenet. Cell Genet. 46:644,1987.
733 Lathrop et al., (HGM9) Cytogenet. Cell Genet. 46:644,1987.
734 Lathrop et al., Am. J. Hum. Genet. 42:38-44,1988.
735 Lathrop et al., Genomics 2:157-164,1988.
736 Lathrop et al., Genomics 3:361-366,1988.
737 Latt et al., (HGM9) Cytogenet. Cell Genet. 46:644,1987.
738 Lavergne et al., Nucleic Acids Res. 15:9099,1987.
739 Lavergne et al., Nucleic Acids Res. 16:2742,1988.
740 Law et al., Lancet i:1301-1303,1986.
741 Law et al., Proc. Natl. Acad. Sci. USA 83:6920-6924,1986.

742 Law et al., Proc. Natl. Acad. Sci. USA 84:2877-2881,1987.
743 Leach et al., Proc. Natl. Acad. Sci. USA 83:3909-3913,1986.
744 Ledley et al., Am. J. Hum. Genet. 42:839-846,1988.
745 Lee and Anvret, Nucleic Acids Res. 15:6307,1987.
746 Lee and Anvret, Nucleic Acids Res. 17:2147,1989.
747 Lee et al., Am. J. Hum. Genet. 36:145S,1984.
748 Lee et al., Cancer Res. 48:4045-4048,1988.
749 Lee et al., Hum. Genet. 76:33-36,1987.
750 Lee et al., Hum. Genet. 79:379-381,1988.
751 Lee et al., Proc. Natl. Acad. Sci. USA 84:4403-4407,1987.
752 Lee, personal communication,1989.
753 Lefranc and Rabbitts, Nature 316:464-466,1985.
754 Lefranc and Rabbitts, Nucleic Acids Res. 12:1303-1311,1984.
755 Lefranc et al., Nature 300:760-762,1982.
756 Lefranc et al., Proc. Natl. Acad. Sci. USA
          83:9596-9600,1986.
757 Leib-Mosch et al., Nucleic Acids Res. 17:2367,1989.
758 Leitersdorf and Hobbs, Nucleic Acids Res. 15:2782,1987.
759 Leitersdorf and Hobbs, Nucleic Acids Res. 16:7215,1988.
760 Lench and Scambler, Nucleic Acids Res. 16:11854,1988.
761 Lentes et al., Nucleic Acids Res. 16:2359,1988.
762 Lentes, personal communication,1988.
763 Leppert et al., (HGM9) Cytogenet. Cell Genet. 46:648,1987.
764 Leppert et al., (HGM9) Cytogenet. Cell Genet. 46:648,1987.
765 Leppert et al., (HGM9) Cytogenet. Cell Genet. 46:727,1987.
766 Leppert et al., Am. J. Hum. Genet. 37:A164,1985.
767 Leppert et al., Science 238:1411-1413,1987.
768 Lerner et al., (HGM10) Cytogenet. Cell Genet. 51:1032,1989.
769 Leroy et al., (HGM8) Cytogenet. Cell Genet. 40:680,1985.
770 Leroy et al., J. Genet. Hum. 34:139,1986.
771 Leveillard et al., Nucleic Acids Res. 16:11852,1988.
772 Leveillard et al., Nucleic Acids Res. 16:2744,1988.
773 Leveillard et al., Nucleic Acids Res. 17:1272,1989.
774 Leveillard et al., Nucleic Acids Res. 17:2875,1989.
775 Leveillard et al., Nucleic Acids Res. 17:5419,1989.
776 Leveillard et al., Nucleic Acids Res. 17:5418,1989.
777 Levilliers et al., (HGM9) Cytogenet. Cell Genet.
          46:650,1987.
778 Lewis et al., Nucleic Acids Res. 15:6760,1987.
779 Leysens et al., Nucleic Acids Res. 17:3621,1989.
780 Leysens et al., Nucleic Acids Res. 17:5417,1989.
781 Li et al., Hum. Hered. 38:273-276,1988.
782 Li et al., J. Immunol. 140:1300-1303,1988.
783 Li et al., Nucleic Acids Res. 16:11856,1988.
784 Li et al., Nucleic Acids Res. 16:2358,1988.
785 Li et al., Nucleic Acids Res. 17:3330,1989.
786 Li et al., Nucleic Acids Res., in press, 1989.
787 Lichter-Konecki et al., Hum. Genet. 78:347-352,1988.
788 Lidereau et al., J. Natl. Cancer Inst. 77:697-701,1986.
789 Lidsky et al., Am. J. Hum. Genet. 37:619-634,1985.
790 Lindlof et al., Genomics 1:87-92,1987.
791 Lindstedt and Anvret, Nucleic Acids Res. 17:2882,1989.
792 Linnenbach et al., Proc. Natl. Acad. Sci. USA 85:74-78,1988.
793 Liou et al., Nucleic Acids Res. 15:3196,1987.
794 Litt and Buder, Nucleic Acids Res. 17:465,1989.
795 Litt and Jorde, Am. J. Hum. Genet. 39:166-178,1986.
796 Litt and Luty, Am. J. Hum. Genet. 44:397-401,1989.
797 Litt and White, Proc. Natl. Acad. Sci. USA
          82:6206-6210,1985.
798 Litt et al., Am. J. Hum. Genet. 38:288-296,1986.
799 Litt et al., Am. J. Hum. Genet. 42:327-334,1988.
800 Litt et al., Genomics, in press, 1988.
801 Litt et al., Hum. Genet. 73:340-345,1986.
802 Litt et al., Nucleic Acids Res. 15:2783,1987.
803 Litt et al., Nucleic Acids Res. 16:6251,1988.
804 Litt et al., Nucleic Acids Res. 17:2371,1989.
805 Litt et al., Nucleic Acids Res. 17:2883,1989.
806 Litt, Nucleic Acids Res. 14:4378,1986.
807 Litt, personal communication,1988.
808 Liu et al., (HGM10) Cytogenet. Cell Genet. 51:1034,1989.
809 Liu et al., (HGM9) Cytogenet. Cell Genet. 46:652,1987.
810 Llewellyn et al., Lancet ii:706-708,1987.
811 Llewellyn et al., Nucleic Acids Res. 15:1349,1987.
812 Lloyd et al., Gene 41:233-239,1986.
813 Lobaccaro et al., Nucleic Acids Res. 16:11851,1988.
814 Lobos and Devor, Nucleic Acids Res. 15:6767,1987.
815 Logan and Joyner, Nucleic Acids Res. 17:2877,1989.
816 Logan and Joyner, Nucleic Acids Res. 17:2878,1989.
817 Lothe et al., Ann. Hum. Genet. 50:361-367,1986.
818 Lubahn et al., Am. J. Hum. Genet. 40:527-536,1987.

819 Lubsen et al., Proc. Natl. Acad. Sci. USA 84:489-492,1987.
820 Lucarelli et al., Hum. Genet. 78:291-292,1988.
821 Lucotte and Ngo, Nucleic Acids Res. 13:8285,1985.
822 Ludecke et al., (HGM10) Cytogenet. Cell Genet. 51:1035,1989.
823 Luster et al., Proc. Natl. Acad. Sci. USA 84:2868-2871,1987.
824 Luty et al., (HGM10) Cytogenet. Cell Genet. 51:1036,1989.
825 Luty et al., Nucleic Acids Res. 16:6250,1988.
826 Luty et al., Nucleic Acids Res. 16:7210, 1988.
827 Lyons et al., Genetics 110:s31,1985.
828 MacDermot et al., Hum. Genet. 77:263-266,1987.
829 MacDonald et al., Genomics 1:29-34,1987.
830 MacDonald et al., Hum. Genet. 77:233-235,1987.
831 MacDonald et al., Somatic Cell Mol. Genet. 13:569-574,1987.
832 MacKenzie et al., (HGM10) Cytogenet. Cell Genet.
          51:1036,1989.
833 Maeda et al., Nature 309:131-135,1984.
834 Mahtani et al., (HGM10) Cytogenet. Cell Genet. 51:1038,1989.
835 Malhotra and Concannon, Nucleic Acids Res. 17:2373,1989.
836 Mankoo and Dalgleish, Nucleic Acids Res. 16:1643,1988.
837 Marcais et al., (HGM10) Cytogenet. Cell Genet. 51:1040,1989.
838 Marchetti et al., (HGM8) Cytogenet. Cell Genet. 40:691,1985.
839 Marchetti et al., Nucleic Acids Res. 17:3329,1989.
840 Mariman et al., Nucleic Acids Res. 15:5502,1987.
841 Marlhens et al., Nucleic Acids Res. 15:1348,1987.
842 Martell et al., C. R. Acad. Sci. Paris, Ser. III
          304:105-110,1987.
843 Martin et al., Nucleic Acids Res. 15:9104,1987.
844 Martin et al., Nucleic Acids Res. 16:3121,1988.
845 Martin et al., Nucleic Acids Res. 16:3596,1988.
846 Martin et al., Nucleic Acids Res. 16:4190,1988.
847 Martin et al., Nucleic Acids Res. 16:5220,1988.
848 Marynen et al., Nucleic Acids Res. 13:8287,1985.
849 Masharani and Frossard, Hum. Genet. 80:308,1988.
850 Masharani and Frossard, Nucleic Acids Res. 16:2357,1988.
851 Masharani et al., Hum. Genet. 80:307,1988.
852 Masharani et al., Nucleic Acids Res. 16:6253,1988.
853 Masharani et al., Nucleic Acids Res. 16:7757,1988.
854 Masharani, Nucleic Acids Res. 17:467,1989.
855 Maslen et al., Genomics 2:66-75,1988.
856 Maslen et al., Nucleic Acids Res. 16:8195,1988.
857 Massaro et al., Nucleic Acids Res. 17:2155,1989.
858 Mathew et al., (HGM10) Cytogenet. Cell Genet. 51:1040,1989.
859 Mathew et al., Nature 328:524-526,1987.
860 Mathew et al., Nature 328:527-528,1987.
861 Mathieu-Mahul et al., Hum. Genet. 71:41-44,1985.
862 Mathy et al., Hum. Genet. 75:359-361,1987.
863 Mattei et al., Hum. Genet. 69:268-271,1985.
864 Matteson et al., Hum. Genet. 69:263-267,1985.
865 Mauchauffe et al., Nucleic Acids Res. 15:3190,1987.
866 Max et al., Proc. Natl. Acad. Sci. USA 83:5592-5596,1986.
867 McBride et al., (HGM9) Cytogenet. Cell Genet.
          46:659-660,1987.
868 McBride et al., J. Biol. Chem. 262:11163-11166,1987.
869 McBride et al., Nucleic Acids Res. 15:10071,1987.
870 McBride et al., Proc. Natl. Acad. Sci. USA 84:503-507,1987.
871 McClure et al., Am. J. Hum. Genet. 39:A162,1986.
872 McDermid et al., (HGM8) Cytogenet. Cell Genet.
          40:695-696,1985.
873 McDermid et al., Nucleic Acids Res. 15:5498,1987.
874 Meera Khan et al., Hum. Genet. 79:183-185,1988.
875 Meisler et al., Cytogenet. Cell Genet. 44:175-176,1987.
876 Meng and Muller, Nucleic Acids Res., in press, 1989.
877 Menlove et al., (HGM8) Cytogenet. Cell Genet.
          40:697-698,1985.
878 Mergia et al., Biochem. Biophys. Res. Comm.
          138:644-651,1986.
879 Merry et al., (HGM10) Cytogenet. Cell Genet. 51:1043,1989.
880 Middlesworth et al., Nucleic Acids Res. 13:5723,1985.
881 Middleton et al., Nucleic Acids Res. 14:1925,1986.
882 Middleton et al., Nucleic Acids Res. 16:3123,1988.
883 Middleton-Price et al., Ann. Hum. Genet. 52:189-195,1988.
884 Middleton-Price et al., Genomics 4:192-197,1989.
885 Mietus-Snyder et al., (HGM10) Cytogenet. Cell Genet.
          51:1044,1989.
886 Migone et al., Proc. Natl. Acad. Sci. USA 80:467-471,1983.
887 Miki et al., Genomics 3:78-81,1988.
888 Miki et al., Nucleic Acids Res. 16:1649,1988.
889 Miles et al., Nucleic Acids Res. 16:5783-5795,1988.
890 Millington-Ward et al., (HGM8) Cytogenet. Cell Genet.
          40:699,1985.

891 Milner et al., Nucleic Acids Res. 17:4002,1989.
892 Mitchell et al., (HGM9) Cytogenet. Cell Genet. 46:662-663,1987.
893 Mitchell et al., Nucleic Acids Res. 17:2876,1989.
894 Modi et al., (HGM10) Cytogenet. Cell Genet. 51:1046,1989.
895 Modi et al., (HGM10) Cytogenet. Cell Genet. 51:1046,1989.
896 Moisan et al., (HGM8) Cytogenet. Cell Genet. 40:701-702,1985.
897 Moisan et al., Hum. Genet. 79:168-171,1988.
898 Monaco et al., Hum. Genet. 75:221-227,1987.
899 Moore et al., Hum. Genet. 72:297-302,1986.
900 Morle et al., (HGM7) Cytogenet. Cell Genet. 37:544,1984.
901 Mornet and White, (HGM10) Cytogenet. Cell Genet. 51:1047,1989.
902 Mornet et al., Hum. Genet. 74:402-408,1986.
903 Morris and Griffiths, Biochem. Biophys. Res. Comm. 150:219-224,1988.
904 Morris et al., (HGM10) Cytogenet. Cell Genet. 51:1047,1989.
905 Moss et al., Nucleic Acids Res. 14:9927-9932,1986.
906 Motomura et al., Nucleic Acids Res. 17:2369,1989.
907 Mukai and Dryja, Cancer Genet. Cytogenet. 22:45-53,1986. [& AJHG 37:A33,1985. (abs.)]
908 Muller and Grimm, Hum. Genet. 74:181-183,1986.
909 Muller et al., Cytogenet. Cell Genet. 45:16-20,1987.
910 Mulligan et al., Am. J. Hum. Genet. 37:463-472,1985.
911 Mulligan et al., Hum. Genet. 75:381-383,1987.
912 Mulligan et al., Nucleic Acids Res. 17:4421,1989.
913 Munke et al., Am. J. Hum. Genet. 42:542-549,1988.
914 Murday et al., (HGM10) Cytogenet. Cell Genet. 51:1049,1989.
915 Murphy et al., Nucleic Acids Res. 13:3015,1985.
916 Murphy et al., Nucleic Acids Res. 14:4381,1986.
917 Murphy et al., Nucleic Acids Res. 15:6311,1987.
918 Murphy et al., Nucleic Acids Res. 15:7212,1987.
919 Murray et al., (HGM8) Cytogenet. Cell Genet. 40:707-708,1985.
920 Murray et al., Am. J. Hum. Genet. 36:176S,1984.
921 Murray et al., Am. J. Hum. Genet. 40:338-350,1987.
922 Murray et al., Am. J. Hum. Genet. 42:490-497,1988.
923 Murray et al., Cytogenet. Cell Genet. 47:149-151,1988.
924 Murray et al., Hum. Genet. 76:274-277,1987.
925 Murray et al., Nucleic Acids Res. 13:6794,1985.
926 Murray et al., Nucleic Acids Res. 14:5117,1986.
927 Murray et al., Nucleic Acids Res. 14:7136,1986.
928 Murray et al., Nucleic Acids Res. 15:6764,1987.
929 Murray et al., Nucleic Acids Res., in press, 1987.
930 Murray et al., Proc. Natl. Acad. Sci. USA 80:5951-5955,1983.
931 Murray et al., Proc. Natl. Acad. Sci. USA 81:3486-3490,1984.
932 Myal et al., Nucleic Acids Res. 17:5897,1989.
933 Myers et al., Nucleic Acids Res. 16:3591,1988.
934 Myers et al., Nucleic Acids Res. 16:784,1988.
935 Myers et al., Nucleic Acids Res. 16:785,1988.
936 Myers et al., Nucleic Acids Res. 16:9883,1988.
937 Nafa et al., Nucleic Acids Res. 17:1276,1989.
938 Nagarajan et al., Nucleic Acids Res. 17:2154,1989.
939 Nakahori et al., Nucleic Acids Res. 17:2152,1989.
940 Nakamura and White, Nucleic Acids Res. 16:9369,1988.
941 Nakamura et al., (HGM9) Cytogenet. Cell Genet. 46:667,1987.
942 Nakamura et al., Am. J. Hum. Genet. 43:638-644,1988.
943 Nakamura et al., Am. J. Hum. Genet. 43:854-859,1988.
944 Nakamura et al., Am. J. Hum. Genet. 44:751-755,1989.
945 Nakamura et al., Genomics 2:302-309,1988.
946 Nakamura et al., Genomics 3:342-346,1988.
947 Nakamura et al., Genomics 3:389-392,1988.
948 Nakamura et al., Nucleic Acids Res. 15:10073,1987.
949 Nakamura et al., Nucleic Acids Res. 15:10079,1987.
950 Nakamura et al., Nucleic Acids Res. 15:10080,1987.
951 Nakamura et al., Nucleic Acids Res. 15:10607,1987.
952 Nakamura et al., Nucleic Acids Res. 15:10608,1987.
953 Nakamura et al., Nucleic Acids Res. 15:2537-2547,1987.
954 Nakamura et al., Nucleic Acids Res. 15:3594,1988.
955 Nakamura et al., Nucleic Acids Res. 15:9620,1987.
956 Nakamura et al., Nucleic Acids Res. 15:9622,1987.
957 Nakamura et al., Nucleic Acids Res. 16:10402,1988.
958 Nakamura et al., Nucleic Acids Res. 16:10405,1988.
959 Nakamura et al., Nucleic Acids Res. 16:10407,1988.
960 Nakamura et al., Nucleic Acids Res. 16:10408,1988.
961 Nakamura et al., Nucleic Acids Res. 16:10409,1988.
962 Nakamura et al., Nucleic Acids Res. 16:10947,1988.
963 Nakamura et al., Nucleic Acids Res. 16:1228,1988.
964 Nakamura et al., Nucleic Acids Res. 16:1229,1988.

965 Nakamura et al., Nucleic Acids Res. 16:3119,1988.
966 Nakamura et al., Nucleic Acids Res. 16:3120,1988.
967 Nakamura et al., Nucleic Acids Res. 16:3122,1988.
968 Nakamura et al., Nucleic Acids Res. 16:3590,1988.
969 Nakamura et al., Nucleic Acids Res. 16:3592,1988.
970 Nakamura et al., Nucleic Acids Res. 16:3593,1988.
971 Nakamura et al., Nucleic Acids Res. 16:3595,1988.
972 Nakamura et al., Nucleic Acids Res. 16:3597,1988.
973 Nakamura et al., Nucleic Acids Res. 16:3598,1988.
974 Nakamura et al., Nucleic Acids Res. 16:374,1988.
975 Nakamura et al., Nucleic Acids Res. 16:375,1988.
976 Nakamura et al., Nucleic Acids Res. 16:376,1988.
977 Nakamura et al., Nucleic Acids Res. 16:377,1988.
978 Nakamura et al., Nucleic Acids Res. 16:379,1988.
979 Nakamura et al., Nucleic Acids Res. 16:381,1988.
980 Nakamura et al., Nucleic Acids Res. 16:4186,1988.
981 Nakamura et al., Nucleic Acids Res. 16:4187,1988.
982 Nakamura et al., Nucleic Acids Res. 16:4191,1988.
983 Nakamura et al., Nucleic Acids Res. 16:4192,1988.
984 Nakamura et al., Nucleic Acids Res. 16:4743,1988.
985 Nakamura et al., Nucleic Acids Res. 16:4747,1988.
986 Nakamura et al., Nucleic Acids Res. 16:5218,1988.
987 Nakamura et al., Nucleic Acids Res. 16:5219,1988.
988 Nakamura et al., Nucleic Acids Res. 16:5222,1988.
989 Nakamura et al., Nucleic Acids Res. 16:5223,1988.
990 Nakamura et al., Nucleic Acids Res. 16:5702,1988.
991 Nakamura et al., Nucleic Acids Res. 16:5703,1988.
992 Nakamura et al., Nucleic Acids Res. 16:5705,1988.
993 Nakamura et al., Nucleic Acids Res. 16:5706,1988.
994 Nakamura et al., Nucleic Acids Res. 16:5707,1988.
995 Nakamura et al., Nucleic Acids Res. 16:5708,1988.
996 Nakamura et al., Nucleic Acids Res. 16:6254,1988.
997 Nakamura et al., Nucleic Acids Res. 16:6255,1988.
998 Nakamura et al., Nucleic Acids Res. 16:6257,1988.
999 Nakamura et al., Nucleic Acids Res. 16:6258,1988.
1000 Nakamura et al., Nucleic Acids Res. 16:778,1988.
1001 Nakamura et al., Nucleic Acids Res. 16:779,1988.
1002 Nakamura et al., Nucleic Acids Res. 16:780,1988.
1003 Nakamura et al., Nucleic Acids Res. 16:782,1988.
1004 Nakamura et al., Nucleic Acids Res. 16:786,1988.
1005 Nakamura et al., Nucleic Acids Res. 16:9354,1988.
1006 Nakamura et al., Nucleic Acids Res. 16:9355,1988.
1007 Nakamura et al., Nucleic Acids Res. 16:9356,1988.
1008 Nakamura et al., Nucleic Acids Res. 16:9359,1988.
1009 Nakamura et al., Nucleic Acids Res. 16:9361,1988.
1010 Nakamura et al., Nucleic Acids Res. 16:9362,1988.
1011 Nakamura et al., Nucleic Acids Res. 16:9363,1988.
1012 Nakamura et al., Nucleic Acids Res. 16:9364,1988.
1013 Nakamura et al., Nucleic Acids Res. 16:9365,1988.
1014 Nakamura et al., Nucleic Acids Res. 16:9367,1988.
1015 Nakamura et al., Nucleic Acids Res. 16:9368,1988.
1016 Nakamura et al., Nucleic Acids Res. 16:9370,1988.
1017 Nakamura et al., Nucleic Acids Res. 16:9881,1988.
1018 Nakamura et al., Nucleic Acids Res. 16:9882,1988.
1019 Nakamura et al., Nucleic Acids Res. 16:9884,1988.
1020 Nakamura et al., Nucleic Acids Res. 16:9886,1988.
1021 Nakamura et al., Nucleic Acids Res. 16:9888,1988.
1022 Nakamura et al., Science 235:1616-1622,1987.
1023 Nakamura, personal communication,1988.
1024 Naylor et al., (HGM10) Cytogenet. Cell Genet. 51:1051,1989.
1025 Naylor et al., (HGM7) Cytogenet. Cell Genet. 37:553,1984.
1026 Naylor et al., (HGM8) Cytogenet. Cell Genet. 40:710,1985.
1027 Naylor et al., Proc. Natl. Acad. Sci. USA 81:2447-2451,1984.
1028 Nemer et al., Nucleic Acids Res. 14:8697,1986.
1029 Newgard et al., Am. J. Hum. Genet. 40:351-364,1987.
1030 Newton et al., Nucleic Acids Res. 16:8233-8243,1988.
1031 Ngo et al., Am. J. Hum. Genet. 38:407-418,1986.
1032 Nicholls et al., J. Med. Genet. 24:39-46,1987.
1033 Nielsen et al., Hum. Genet. 78:179-182,1988.
1034 Nigro et al., Am. J. Hum. Genet. 40:115-125,1987.
1035 Nimmo et al., Nucleic Acids Res. 15:3940,1987.
1036 Nishimura et al., Genomics 3:393-395,1988.
1037 Nishino and Lynch, Nucleic Acids Res. 14:4697,1986.
1038 Nishioka and Lamothe, Am. J. Med. Genet. 27:711-717,1987.
1039 Nishisho et al., Jpn. J. Hum. Genet. 31:249-258,1986.
1040 Noll and Collins, Proc. Natl. Acad. Sci. USA 84:3339-3343,1987.
1041 Norby and Schwartz, Clin. Genet. 31:192-197,1987.
1042 Northrup et al., Nucleic Acids Res. 17:1784,1989.
1043 Nozari et al., Analyt. Biochem. 172:810-184,1988.

1044 Nukiwa et al., J. Biol. Chem. 261:15989-15994,1986.
1045 O'Connell et al., (HGM9) Cytogenet. Cell Genet. 46:672,1987.
1046 O'Connell et al., (HGM9) Cytogenet. Cell Genet. 46:672,1987.
1047 O'Connell et al., (HGM9) Cytogenet. Cell Genet. 46:673,1987.
1048 O'Connell et al., Am. J. Hum. Genet. 37:A169,1985.
1049 O'Connell et al., Am. J. Hum. Genet. 44:51-57,1989.
1050 O'Connell et al., Genomics 1:93-102,1987.
1051 O'Connell et al., Genomics 3:367-372,1988.
1052 O'Dowd et al., Nucleic Acids Res. 15:3194,1987.
1053 O'Malley and Rotwein, Nucleic Acids Res. 16:4437-4446,1988.
1054 Oberle et al., Hum. Genet. 77:60-65,1987.
1055 Oberle et al., N. Engl. J. Med. 312:682-686,1985.
1056 Oberle et al., Nucleic Acids Res. 16:8749,1988.
1057 Oberle et al., Proc. Natl. Acad. Sci. USA 82:2824-2828,1985.
1058 Oettgen et al., Nucleic Acids Res. 14:5571,1986.
1059 Oettgen et al., Nucleic Acids Res. 14:7138,1986.
1060 Ogilvie et al., Nucleic Acids Res. 15:4699,1987.
1061 Ohlsson et al., Proc. Natl. Acad. Sci. USA
        82:4473-4476,1985.
1062 Okamoto et al., Nucleic Acids Res. 16:2363,1988.
1063 Old and Wainscoat, Brit. J. Haematol. 53:337-341,1983.
1064 Olek et al., (HGM8) Cytogenet. Cell Genet. 40:717,1985.
1065 Oliviero et al., Hum. Genet. 70:66-70,1985.
1066 Ordovas et al., N. Engl. J. Med. 314:671-677,1986.
1067 Orzechowski et al., Nucleic Acids Res. 15:6310,1987.
1068 Overhauser et al., Am. J. Hum. Genet. 39:562-572,1986.
1069 Overhauser et al., Nucleic Acids Res. 15:1345,1987.
1070 Overhauser et al., Nucleic Acids Res. 15:4617-4627,1987.
1071 Owerbach and Aagaard, Gene 32:475-479,1984.
1072 Ozelius et al., Genomics 3:53-58,1988.
1073 Page et al., Genomics 1:243-256,1987.
1074 Page et al., Nature 311:119-123,1984.
1075 Page et al., Nature 328:437-440,1987.
1076 Page et al., Proc. Natl. Acad. Sci. USA 79:5352-5356,1982.
1077 Pakstis et al., (HGM10) Cytogenet. Cell Genet. 51:1057,1989.
1078 Palotie et al., (HGM10) Cytogenet. Cell Genet. 51:1057,1989.
1079 Pals et al., Genomics 4:137-145,1989.
1080 Palsdottir et al., Nature 306:615-616,1983.
1081 Palsdottir et al., Nucleic Acids Res. 15:2395,1987.
1082 Pandolfo and Smith, Nucleic Acids Res. 16:11857,1988.
1083 Pandolfo et al., Nucleic Acids Res. 16:7213,1988.
1084 Paolella et al., Hum. Genet. 77:115-117,1987.
1085 Paolella et al., Nucleic Acids Res. 14:5568,1986.
1086 Parmentier and Vassart, Nucleic Acids Res. 16:9373,1988.
1087 Partanen and Koskimies, Scand. J. Immunol. 28:313-316,1988.
1088 Partanen et al., Hum. Hered. 37:241-249,1987.
1089 Pasquinelli et al., J. Virol. 62:629-632,1988.
1090 Patel et al., Nucleic Acids Res. 16:5700,1988.
1091 Patel et al., Somatic Cell Mol. Genet. 10:483-493,1984.
1092 Patterson et al., Am. J. Hum. Genet. 43:684-688,1988.
1093 Patterson et al., Nucleic Acids Res. 15:2639-2651,1987.
1094 Paul et al., Hum. Genet. 75:264-268,1987.
1095 Pedersen and Berg, Clin. Genet. 34:306-312,1988.
1096 Pelicci et al., Science 237:1051-1055,1987.
1097 Pellegrino et al., Nucleic Acids Res. 16:7758,1988.
1098 Perryman et al., Nucleic Acids Res. 16:8744,1988.
1099 Petit et al., Cell 49:595-602,1987.
1100 Phelan et al., Am. J. Hum. Genet. 43:511-519,1988.
1101 Pierotti et al., Cytogenet. Cell Genet. 43:174-180,1986.
1102 Pierotti et al., Nucleic Acids Res. 14:4379,1986.
1103 Pikkarainen et al., Nucleic Acids Res. 17:4424,1989.
1104 Poller et al., Nucleic Acids Res. 17:2151,1989.
1105 Powell et al., FEBS Lett. 175:37-40,1984.
1106 Prchal et al., Proc. Natl. Acad. Sci. USA 84:7468-7472,1987.
1107 Priestley et al., Nucleic Acids Res. 13:6790,1985.
1108 Priestley et al., Nucleic Acids Res. 13:6793,1985.
1109 Prior et al., Nucleic Acids Res. 17:2370,1989.
1110 Prochownik et al., J. Biol. Chem. 258:8389-8394,1983.
1111 Quadt et al., Nucleic Acids Res. 14:7139,1986.
1112 Quan et al., Am. J. Hum. Genet. 37:A171,1985.
1113 Quan et al., Nucleic Acids Res. 13:8288,1985.
1114 Radford et al., (HGM9) Cytogenet. Cell Genet. 46:678,1987.
1115 Radice et al., Nucleic Acids Res. 16:9062,1988.
1116 Radice et al., Oncogene 2:91-95,1987.
1117 Raeymaekers et al., Hum. Genet. 78:76-78,1988.
1118 Raeymaekers et al., Nucleic Acids Res. 16:2738,1988.
1119 Raeymaekers et al., Nucleic Acids Res. 16:8196,1988.
1120 Raeymaekers et al., Nucleic Acids Res. 17:1278,1989.
1121 Raimondi et al., Nucleic Acids Res. 15:7653,1987.
1122 Ramesh et al., Am. J. Hum. Genet. 42:365-372,1988.
1123 Ramesh et al., Hum. Genet. 76:121-126,1986.
1124 Ramirez et al., Nucleic Acids Res. 7:1147-1162,1979.
1125 Ramsay et al., J. Med. Genet. 23:145-150,1986.
1126 Ramsay et al., Nucleic Acids Res. 17:1793,1989.
1127 Ramsay, personal communication,1989.
1128 Ray and Cooke, Nucleic Acids Res. 16:7211,1988.
1129 Ray et al., Nature 318:672-675,1985.
1130 Ray et al., Nucleic Acids Res. 16:2361,1988.
1131 Read et al., Hum. Genet. 80:152-156,1988.
1132 Reeders et al., (HGM9) Cytogenet. Cell Genet. 46:680,1987.
1133 Reeders et al., Nature 317:542-544,1985.
1134 Rees et al., Hum. Genet. 72:168-171,1986.
1135 Rees et al., Lancet i:444-446,1983.
1136 Reeve et al., Nature 309:174-176,1984.
1137 Reilly et al., Am. J. Hum. Genet. 42:748-755,1988.
1138 Rekila et al., Hum. Genet. 80:193,1988.
1139 Retief et al., (HGM10) Cytogenet. Cell Genet. 51:1065,1989.
1140 Rey-Campos et al., J. Exp. Med. 166:246-252,1987.
1141 Rich et al., Nucleic Acids Res. 16:8740,1988.
1142 Richards et al., Proc. Natl. Acad. Sci. USA
        85:6437-6441,1988.
1143 Rider et al., Nucleic Acids Res. 15:863,1987.
1144 Rimokh et al., Nucleic Acids Res. 17:823,1989.
1145 Ritty et al., personal communication,1989.
1146 Robbins et al., Nucleic Acids Res. 15:8122,1987.
1147 Roberts et al., Nucleic Acids Res. 17:811,1989.
1148 Robertson et al., Nucleic Acids Res. 17:3327,1989.
1149 Robinson and Kindt, Immunogenetics 24:259-266,1986.
1150 Robinson and Kindt, Proc. Natl. Acad. Sci. USA
        82:3804-3808,1985.
1151 Robinson and Kindt, Proc. Natl. Acad. Sci. USA
        84:9089-9093,1987.
1152 Rogne et al., Hum. Genet. 72:68-71,1986.
1153 Romeo et al., Lancet ii:8-10,1988.
1154 Romeo et al., Proc. Natl. Acad. Sci. USA 84:2829-2832,1987.
1155 Rommens et al., Am. J. Hum. Genet. 43:645-663,1988.
1156 Roschmann et al., Nucleic Acids Res. 15:1883,1987.
1157 Roses et al., Nucleic Acids Res. 14:5569,1986.
1158 Ross et al., (HGM10) Cytogenet. Cell Genet. 51:1069,1989.
1159 Rotwein et al., Am. J. Hum. Genet. 39:291-299,1986.
1160 Rotwein et al., J. Biol. Chem. 261:4828-4832,1986.
1161 Rotwein et al., N. Engl. J. Med. 308:65-71,1983.
1162 Rouleau et al., Nucleic Acids Res. 16:11848,1988.
1163 Rouleau et al., Nucleic Acids Res. 16:1646,1988.
1164 Rouleau, personal communication,1988.
1165 Rouyer et al., (HGM10) Cytogenet. Cell Genet. 51:1070,1989.
1166 Rouyer et al., Cold Spring Harbor Symp. Quant. Biol.
        51:221-228,1986.
1167 Rouyer et al., Nature 319:291-295,1986.
1168 Rozen et al., Nature 313:815-817,1985.
1169 Rubin et al., Nucleic Acids Res. 16:8791,1988.
1170 Sacchi et al., Nucleic Acids Res. 14:9545,1986.
1171 Sadler, personal communication,1988.
1172 Sam et al., Nucleic Acids Res. 16:8748,1988.
1173 Sanna et al., Nucleic Acids Res. 14:6776,1986.
1174 Sasaki et al., Nucleic Acids Res. 15:6309,1987.
1175 Scambler and Lench, Nucleic Acids Res. 15:8121,1987.
1176 Scambler and Williamson, Nucleic Acids Res. 13:6788,1985.
1177 Scambler et al., Am. J. Hum. Genet. 38:567-572,1986.
1178 Scambler et al., Am. J. Hum. Genet. 41:925-932,1987.
1179 Scambler et al., Hum. Genet. 76:278-282,1987.
1180 Scambler et al., Nucleic Acids Res. 13:3016,1985.
1181 Scambler et al., Nucleic Acids Res. 13:6787,1985.
1182 Scambler et al., Nucleic Acids Res. 14:1927,1986.
1183 Scambler et al., Nucleic Acids Res. 14:1951-1956,1986.
1184 Scambler et al., Nucleic Acids Res. 14:4382,1986.
1185 Scambler et al., Nucleic Acids Res. 15:3639-3652,1987.
1186 Scarpati et al., Nucleic Acids Res. 15:9098,1987.
1187 Scheffer et al., Hum. Genet. 74:249-255,1986.
1188 Scheffer et al., Hum. Genet. 77:335-337,1987.
1189 Scheffer et al., Nucleic Acids Res. 14:3148,1986.
1190 Scheffer et al., Nucleic Acids Res. 14:4374,1986.
1191 Schellenberg et al., J. Neurogenet. 4:97-108,1987.
1192 Schepens et al., Nucleic Acids Res. 15:3192,1987.
1193 Schepens et al., Nucleic Acids Res. 15:3193,1987.
1194 Schmeckpeper et al., Nucleic Acids Res. 13:5724,1985.
1195 Schmidt et al., Nucleic Acids Res. 15:7652,1987.
1196 Schmidt et al., Science 224:1104-1106,1984.
1197 Schmidtke et al., Hum. Genet. 67:428-431,1984.
1198 Schmidtke et al., J. Med. Genet. 23:217-219,1986.

1199 Schneid et al., Nucleic Acids Res. 16:9059,1988.
1200 Schneid et al., Nucleic Acids Res. 17:466,1989.
1201 Schneider et al., J. Clin. Invest. 78:650-657,1986.
1202 Schnur et al., Am. J. Hum. Genet. 44:248-254,1989.
1203 Schumm et al., Am. J. Hum. Genet. 42:143-159,1988.
1204 Schwartz and Barjon, Nucleic Acids Res. 15:862,1987.
1205 Schwartz et al., (HGM10) Cytogenet. Cell Genet.
        51:1075,1989.
1206 Schwartz et al., (HGM8) Cytogenet. Cell Genet. 40:740,1985.
1207 Schwartz et al., Clin. Genet. 29:472-473,1986.
1208 Schwartz et al., Cytogenet. Cell Genet. 43:117-120,1986.
1209 Schwartz et al., Hum. Genet. 78:156-160,1988.
1210 Schwartz et al., Nucleic Acids Res. 16:5225,1988.
1211 Schwartz, personal communication,1987.
1212 Scott et al., Lancet i:771-773,1985.
1213 Sebastio et al., Nucleic Acids Res. 13:5404,1985.
1214 Sefiani et al., (HGM10) Cytogenet. Cell Genet. 51:1076,1989.
1215 Seidegard et al., Proc. Natl. Acad. Sci. USA
        85:7293-7297,1988.
1216 Seilhamer et al., DNA 3:309-317,1984.
1217 Seizinger et al., Nature 332:268-269,1988.
1218 Seizinger et al., Science 236:317-319,1987.
1219 Semenza et al., Nucleic Acids Res. 15:6768,1987.
1220 Shaw and Bell, Nucleic Acids Res. 13:8661,1985.
1221 Shaw et al., (HGM8) Cytogenet. Cell Genet. 40:741,1985.
1222 Shaw et al., Hum. Genet. 74:262-266,1986.
1223 Shaw et al., Hum. Genet. 74:267-269,1986.
1224 Shiang et al., Genomics 4:82-86,1989.
1225 Shiloh et al., Nucleic Acids Res. 13:5403,1985.
1226 Shows et al., Diabetes 36:546-549,1987.
1227 Siddique et al., Nucleic Acids Res. 17:1785,1989.
1228 Silva et al., Nucleic Acids Res. 15:3845-3857,1987.
1229 Simmler et al., EMBO J. 6:963-969,1987.
1230 Simmler et al., Nature 317:692-697,1985.
1231 Simon et al., Nucleic Acids Res. 15:373,1987.
1232 Simpson et al., Nature 328:528-530,1987.
1233 Singh et al., (HGM10) Cytogenet. Cell Genet. 51:1080,1989.
1234 Singh et al., (HGM10) Cytogenet. Cell Genet. 51:1080,1989.
1235 Singh et al., (HGM9) Cytogenet. Cell Genet. 46:693,1987.
1236 Siniscalco et al., (HGM10) Cytogenet. Cell Genet.
        51:1081,1989.
1237 Skare et al., Nucleic Acids Res., in press, 1989.
1238 Skare et al., Proc. Natl. Acad. Sci. USA 84:2015-2018,1987.
1239 Skoda et al., J. Biol. Chem. 263:1549-1554,1988.
1240 Skolnick et al., (HGM7) Cytogenet. Cell Genet.
        37:210-273,1984.
1241 Skraastad et al., Am. J. Hum. Genet. 44:560-566,1989.
1242 Smead et al., Nucleic Acids Res., in press, 1989.
1243 Smeets et al., Hum. Genet. 80:49-52,1988.
1244 Smeets et al., Nucleic Acids Res. 15:8120,1987.
1245 Smeets et al., Nucleic Acids Res. 17:3323,1989.
1246 Smeets et al., Nucleic Acids Res. 17:3324,1989.
1247 Smeets et al., Nucleic Acids Res. 17:3325,1989.
1248 Smeets et al., Nucleic Acids Res. 17:3627,1989.
1249 Smeets et al., Nucleic Acids Res. 17:3628,1989.
1250 Smit et al., Hum. Hered. 38:277-282,1988.
1251 Smith et al., (HGM8) Cytogenet. Cell Genet. 40:748-749,1985.
1252 Smith et al., (HGM8) Cytogenet. Cell Genet. 40:748,1985.
1253 Smith et al., Am. J. Hum. Genet. 42:335-344,1988.
1254 Smith et al., Nucleic Acids Res. 15:6764,1987.
1255 Smith et al., Nucleic Acids Res. 17:5878,1989.
1256 Snell et al., (HGM10) Cytogenet. Cell Genet. 51:1083,1989.
1257 So et al., Immunogenetics 25:141-144,1987.
1258 Solomon et al., Nature 328:616-619,1987.
1259 Sood et al., Nucleic Acids Res. 17:4422,1989.
1260 Soria et al., Proc. Natl. Acad. Sci. USA 86:587-591,1989.
1261 Spence et al., Am. J. Hum. Genet. 39:729-734,1986.
1262 Spielman et al., Proc. Natl. Acad. Sci. USA
        81:3461-3465,1984.
1263 Spritz et al., Nucleic Acids Res. 16:9890,1988.
1264 Spurr et al., (HGM10) Cytogenet. Cell Genet. 51:1085,1989.
1265 Spurr et al., Ann. Hum. Genet. 50:145-152,1986.
1266 Spurr et al., Genomics 2:119-127,1988.
1267 Spurr et al., Hum. Genet. 78:333-337,1988.
1268 Spurr et al., Nucleic Acids Res. 15:5901,1987.
1269 Squire et al., Proc. Natl. Acad. Sci. USA 83:6573-6577,1986.
1270 Srivatsan et al., Cancer Res. 46:6174-6179,1986.
1271 St Clair et al., Nature 339:305-309,1989.
1272 St George-Hyslop et al., Science 235:885-890,1987.
1273 Stafford et al., Proc. Natl. Acad. Sci. USA 85:880-884,1988.

1274 Stambrook et al., Somatic Cell Mol. Genet. 10:359-367,1984.
1275 Stanier et al., Hum. Genet. 80:309-310,1988.
1276 Stapleton, Nucleic Acids Res. 116:2735,1988.
1277 Starr and Wood, Genome 29:201-205,1987.
1278 Starr et al., Nucleic Acids Res. 15:2784,1987.
1279 Sten-Linder et al., Nucleic Acids Res. 17:1277,1989.
1280 Stephens et al., (HGM9) Cytogenet. Cell Genet. 46:699,1987.
1281 Stephens et al., Genomics 1:353-357,1987.
1282 Stetler et al., Proc. Natl. Acad. Sci. USA
        82:8100-8104,1985.
1283 Stewart et al., Genomics, in press,1989.
1284 Stewart et al., Nucleic Acids Res. 13:4125-4132,1985.
1285 Stewart et al., Nucleic Acids Res. 13:7168,1985.
1286 Stewart et al., Nucleic Acids Res. 15:3939,1987.
1287 Steyn et al., Nucleic Acids Res. 15:4702,1987.
1288 Stocks et al., Am. J. Hum. Genet. 41:106-118,1987.
1289 Stoker et al., Nucleic Acids Res. 13:4613-4622,1985.
1290 Stolz et al., (HGM10) Cytogenet. Cell Genet. 51:1086,1989.
1291 Strachan et al., Lancet ii:1272-1273,1987.
1292 Strom, Nucleic Acids Res. 16:9077,1988.
1293 Suthers et al., (HGM10) Cytogenet. Cell Genet. 51:1087,1989.
1294 Suthers et al., Nucleic Acids Res. 16:11389,1988.
1295 Swallow et al., Ann. Hum. Genet. 51:289-294,1987.
1296 Swallow et al., Nature 328:82-84,1987.
1297 Sykes et al., Hum. Genet. 70:35-37,1985.
1298 Sykes et al., Lancet ii:69-72,1986.
1299 Taggart et al., Am. J. Hum. Genet. 38:848-854,1986.
1300 Taggart et al., Proc. Natl. Acad. Sci. USA
        82:6240-6244,1985.
1301 Takeda et al., Nucleic Acids Res. 14:6777,1986.
1302 Talmud et al., (HGM8) Cytogenet. Cell Genet. 40:760,1985.
1303 Tanzi et al., Nature 329:156-157,1987.
1304 Tanzi et al., Science 235:880-884,1987.
1305 Tanzi et al., Science 238:666-669,1987.
1306 Tasset et al., Am. J. Hum. Genet. 42:854-866,1988.
1307 Tateishi et al., Nucleic Acids Res. 14:1926,1986.
1308 Taylor et al., Hum. Genet. 79:273-276,1988.
1309 Taylor et al., Nucleic Acids Res. 16:6259,1988.
1310 Taylor et al., Nucleic Acids Res. 16:7217,1988.
1311 Taylor et al., Nucleic Acids Res. 17:6426,1989.
1312 Tesch et al., Nucleic Acids Res. 16:4193,1988.
1313 Thibodeau et al., (HGM10) Cytogenet. Cell Genet.
        51:1089,1989.
1314 Thomas et al., (HGM9) Cytogenet. Cell Genet. 46:704,1987.
1315 Thompson et al., Am. J. Hum. Genet. 42:113-124,1988.
1316 Tikka et al., Nucleic Acids Res. 15:5497,1987.
1317 Tinklenberg et al., Nucleic Acids Res. 16:10948,1988.
1318 Torroni et al., Nucleic Acids Res. 16:9061,1988.
1319 Trent et al., Am. J. Hum. Genet. 39:350-360,1986.
1320 Tsavaler et al., Proc. Natl. Acad. Sci. USA
        84:4529-4532,1987.
1321 Tsavaler et al., Proc. Natl. Acad. Sci. USA
        85:7680-7684,1988.
1322 Tsipouras et al., (HGM8) Cytogenet. Cell Genet.
        40:762-763,1985.
1323 Tsipouras et al., Am. J. Hum. Genet. 35:182A,1983.
1324 Tsipouras et al., Am. J. Hum. Genet. 36:1172-1179,1984.
1325 Tsipouras et al., Genomics 3:275-277,1988.
1326 Tsui et al., (HGM8) Cytogenet. Cell Genet. 40:763-764,1985.
1327 Tsui et al., (HGM9) Cytogenet. Cell Genet. 46:706,1987.
1328 Tsui et al., Am. J. Hum. Genet. 39:720-728,1986.
1329 Tsui et al., Science 230:1054-1057,1985.
1330 Tsujimoto et al., Proc. Natl. Acad. Sci. USA
        84:1329-1331,1987.
1331 Tuan et al., Proc. Natl. Acad. Sci. USA 80:6937-6941,1983.
1332 Turnbull et al., Immunogenetics 25:184-192,1987.
1333 Tyler-Smith et al., J. Mol. Biol. 203:837-848,1988.
1334 Upadhyaya et al., Genomics 1:358-360,1987.
1335 Upadhyaya et al., Hum. Genet. 74:391-398,1986.
1336 Uring-Lambert et al., FEBS Lett. 217:65-68,1987.
1337 Uzan et al., Biochem. Biophys. Res. Comm. 119:273-281,1984.
1338 Vaisanen et al., Hum. Hered. 38:65-71,1988.
1339 Van Broeckhoven et al., Nature 329:153-155,1987.
1340 Van Camp et al., (HGM10) Cytogenet. Cell Genet.
        51:1095,1989.
1341 van Camp et al., Nucleic Acids Res. 17:4420,1989.
1342 Van Cong et al., (HGM9) Cytogenet. Cell Genet. 46:670,1987.
1343 van Ommen et al., Genomics 1:329-336,1987.
1344 van Tuinen and Ledbetter, (HGM9) Cytogenet. Cell Genet.
        46:708-709,1987.

1345 van Tuinen et al., Am. J. Hum. Genet. 43:587-596,1988.
1346 Vanin, Nucleic Acids Res., in press, 1989.
1347 Vella et al., Hum. Genet. 69:275-276,1985.
1348 Verga et al., Nucleic Acids Res. 17:4007,1989.
1349 Verga et al., Nucleic Acids Res. 17:4008,1989.
1350 Verga et al., Nucleic Acids Res. 17:4009,1989.
1351 Verga et al., Nucleic Acids Res. 17:4010,1989.
1352 Verga et al., Nucleic Acids Res. 17:4011,1989.
1353 Verweij et al., Nucleic Acids Res. 13:8289,1985.
1354 Vincent et al., (HGM10) Cytogenet. Cell Genet. 51:1099,1989.
1355 Vogelstein et al., Cancer Res. 47:4806-4813,1987.
1356 Vortkamp et al., (HGM9) Cytogenet. Cell Genet. 46:709-710,1987.
1357 Wadey and Cowell, Nucleic Acids Res. 17:3332,1989.
1358 Wadey et al., (HGM10) Cytogenet. Cell Genet. 51:1100,1989.
1359 Wagner et al., Nucleic Acids Res. 17:3328,1989.
1360 Wainscoat et al., Am. J. Hum. Genet. 35:1086-1089,1983.
1361 Wainscoat et al., Hum. Genet. 74:90-92,1986.
1362 Wainscoat et al., Lancet :1299-1302,1984.
1363 Wainscoat et al., Nature 319:491-493,1986.
1364 Wainwright et al., Am. J. Hum. Genet. 41:944-947,1987.
1365 Wainwright et al., Nucleic Acids Res. 13:4610,1985.
1366 Wainwright et al., Nucleic Acids Res. 14:3149,1986.
1367 Wake et al., Nucleic Acids Res. 16:3124,1988.
1368 Walker et al., Nucleic Acids Res. 16:9063,1988.
1369 Wallis and Nakamura, Nucleic Acids Res., in press, 1989.
1370 Wallis et al., Genomics 3:299-301,1988.
1371 Wallis et al., Hum. Genet. 68:286-289,1984.
1372 Wallis et al., J. Med. Genet. 23:411-416,1986.
1373 Walter and Cox, Am. J. Hum. Genet. 42:446-451,1988.
1374 Walter et al., Nucleic Acids Res. 15:4697,1987.
1375 Wandersee et al., personal communication,1989.
1376 Wang and Rogler, Cytogenet. Cell Genet. 48:72-78,1988.
1377 Wapenaar et al., (HGM10) Cytogenet. Cell Genet. 51:1102,1989.
1378 Wapenaar et al., Genomics 2:101-108,1988.
1379 Wappenschmidt et al., Nucleic Acids Res. 14:9221,1986.
1380 Wappenschmidt et al., Nucleic Acids Res. 15:2398,1987.
1381 Wappenschmidt et al., Nucleic Acids Res. 15:2399,1987.
1382 Wappenschmidt et al., Nucleic Acids Res. 15:7658,1987.
1383 Warnich et al., Nucleic Acids Res. 14:1920,1986.
1384 Warnich et al., Nucleic Acids Res. 17:468,1989.
1385 Warnich et al., Nucleic Acids Res. 17:469,1989.
1386 Warnich et al., Nucleic Acids Res. 17:474,1989.
1387 Warren et al., Genomics 1:307-312,1987.
1388 Warren et al., Nucleic Acids Res. 16:9060,1988.
1389 Warren et al., Science 237:652-654,1987.
1390 Wasmuth et al., Nature 332:734-736,1988.
1391 Watkins et al., (HGM10) Cytogenet. Cell Genet. 51:1103,1989.
1392 Watkins et al., (HGM7) Cytogenet. Cell Genet. 37:602,1984.
1393 Watkins et al., (HGM8) Cytogenet. Cell Genet. 40:773-774,1985.
1394 Watkins et al., Cytogenet. Cell Genet. 42:214-218,1986.
1395 Watkins et al., DNA 6:205-212,1987.
1396 Watkins et al., Nucleic Acids Res. 13:6075-6088,1985.
1397 Waye and Willard, Nucleic Acids Res. 14:6915-6927,1986.
1398 Waye et al., (HGM9) Cytogenet. Cell Genet. 46:712-713,1987.
1399 Waye et al., (HGM9) Cytogenet. Cell Genet. 46:712,1987.
1400 Waye et al., Genomics 1:43-51,1987.
1401 Waye et al., Hum. Genet. 77:151-156,1987.
1402 Waye et al., Nucleic Acids Res. 16:2362,1988.
1403 Weaver and Knowlton, Nucleic Acids Res. 17:6429,1989.
1404 Weaver et al., Nucleic Acids Res. 16:11386,1988.
1405 Webb et al., Hum. Genet. 81:157-160,1989.
1406 Weber and May, Am. J. Hum. Genet. 44:388-396,1989.
1407 Weber et al., Hum. Genet. 81:54-56,1988.
1408 Weiffenbach et al., (HGM10) Cytogenet. Cell Genet. 51:1104,1989.
1409 Weiffenbach, personal communication, 1989.
1410 Weisgraber et al., Biochem. Biophys. Res. Comm. 157:1212-1217,1988.
1411 Weiss et al., Nucleic Acids Res. 15:860,1987.
1412 Weller et al., EMBO J. 3:439-446,1984.
1413 Weremowicz et al., (HGM10) Cytogenet. Cell Genet. 51:1107,1989.
1414 Westphal et al., Genomics 1:313-319,1987.
1415 Westphal et al., Hum. Genet. 79:260-264,1988.
1416 Wevrick et al., (HGM10) Cytogenet. Cell Genet. 51:1107,1989.
1417 Wexler et al., Am. J. Hum. Genet. 39:A248,1986.
1418 White et al., Genomics 1:364-367,1987.
1419 White et al., Nature 318:382-384,1985.
1420 Whitehead et al., N. Engl. J. Med. 310: 88-91,1984.
1421 Wieacker et al., (HGM7) Cytogenet. Cell Genet. 37:606-607,1984.
1422 Wieacker et al., Hum. Genet. 76:248-252,1987.
1423 Wiggs et al., N. Engl. J. Med. 318:151-157,1988.
1424 Wilichowski et al., Hum. Genet. 75:32-40,1987.
1425 Willard et al., (HGM8) Cytogenet. Cell Genet. 40:360-489,1985.
1426 Willard et al., Proc. Natl. Acad. Sci. USA 83:5611-5615,1986.
1427 Williams et al., Hum. Genet. 71:227-230,1985.
1428 Williams et al., Lancet ii:102-103,1988.

1429 Williams et al., Oncogene 3:345-348,1988.
1430 Willison et al., (HGM8) Cytogenet. Cell Genet. 40:779-780,1985.
1431 Wilson et al., J. Exp. Med. 164:50-59,1986.
1432 Winqvist et al., (HGM10) Cytogenet. Cell Genet. 51:1108,1989.
1433 Winship et al., Nucleic Acids Res. 12:8861-8872,1984.
1434 Wion et al., Nucleic Acids Res. 14:4535-4542,1986.
1435 Wirth and Gal, Nucleic Acids Res. 17:3326,1989.
1436 Wohn et al., Nucleic Acids Res. 15:2396,1987.
1437 Wolff et al., Nucleic Acids Res. 16:9885,1988.
1438 Wong et al., (HGM9) Cytogenet. Cell Genet. 46:719,1987.
1439 Wong et al., Ann. Hum. Genet. 51:269-288,1987.
1440 Wong et al., J. Exp. Med. 164:1531-1546,1986.
1441 Wong et al., J. Exp. Med. 169:847-863,1989.
1442 Wong et al., Proc. Natl. Acad. Sci. USA 82:7711-7715,1985.
1443 Wong et al., Proc. Natl. Acad. Sci. USA 86:1914-1918,1989.
1444 Woo et al., Lancet ii:767-769,1987.
1445 Woo et al., Nature 306:151-155,1983.
1446 Woo, Am. J. Hum. Genet. 43:781-783,1988.
1447 Wood et al., Am. J. Hum. Genet. 39:744-750,1986.
1448 Wood et al., Cytogenet. Cell Genet. 42:113-118,1986.
1449 Wood et al., Nucleic Acids Res. 17:2368,1989.
1450 Woods et al., J. Clin. Invest. 74:634-638,1984.
1451 Wright et al., Nucleic Acids Res. 17:824,1989.
1452 Wright et al., Nucleic Acids Res. 17:825,1989.
1453 Wrogemann et al., Nucleic Acids Res. 14:4377,1986.
1454 Wrogemann et al., Nucleic Acids Res. 14:5572,1986.
1455 Wu et al., (HGM10) Cytogenet. Cell Genet. 51:1110,1989.
1456 Wu et al., Cytogenet. Cell Genet. 48:246-247,1988.
1457 Wu et al., Nucleic Acids Res. 15:1882,1987.
1458 Wu et al., Nucleic Acids Res. 15:3191,1987.
1459 Wu et al., Nucleic Acids Res. 16:11855,1988.
1460 Wu et al., Nucleic Acids Res. 16:1647,1988.
1461 Wu et al., Nucleic Acids Res. 17:6433,1989.
1462 Wu et al., Nucleic Acids Res., in press, 1989.
1463 Wullich et al., Nucleic Acids Res. 17:3331,1989.
1464 Wyllie et al., Nucleic Acids Res. 16:5224,1988.
1465 Wyman et al., Am. J. Hum. Genet. 36:159S,1984.
1466 Xiang et al., Nucleic Acids Res. 15:7654,1987.
1467 Xiang et al., Nucleic Acids Res. 15:7655,1987.
1468 Xiang et al., Nucleic Acids Res. 15:9101,1987.
1469 Xiang et al., Nucleic Acids Res. 16:3599,1988.
1470 Xiao et al., (HGM9) Cytogenet. Cell Genet. 46:721,1987.
1471 Xiao et al., Nucleic Acids Res. 15:5499,1987.
1472 Xiao et al., Nucleic Acids Res. 15:5500,1987.
1473 Xu and Galibert, Proc. Natl. Acad. Sci. USA 83:3447-3450,1986.
1474 Xu et al., (HGM10) Cytogenet. Cell Genet. 51:1112,1989.
1475 Xu et al., Genomics 2:209-214,1988.
1476 Yamakawa et al., Hum. Genet. 80:1-5,1988.
1477 Yamakawa et al., Nucleic Acids Res. 15:7659,1987.
1478 Yang et al., Nucleic Acids Res. 13:8007-8017,1985.
1479 Yang-Feng et al., Am. J. Hum. Genet. 39:79-87,1986.
1480 Yang-Feng et al., Genomics 2:128-138,1988.
1481 Yates et al., Genomics 1:52-59,1987.
1482 Yokota et al., Science 231:261-265,1986.
1483 Yoshida et al., Am. J. Hum. Genet. 42:872-876,1988.
1484 Yoshimoto et al., Nucleic Acids Res. 16:11850,1988.
1485 Yoshioka et al., Mol. Biol. Med. 3:319-328,1986.
1486 Yoshioka et al., Nucleic Acids Res. 14:3147,1986.
1487 Youngman et al., (HGM9) Cytogenet. Cell Genet. 46:724-725,1987.
1488 Youngman et al., Hum. Genet. 73:333-339,1986.
1489 Youngman et al., Nucleic Acids Res. 16:1648,1988.
1490 Youssoufian et al., Nucleic Acids Res. 15:6312,1987.
1491 Yu and Campbell, Biochem. Soc. Trans. 15:654-655,1987.
1492 Yu and Campbell, Immunogenetics 25:383-390,1987.
1493 Yu et al., EMBO J. 5:2873-2881,1986.
1494 Yurov et al., (HGM9) Cytogenet. Cell Genet. 46:725,1987.
1495 Yurov et al., (HGM9) Cytogenet. Cell Genet. 46:725,1987.
1496 Zabel et al., (HGM10) Cytogenet. Cell Genet. 51:1115,1989.
1497 Zabel et al., (HGM7) Cytogenet. Cell Genet. 37:615,1984.
1498 Zabel et al., Proc. Natl. Acad. Sci. USA 80:6932-6936,1983.
1499 Zbar et al., Nature 327:721-724,1987.
1500 Zhang et al., Nucleic Acids Res. 16:2739,1988.
1501 Zhang et al., Nucleic Acids Res. 17:1788,1989.
1502 Zhang et al., Nucleic Acids Res. 17:1791,1989.
1503 Zoghbi et al., (HGM10) Cytogenet. Cell Genet. 51:1116,1989.
1504 Zoghbi et al., Am. J. Hum. Genet. 42:877-883,1988.
1505 Zoghbi et al., Genomics 3:396-398,1988.
1506 Dr. Michael Dean (pers. comm.)
1507 Dr. Peter O'Connell (pers. comm.)
1508 Dr. James Webber (pers. comm.)
1509 Dr. Barbara Weiffenbach (pers. comm.)
1510 Dr. Andreas Weith (pers. comm.)

The Human Mitochondrial DNA

Date compiled: June 16, 1989

Douglas C. Wallace
Emory University School of Medicine
Department of Biochemistry
Atlanta, Georgia   30322

The Mitochondrial DNA Functional Map

The entire human mitochondrial DNA (mtDNA) sequence has been determined (2).   Functions and gene products have been assigned to all of the structural genes (Figure 1, Table I, Table II). The mtDNA strands have an asymmetric distribution of Gs and Cs generating heavy (H)- and light (L)-strands.   In Figure 1 the loci encoded by the L-strand are shown in the inner complete circle and the loci of the H-strand in the outer complete circle.

Each strand is transcribed from one predominant promoter, P_L and P_H1, located in the displacement (D)-loop.   The D-loop is a triple stranded region generated by the synthesis of a short piece of H-strand DNA, the 7SDNA.   Both P_L and P_H1 are bidirectional (29) and associated with an upstream binding sites (*) for the bidirectional, mitochondrial transcription factor (45). RNA synthesis proceeds around the circle in both directions. A bidirectional attenuator sequence (t) within the MTTL1 gene [L(UUR)] limits L-strand synthesis and maintains a high ratio of rRNA to mRNA transcripts from the H-strand (34,35). The mature RNAs (light grey arcs), numbered 1 to 17, are generated by cleavage of the transcript at the tRNAs (5,6,68,75,76).

The D-loop 7SDNA is initiated at four major and three minor sites. Three of these correspond to L-strand transcription stop sequences at the conserved sequence blocks (CSB) I to III. The most prevalent 7SDNA start is at CSB-II (MTCSB2). Primers for 7SDNA synthesis at this site are generated by the cleavage of the L-strand transcript by RNAse MRP which includes a nuclear encoded RNA which appears to guide the cleavage of the RNA at the CSB-II (26,27,28).   All 7SDNA molecules end at nucleotides just past the termination associated sequence (TAS) (40).   H-strand replication starts at the 7SDNA and proceeds around the L-strand, displacing the single-stranded, parental H-strand.   When the L-strand origin is exposed, replication is initiated with a L-strand specific primase containing the cytosol 5.8S rRNA (115).   L-strand replication then proceeds back along the free H-strand.

The polypeptides of the mtDNA are all subunits of the mitochondrial energy generating pathway, oxidative phosporylation. The mitochondrial mRNAs are translated within the mitochondrion using mtDNA encoded rRNAs and tRNAs on chloramphenicol-sensitive ribosomes.   The mammalian mtDNAs share a unique genetic code

This data was first published in Cytogenetics and Cell Genetics, volume 51, number (1-4), pp 612-621, 1989. The Report of the Committee on Human Mitochondrial DNA . Published by S. Karger Publishers.

where UGA = tryptophan, AGA and AGG = stop, and AUA = methionine (2,3,7,8,104).

The Mitochondrial DNA Sequence Variation

The mtDNA sequence evolves 6 to 17 times faster than comparable nuclear DNA gene sequences (17,19,65,72,113). This has generated a high frequency of restriction fragment length polymorphisms (RFLPs) (Table III). Polymorphic sites for AvaII, BamHI, HaeII, HincII, HpaI and MspI have been found to correlate with the ethnic and geographic origin of the samples (10,13,14,15, 39, 61, 105,108). Analysis of polymorphic restriction endonuclease sites has permitted construction of human mtDNA phylogenies (10,13,14,21,58,61,105,108).

Most RFLPs result from synonymous nucleotide substitutions. However, naturally occurring mtDNA mutations which affect the mtDNA shape (conformational variants), length (length variants) and amino acid sequence (amino acid variants) have also been reported. Nucleotide substitutions in cultured cell mtDNAs which impart resistance to chloramphenicol (Table IV) have also been identified.

The Mitochondrial DNA Mutations in Clinical Disease

Mitochondrial DNA nucleotide substitutions, deletions and insertions have been associated with a spectrum of clinical disorders (106, 107) (Figure 2, Table V). The human mtDNA is maternally inherited (23,47) and several maternally inherited diseases have been described. A specific mtDNA nucleotide substitution has been shown to cause the maternally inherited disease, Leber's Hereditary Optic Neuropathy (111) and comparison of linked polymorphisms indicates that the Leber's mutation has arisen more than once (94,101).

Many insertion-deletion mutations have been identified (Figure 2, Table V) and appear to be spontaneous mutations that occur during development. One common deletion occurs between a pair of 13 base pair direct repeats and removes 4977 nps of the mtDNA. This deletion has been found in approximately one third of all deletions characterized (86,87,90,93,109). MtDNA carrying both deletions and duplications (83) may predominate because they impart a replicative advantage, the deletions shortening the template and the duplications doubling the number of origins.

One family with autosomal dominant ocular myopathy has been described in which affected family members have multiple different muscle mtDNA deletions (118). This family may have a nuclear mutation that predisposes individuals to mtDNA deletions.

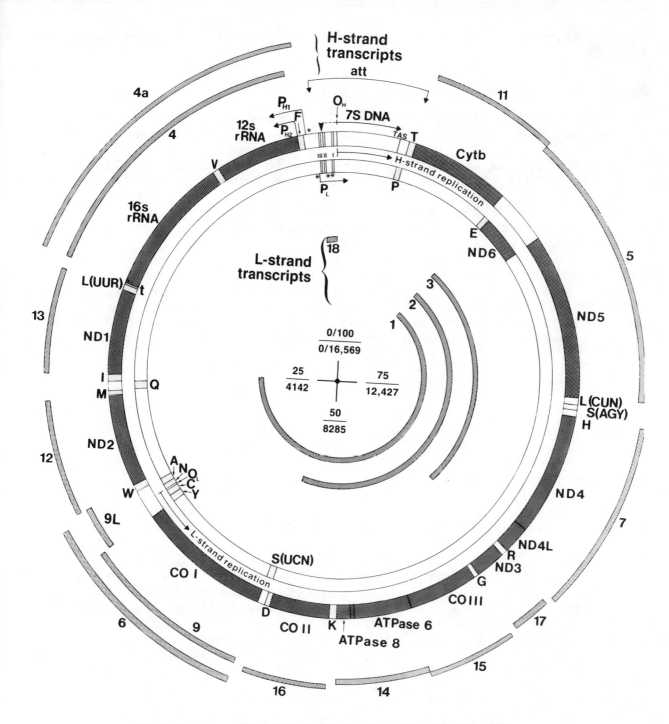

FIGURE 1:   MITOCHONDRIAL DNA FUNCTION MAP

Inner and outer double circles are the C-rich light (L)-strand
and the G-rich heavy (H)-strand, respectively.   Dark grey regions
are rRNA and polypeptide genes.   The tRNA genes are indicated by
light blocks between the larger genes.   The loci designations are
defined and their nucleotide positions are given in Table I.   The
inner and outer light arcs are the stable, processed, L-strand
and H-strand transcripts, respectively. O_H and O_L are the origins of H- and L- strand replication. P_H and P_L are the H- and
L- strand promoters. In the 7SDNA D-loop region, I to III are
conserved sequence blocks (CSBs), the *s indicate transcription
factor binding sites and the heavy arrow between II and III
indicates the L-strand transcript processing site for generating
H-strand replication primers. The "t" within MTTL1 is the bidirectional transcription termination factor.

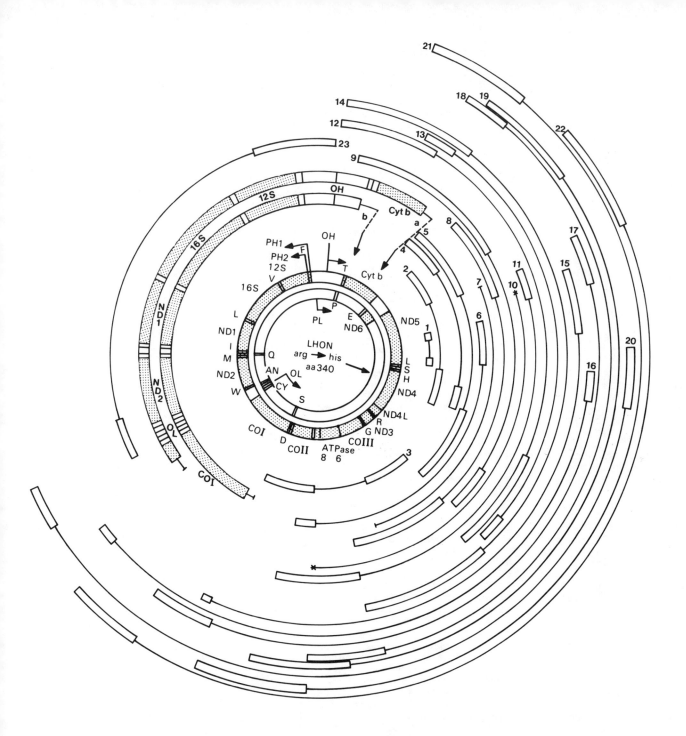

FIGURE 2:   MITOCHONDRIAL DNA MORBID MAP

The positions of genetic disease mutations are shown relative to
the mtDNA function map (inner circles) and their nucleotide
positions and citations are given in Table V.   The position of
the primary Leber's Hereditary Optic Neuropathy (LHON) mutation
at np 11778 is indicated in the center of the circle. The arcs
numbered 1-23 indicate the regions of the mtDNA that were lost in
various deletions.   The open bars at the ends of the arcs are
regions of uncertainty.   Deletion 1 was found in a patient with
Myoclonic Epilepsy and Ragged Red Fiber Disease (MERRF) together
with stroke-like symptoms.   Deletions 2-23 were found in ocular
myopathy patients with symptoms ranging in severity from Chronic
External Ophthalmoplegia Plus (CEOP) to Kearns-Sayre (KS) Syn-
dromes.   Deletion 10 was found in approximately one third of all
ocular myopathy cases and was also found in a Pearson's
Marrow/Pancreas syndrome patient.   The * at the ends of deletion
10 represent the associated 13 base pair direct repeat.   The two
partial mtDNA maps labeled a and b to the left of the function
map indicate the regions that were tandemly duplicated in patients with ocular myopathy associated with diabetes mellitus.
The insertion sites around MTCYB (cytb) are indicated by arrows.

TABLE I: MITOCHONDRIAL DNA FUNCTION LOCATIONS

MAP LOCUS	MAP POSITION (np)	DESIGNATION	NOTES	REFERENCES
MTTAS	16157-16172	TAS	termination seq.	40
MTOHR	191+110,174, 169,219,310,441	O_H	H-strand Origin	25,38,97
MT7SDNA	O_L-(16104-16106)	7S	7S DNA	40
MTCSB1	213-235	I	CSB-I	25
MTTFX	233-260	*	mtTF binding site	45
MTTFY	276-303	*	mtTF binding site	45
MTCSB2	299-315	II	CSB-II	25
MTHPR	317-321	arrow	replication primer	26,27
MTCSB3	346-363	III	CSB-III	25
MTLSP	407±1 392-445	P_L	L-strand promoter	24,37,54, 66,102
MTTFL	418-445	*	mtTF binding site	45
MTTFH	523-550	*	mtTF binding site	45
MTHSP1	559±2,561±2 (545-567)	P_{H1}	Major H-strand promoter	12,24,54, 66,117
MTTF	577-647	P	tRNA Phenylalanine	2
MTHSP2	645	P_{H2}	Minor H-strand promoter	66,67, 117
MTRNR1	648-1601	12S	12S rRNA	2
MTTV	1602-1670	V	tRNA Valine	2
MTRNR2	1671-3229	16S	16S rRNA	2
MTRNR3	3206-3229		5S-like sequence	73
MTTER	3237-3249	t	Transcrip.terminator	34,35
MTTL1	3230-3304	L(UUA/G)	tRNA Leucine1	2
MTND1	3307-4262	ND1	NADH Dehydrogenase1	12,32,42
MTTI	4263-4331	I	tRNA Isoleucine	2
MTTQ	4329-4400	Q	tRNA Glutamine	2
MTTM	4402-4469	M	tRNA Methionine	2
MTND2	4470-5511	ND2	NADH Dehydrogenase2	22,32
MTTW	5512-5576	W	tRNA Tryptophan	2
MTTA	5587-5655	A	tRNA Alanine	2
MTTN	5657-5729	N	tRNA Asparagine	2
MTOLR	5721-5781 (5761,5798)	O_L	L-strand Origin	2,55,97
MTTC	5761-5826	C	tRNA cysteine	2
MTTY	5826-5891	Y	tRNA Tyrosine	2
MTCO1	5904-7444	COI	Cyt c Oxidase I	2
MTTS1	7445-7516	S(UCN)	tRNA Serine1	2
MTTD	7518-7585	D	tRNA Aspartic acid	2
MTCO2	7586-8262	COII	Cyt c Oxidase II	2
MTTK	8295-8364	K	tRNA Lysine	2,63
MTATP8	8366-8572	ATPase8	ATP Synthase8	2,63
MTATP6	8527-9207	ATPase6	ATP Synthase6	2
MTCO3	9207-9990	COIII	Cyt c Oxidase III	2
MTTG	9991-10058	G	tRNA Glycine	2
MTND3	10059-10404	ND3	NADH Dehydrogenase3	2,32
MTTR	10405-10469	R	tRNA Arginine	2
MTND4L	10470-10766	ND4L	NADH Dehydrogenase4L	2,32
MTND4	10760-12137	ND4	NADH Dehydrogenase4	2,32
MTTH	12138-12206	H	tRNA Histidine	2
MTTS2	12207-12265	S(AGY)	tRNA Serine2	2
MTTL2	12266-12336	L(CUN)	tRNA Leucine2	2
MTND5	12337-14148	ND5	NADH Dehydrogenase5	2,32
MTND6	14149-14673	ND6	NADH Dehydrogenase6	2,31
MTTE	14674-14742	E	tRNA Glutamic acid	2
MTCYB	14747-15887	Cytb	Cytochrome b	2
MTTT	15888-15953	T	tRNA Threonine	2
MTTP	15955-16023	P	tRNA Proline	2
MTATT	˜15925-499	att	membrane attachment site	1,2

Legend: The map positions correspond to the nucleotide pair (np) numbers determined from the DNA sequence (2). Locus names are the official designations delineated by the given nucleotide numbers. The map designations are used to indicate the position of the locus on the map. Notes further define each locus: TAS = termination associated sequence, CSB = conserved sequence block, mtTF = mitochondrial transcription factor, P = either purine, Q = either pyrimidine, N = any base.

TABLE II: MITOCHONDRIAL DNA POLYPEPTIDE ASSIGNMENTS

LOCUS(a)	PREDICTED Mol. Wt.(b)	TRIS-GLYCINE (c)		SDS-PAGE MOBILITY PO4= -UREA (d)	
		No.	MWt.	No.	MWt.
MTND5	66,600	1	43.5		
MTCO1	57,000	2	39.5	2,3	45
MTND4	51,400	3	36.5	4,5	39,36
MTCYB	42,700	6	27.5	7-10	29
MTND2	38,900	4	31.5	11	25
MTND1	35,600	5	29.5	12	24
MTCO3	30,000	8	22.5	15,16	18
MTCO2	25,500	7	23.6	13,14	20
MTATP6	24,800	9	21.6	17,18	16
MTND6	18,600	10	16.7		
MTND3	13,200	12	13.5	23,24	6
MTND4L	10,700	11?	14.8	26	3.5
MTATP8	7,900	13+13a	9.8	25	4.5

Notes: (a) As described in Table 1. (b) From (2). (c) Labeling of mitochondrial proteins and gel systems (78, 79). Mitochondrial proteins are numbered from slowest, #1, to fastest, #13a, mobility. Apparent molecular weights are from (79,112). Assignments of MTND6 (ND6) and MTND3 (ND3) (77); remaining assignments (112). (d) Labeling of mitochondrial proteins, gel systems, protein band numbering, and apparent molecular weights (30). Gene assignments are MTCO1 (COI), MTCO2 (COII), and MTCO3 (COIII) (30, 49); MTCO2 (COII) and MTATP8 (ATPase 8) (64); MTND1 (ND1), MTATP6 (ATPase6), and MTND3 (ND3) (33) MTND4 (ND4), MTCYB (Cytb), MTND2 (ND2), and MTND4L (ND4L) (32) and MTND6 (ND6) (31, 33).

TABLE III: MITOCHONDRIAL DNA RESTRICTION SITES AND POLYMORPHISMS

```
Acc II     CGCG     *78 *898 *1473 *2570 *9381 *10200 (13938/9)
Alu I**    AGCT     (259) (1240) (1403) (*1610) (*1917) (*2208)
                    (2223) (2384) (2734) (*3537) (*4411) (*4631)
                    (*4769) (*5176) (*5978) (*5996) (*6022) (7025)
                    (*7055) (*7474) (*8074) (9009) (10028) (10352)
                    (10413) (10694) (*11350) (11806) (*12560)
                    (12990) (13068) (*14015) *14304 (14322) (14509)
                    (15606) (16254)
Ava II     GGRCC    *657 (748) *1169 *2268 *2621 *2776 (3876) (4276)
                    (4280) (4346) (4481) (4776) (5165) (5226) (5264)
                    (5984) (6384) (6726) (7779) (8229) (8249) (8270)
                    (8329) (8342) (8380) (8400) (8723) (10933)
                    (12130) (*12629) (*13367) (14892) (15490)
                    (15870) (15890) (15917) (*16390) (16503)
Bam HI     GGATCC   (13368) (13460) (*14258) (16350) (16391)
Bst EII    GGTNACC  (184) *4820 *6618
Bst N1     CCRGG    (8992) (13705) (˜15045) (˜15728) (˜16502)
Dde I**    CTNAG    (*1637) (*1667) (*1715) (*3930) (*5552) (6356)
                    (*6377) (*7750) (*8515) (10394) (11146) (14385)
                    (*14608) (*15238) (*15250) (*15996) (16482)
Eco RI     GAATTC   (*4121) *5274 *12640
Eco RV     GATATC   *3179 *6734 *12871 (16274)
Fnu 4HI**  GCNGC    (260)
Hae II     PGCGCQ   *3160 (*4529) (4830) (8988) (*9052)
                    (9266/9278/9327/9331) (9689) (11002) *13177
                    (*13594) (14534) (*14858) *15002 (15552/15618)
                    (16274)
Hae III**  GGCC     (663) (816) (*1463) (1484) (1515) (*3315) (3391)
                    (*3427) (3842) (*3849) (4793/4794) (4848) (5261)
                    (*6260) (6333) (6425) (*6957) (7347) (8165)
                    (*8250) (*8391) (*8994) (*9266) (*9294) (*9342)
                    (9438) (*9553) (9714) (*10364) (*10689) (*10725)
                    (11329) (*13051) (*13702) (14279) (14749)
                    (*15047) (*15152) (*15172) (15431) (*15883)
                    (16254) (16398) (16517) (16534)
Hha I      GCGC     (255) *1301 *1472 (1536) *1768 *2941 *3161
                    (*3698) *4530 (4831) (5351) (5538) *5971 (6166)
                    *7598 (7617) (7828) (7898) (8852) (*9053)
                    (*9380) (10066) *11691 (12501) *13178 (*13208)
                    (*13595) (13938) *14859 *15003
Hinc II    GTQPAC   (*1004) (1096) (1118) (1921) *2421 *3592 *5691
                    *5917 (*7853) *7997 *10014 (12115) (12148)
                    (*12406) (*13259) (*13634) (*14199) (16078)
                    (16200)
Hind III   AAGCTT   (4582) *6203 (9736) *11680 *12570
Hinf I**   GANTC    (4092) (*4360) (*5983) (5987) (*6211) (6610)
                    (*6931) (7902) (7970) (8750) (*8783) (9376)
                    (9859) (10806) (12192) (12925) (*13031) (*13103)
                    (*13268) (13605) (*14976) (15005) (*15234)
                    (15723) (*16000) (*16065) (16246) (16389)
                    (16390) (16490)
Hpa I      GTTAAC   (207) (212) (1004) (1921) (3592) (*5691) (6501)
                    (*10014) (11852/11876) (12026) (12114) (*12406)
                    (14408)
Hph I**    GGTGA    (186)
Kpn I      GGTACC   *2573 (*16048) (*16129)
Mbo I      GATC     *1 (8) (340) (719) (*740) *951 *1227 (2390)
                    *2896 *2996 *3063 *3659 *3693 *6460 *6904 *7658
                    (*7859) (8565) (*8592) *8616 *8729 (9150) *10254
                    (10771) (*11922) (12795) (13004) (13116) (13182)
                    *14259 (14279) (*14869) *15060 (15195) (15235)
                    *15357 *15591 (15790) (16215) (16373)
```

TABLE III: MITOCHONDRIAL DNA RESTRICTION SITES AND POLYMORPHISMS

Mnl I**	CCTC	(*144) (16185) (*16187) (*16222) (16278) (16355)
Msp I	CCGG	(64) *104 (496) (707) (712) *931 *3077 *3246
		(*4711) *4846 *5242 (*5742) (5754) *5766 *6262
		(6501) *6571 *6688 *6850 *7204 (7977) (8002)
		(*8112) (*8150) (8649) *9292 (11161) (11164)
		(11440) (11454) *11688 (*12123) (12813) (13070)
		(13100) *13364 *13712 (14205) (14567) (15510)
		(15885) (15912) (*15925) *16453
Pst I	CTGCAG	*6910 (7364) *9020 (16154)
Pvu II	CAGCTG	(981) *2650 (12753)
Rsa I	GTAC	(*1307) (1315) *1476 *1576 *1782 (2120) *2574
		(*2758) (*2849) (*3123) (*3337) (*4464) (4621)
		(4643) (4762) (5009) *5054 (5985) (5987) (6915)
		*6999 *7013 (7241) *7897 *7912 *8012 (8299)
		(8356) *8587 (*8998) (9429) (*9746) *10009
		(10644) *10737 *11447 *11459 *11546 (12345)
		(12810) (13096) *13325 (15282) (*15812) (15897)
		(15949) (15907) (*16049) (16089) (*16096)
		(*16125) *16130 *16156 (*16208) (*16303)
		(*16310) *16329
Sac I	GAGCTC	*36 (*9643)
Sau 96I**	GGNCC	(16516)
Sca I	AGTACT	(4731) *8011 *8586 (*9745) *11446 *11458
Stu I	AGGCCT	*2172 *3145 *3606 *4562 *5225 *5836 (7196) *8571
		*9341 *9437 *12978 (*13701) *13956 (*15046)
Taq I**	TCGA	(134) (270) (*3899) (*3944) (*5269) (6409)
		(7214) (*7335) (*7461) (8022) (9070) (*9751)
		(10084) (*10180) (10893) (*11421) (*13404)
		(13636) (14050) (14158) (*14956) (16127)
Xba I	TCTAGA	*1193 *2952 (*7440) *8286 (*10256)
Xho I	CTCGAG	(9758) (12350) (14157) *14955

Legend: Constant and polymorphic restriction endonuclease sites.
* = sites identified in the published sequence (2) which are
accurate to the nucleotide. The other sites have been assigned
by restriction mapping and are accurate to varying degrees. Codes
for recognition sites are: N=A/G/C/T; R=A/T; P=A/G; Q=C/T. Sites
in parentheses are polymorphic in the human population. The **
notation directly next to an enzyme name indicates that all
nonpolymorphic sites are not listed due to space constraints.

References: Acc II: 59; Alu I**: 21 (**Note: has 64 sites in published sequence); Ava II: 13, 14, 15, 20, 21, 50, 59, 61, 81, 89,
91, 96, 108; Bam HI: 10, 13, 14, 15, 16, 18, 58, 61, 89, 100; Bst
EII: 4; Bst N1: 98; Dde I**: 21, 96 (**Note: has 72 sites in published sequence); Eco RI: 58; Eco RV: 58; Fnu 4HI**: 4, 21
(**Note: has 30 sites in published sequence); Hae II: 13, 14, 15,
58, 61, 89, 91, 103, 108; Hae III**: 4, 21, 59, 81, 96 (**Note:
has 50 sites in published sequence); Hha I: 21, 57, 96; Hinc II:
10, 14, 15, 16, 18, 58, 81, 89, 100, 108; Hind III: 58; Hinf I**:
21, 57, 96 (**Note: has 36 sites in published sequence); Hpa I:
10, 13, 14, 15, 16, 20, 21, 39, 61, 89, 91, 108; Hph I**: 4
(**Note: has 55 sites in published sequence); Kpn I: 4, 16, 48,
58; Mbo I: 4, 16, 20, 21; Mnl I**: 4 (**Note: has 199 sites in
published sequence); Msp I: 13, 14, 15, 16, 18, 20, 21, 51, 61,
91, 100, 108; Pst I: 51, 58; Pvu II: 50, 58; Rsa I: 4, 21, 57,
96; Sac I: 58; Sau 96I**: 4 (**Note: has 32 sites in published
sequence); Sca I: 58; Stu I: 58; Taq I**: 21, 59 (**Note: has 29
sites in published sequence); Xba I: 51, 58; Xho I: 51, 58.

TABLE IV: ADDITIONAL MITOCHONDRIAL DNA VARIANTS

TYPE OF VARIANT np location	LOCUS	NUCLEOTIDE CHANGE	FUNCTIONAL CHANGE	REFERENCE
Conformational Variants				
10398	MTND3	A↔G	Thr↔Ala	95,77
11253	MTND4	T↔C	Ile↔Thr	95,99
Amino Acid Variants				
8701	MTATP6	A↔G	Thr↔Ala	111
9163	MTATP6	G↔A	Val↔Ile	111
10086	MTND3 MV-1/MV-2	A↔G	AspN↔Asp	77
12385	MTND5	C↔T	Pro↔Ser	111
12406	MTND5	A↔G	Ile↔Val	10
Length Variants				
15925-16303	D-loop	-10	—	22
16222	D-loop	+1	—	4,48
309	D-loop	-1-6	—	4,48,52
520	D-loop	-2	—	4,48
37-585	D-loop	±7	—	22
3958-4428	MTND1,MTTI, MTTQ,MTTM	+10	—	22 59
5261-5552	MTND2,MTTY	±7	—	22
5837-6027	MTTC,MTTY	+10	—	59
5877-5978	MTTY,MTCO2	+14	—	22
8272-8289	MTCO2,MTT	±6 or 9	—	22,53,59, 116
8957-9443	MTATP6	-11	—	59
10014-10258	MTTG,MTND3	±80	—	59
10352-10598	MTND3, MTTR,MTND4L	+10	—	22
14608-14802	MTCYB	±7	—	22,59
15756-16000	MTCYB	±4	—	59
15812-16049	MTCYB	+7	—	59
15883-15994	MTCYB,	±6	—	22,59
Drug-Resistant Variants				
1670-3229	MTRNR2	?	ERY-R	41
2991	MTRNR2	T↔C	CAP-R	9,62
2939	MTRNR2	C↔A	CAP-R	9
4747-15886?	MTCYB?	?	ANT-R	70

Legend: Conformational variants change the shape of the mtDNA and alter the mobility of restriction fragments on polyacrylamide gels (95). ERY-R = Erythromycin-resistant, CAP-R = Chlorampheni-col-resistant and ANT-R = Antimycin-resistant.

TABLE V: MITOCHONDRIAL DNA DISEASE MUTATIONS

CLINICAL SYNDROME MAPPING METHOD	NATURE OF MUTATION	MAP LOCATION	REFERENCES
LHON # 1			
P,L,S	G(Arg)→A(His)	11778	44,82,94,111
LHON # 2			
P,L	?	?	101
LHON+IBSN			
P	point mutation	?	74
MERRF			
P	point mutation	?	84,110,114
Cardiac septal defects			
P	?	?	92
MERRF+Strokes			
SB	del 1 (0.4 kb)	12615-13015	88
Ocular Myopathies			
P	?	?	43
SB	del 2 (1.3 kb)	(11.6-12.4)-(13.8-14.6)	69
SB	del 3 (2.0 kb)	(7.4-8.4)-(9.5-10.4)	69
SB	del 4 (2.5 kb)	(11688-12123)-(14258-15047)	80
SB	del 5 (3.4 kb)	(10.4-11.8)-(13.7-15)	69
SB	del 6 (4.2 kb)	(8.4-9.0)-(12.5-13.2)	69
SB	del 7 (4.3 kb)	˜9400-˜13700	60
SB	del 8 (4.4 kb)	(9.0-10.4)-(13.5-14.7)	69
SB	del 9 (4.6 kb)	(10.5-11.1)-(14.1-16.3)	69
SB,S	del 10 (5.0 kb)	(8470-8482)-(13447-13459)	90,93,109
SB	del 11 (5.0kb)	(8005-9020)-(13364-13712)	80
SB	del 12 (5.1kb)	(10.3-11.5)-(15.4-16.6)	69
SB	del 13 (5.5kb)	(10.5-10.8)-(15.3-15.6)	69
SB	del 14 (6.0kb)	(9.0-10.4)-(15.1-16.5)	69
SB	del 15 (5.9kb)	(7.3-7.4)-(12.9-13.5)	56,57
SB	del 16 (6.4kb)	(6.1-6.3)-(12.3-12.5)	69
SB	del 17 (6.4kb)	(6.9-7.5)-(13.4-13.8)	69
SB	del 18 (6.4kb)	(8.4-9.1)-(15.0-15.4)	69
SB	del 19 (6.9kb)	(7.9-8.8)-(14.3-15.3)	69
SB	del 20 (7.0kb)	(5.5-5.9)-(12.3-12.8)	69
SB	del 21 (7.5kd)	(7.5-8.4)-(15-15.9)	69
SB	del 22 (7.6kb)	(6.5-7.0)-(13.7-14.6)	69
SB	del 23 (4.9kb)	(.05-1.1)-(5.2-5.7)	56,57
SB	ins a (8.9kb)	(6033-6203)-(14960-15130)	83
SB	ins b (8.7kb)	(7203-7376)-(˜15890-15950)	83
Pearson's Marrow/Pancreas Syndrome			
SB,S	del 10 (5.0kb)	(8470-8482)-(13447-13459)	85,86
SB	ins (7.0kb)	(5800-16129)	87
Huntington's age of onset			
P	?	?	11,71

Legend: LHON = Leber's Hereditary Optic Neuropathy; LHON + IBSN = Leber's and Infantile Bilateral Striatal Necrosis; MERRF = Myoclonic Epilepsy and Ragged Red Fiber Disease; Ocular Myopathies = Ocular Myopathies with a range of symptoms from Chronic External Ophthalmo-plegia Plus to Kearns-Sayre Syndrome; P = maternally inherited pedigrees; S = nucleotide sequence; SB = Southern blot analysis; point mutation = a single nucleotide change or small deletion/inser-tion; del = deletion with the numbers corresponding to those in the figure; ins = insertion with the letters corresponding to those in the figure, the numbers in the parentheses give the approximate sizes and end points of each del and ins. The 13 nucleotide repeat shown as a * for 10, has been found in about 1/3 of all deletions. The numbers in parentheses in the "Map Location" indicate the limits of each deletion or insertion.

References

1. Albring, M.; Griffith, J.; Attardi, G. Proc. Natl. Acad. Sci. USA 74:1348-1352,1977.

2. Anderson, S.; Bankier, A.T.; Barrell, B.G.; de Bruijn, M.H.L.; Coulson, A.R.; Drouin, J.; Eperon, I.C.; Nierlich, D.P.; Rose, B.A.; Sanger, F.; Schreier, P.H.; Smith, A.J.H.; Staden, R.; Young, I.G. Nature 290:457-465,1981.

3. Anderson, S.; de Bruijn, M.H.L.; Coulson, A.R.; Eperon, I.C.; Sanger, F.; Young, I.G. J. Mol. Biol. 156:683-717,1982.

4. Aquadro, C.F.; Greenberg, B.D. Genetics 103:287-312,1983.

5. Attardi, G.; Chomyn, A.; Montoya, J.; Ojala, D. Cytogenet. Cell Genet. 32:85-98,1982.

6. Attardi, G.; Montoya, J. Methods Enzymol. 97:435-469,1983.

7. Barrell, B.G.; Bankier, A.T.; Drouin, J. Nature 282(5735):189-94,1979.

8. Bibb, M.J.; Van Etten, R.A.; Wright, C.T.; Walberg, M.W.; Clayton, D.A. Cell 26:167-180,1981.

9. Blanc, H.; Adams, C.W.; Wallace, D.C. Nucl. Acids Res. 9:5785-5795,1981.

10. Blanc, H.; Chen, K.H.; D'Amore, M.A.; Wallace, D.C. Amer. J. Human Genet. 35:167-176,1983.

11. Boehnke, M.; Conneally, P.M.; Lange, K. Am. J. Hum. Genet. 35:845-860,1983.

12. Bogenhagen, D.F.; Applegate, E.F.; Yoza, B.K. Cell 36:1105-1113,1984.

13. Bonne'-Tamir, B.; Johnson, J.H.; Natali, A.; Wallace, D.C.; Cavalli-Sforza, L.L. Am. J. Hum. Genet. 38:341-351,1986.

14. Brega, A.; Gardella, R.; Semino, O.; Morpurgo, G.; Astaldi Ricotti, G.B.; Wallace, D.C.; Santachiara Benerecetti, A.S. Amer. J. Hum. Genet. 39:502-512, 1986.

15. Brega, A.; Scozzari, R.; Maccioni, O.; Iodice, C.; Wallace, D.C.; Bianco, I.; Cao, A.; Santachiara Benerecetti, A.S. Ann. Hum. Genet. 50:327-338,1986.

16. Brown, W.M. Proc. Natl. Acad. Sci. USA 77:3605-3609,1980.

17. Brown, W.M.; George, M.,Jr.; Wilson, A.C. Proc. Natl. Acad. Sci. USA 76:1967-1971,1979.

18. Brown, W.M.; Goodman, H.M. In Extrachromosomal DNA. (1979) Cummings, D.J.; Borst, P.; Dawid, I.B.; Weissman, S.M.; Fox, C.F. (eds.) Academic Press, N.Y., pp. 485-499.

19. Brown, W.M.; Prager, E.M.; Wan, A.; Wilson, A.C. J. Mol. Evol. 18:225-239,1982.

20. Cann, R.L.; Brown, W.M.; Wilson, A.C. Genetics 106:479-499,1984.

21. Cann, R.L.; Stoneking, M.; Wilson, W.C. Nature 325:31-36,1987.

22. Cann, R.L.; Wilson, A.C. Genetics 104:699-711,1983.

23. Case, J.T.; Wallace, D.C. Som. Cell Genet. 7:103-108, 1981.

24. Chang, D.D.; Clayton, D.A. Cell 36:635-643,1984.

25. Chang, D.D.; Clayton, D.A. Proc. Natl. Acad. Sci. USA 82:351-355,1985.

26. Chang, D.D.; Clayton, D.A. Science 235:1178-1184,1987a.

27. Chang, D.D.; Clayton, D.A. EMBO J. 6:409-417,1987b.

28. Chang, D.D.; Clayton, D.A. Cell 56:131-139,1989.

29. Chang, D.D.; Hixson, J.E.; Clayton, D.A. Mol. Cell. Biol. 6:294-301,1986.

30. Ching, E.; Attardi, G. Biochem. 21:3188-3195,1982.

31. Chomyn, A.; Cleeter, W.J.; Ragan, C.I.; Riley, M.; Doolittle, R.F.; Attardi, G. Science 234:614-618,1986.

32. Chomyn, A.; Mariottini, P.; Cleeter, M.W.J.; Ragan, C.I.; Matsuno-Yagi, A.; Hatefi, Y.; Doolittle, R.F.; Attardi,G. Nature 314:592-597, 1985.

33. Chomyn, A.; Mariottini, P.; Gonzalez-Cadavid, N.; Attardi, G.; Strong, D.D.; Trovato, D.; Riley, M.; Doolittle, R.F. Proc. Natl. Acad. Sci. USA 80:5535-5539,1983.

34. Christianson, T.W.; Clayton, D.A. Proc. Natl. Acad. Sci. USA 83:6277-6281,1986.

35. Christianson, T.W.; Clayton, D.A. Mol. Cell. Biol. 8:4502-4509,1988.

36. Clayton, D. A. Cell 28:693-705,1982.

37. Clayton, D. A. Ann. Rev. Biochem. 53:573-594,1984.

38. Crews, S.; Ojala, D.; Posakonoy, J.; Nishiguchi, J.; Attardi, G. Nature 277:192-198,1979.

39. Denaro, M.; Blanc, H.; Johnson, M.J.; Chen, K.H.; Wilmsen, E.; Cavalli-Sforza, L.L.; Wallace, D.C. Proc. Natl. Acad. Sci. USA 78:5768-5772,1981.

40. Doda, J.N.; Wright, C.T.; Clayton, D.A. Proc. Natl. Acad. Sci. USA 78:6116-6120,1981.

41. Doersen, G.J.; Stanbridge, E.J. Proc. Natl. Acad. Sci. USA 76:4549-4553,1979.

42. Early, G.P.; Patel, S.D.; Ragan, C.I., Attardi, G. FEB LETTS 219:108-112.

43. Egger, J.; Wilson. J. New Eng. J. Med. 309:142-146,1983.

44. Erickson, R.P. Amer. J. Human Genet. 24:348-349,1972.

45. Fisher, R.P.; Topper, J.N.; Clayton, D.A. Cell 50:247-258,1987.

46. Gaines, G.; Rossi, C.; Attardi, G. Mol. Cell. Biol. 7:925-931,1987.

47. Giles, R.E.; Blanc, H.; Cann, H.M.; Wallace, D.C. Proc. Natl. Acad. Sci. U.S.A. 77:6715-6719,1980.

48. Greenberg, B.D.; Newbold, J.E.; Sugino, A. Gene. 21:33-49,1983.

49. Hare, J.F.; Ching, E.; Attardi, G. Biochem. 19:2023-2030,1980.

50. Harihara, S.; Hirai, M.; Omoto, K. Jpn. J. Hum. Genet. 31:73-83,1986.

51. Harihara, S.; Saitou, N.; Hirai, M.; Gojobori, T.; Park, K.S.; Misawa, S.; Ellepola, S.B.; Ishida, T.; Omoto, K. Am. J. Hum. Genet. 43:134-143,1988.

52. Hauswirth, W.W.; Clayton, D.A. Nucl. Acids Res. 13:8093-8104, 1985.

53. Hertzberg, M.; Mickleson, K.N.P.; Serjeantson, S.W.; Prior, J.F.; Trent, R.J. Am. J. Hum. Genet. 44:504-510,1989.

54. Hixson, J.E.; Clayton, D.A. Proc. Natl. Acad. Sci. USA 82:2660-2664,1985.

55. Hixon, J.E.; Wong, T.W.; Clayton, D.A. J. Biol. Chem. 261:2384-2390,1986.

56. Holt, I.J.; Cooper, J.M.; Morgan-Hughes, J.A.; Harding, A.E. Lancet i:1462,1988.

57. Holt, I.J.; Harding, A.E.; Morgan-Hughes, J.A. Nature 331:717-719,198.

58. Horai, S.; Gojobori, T.; Matsunaga, E. Hum. Genet. 68:324-332,1984.

59. Horai, S.; Matsunaga, E. Hum. Genet. 72:105-117,1986.

60. Johns, D.R.; Drachman, D.B.; Hurko, O. Lancet i:393-394,1989.

61. Johnson, M.J.; Wallace, D.C.; Ferris, S.D.; Rattazzi, M.C.; Cavalli-Sforza, L.L. J. Mol. Evol. 19:255-271,1983.

62. Kearsey, S.E.; Craig, I.W. Nature 290:607-608,1981.

63. Macreadie, I.G.; Novitski, C.E.; Maxwell, R.J.; John U.; Ooi, B.G.; McMullen, G.L.; Lukins, H.B.; Linnane, A. W.; Nagley, P. Nucl. Acids Res. 11:4435-4451,1983.

64. Mariottini, P.; Chomyn, A.; Attardi, G.; Trovato D.; Strong, D.D.; Doolittle, R.F. Cell 32:1269-1277,1983.

65. Miyata, T.; Hayashida, H.; Kikuno, R.; Hasegawa, M.; Ko--bayashi, M.; Koike, K. J. Mol. Evol. 19:28-35,1982.

66. Montoya, J.; Christianson, T.; Levens, D.; Rabinowitz, M.; Attardi, G. Proc. Natl. Acad. Sci. USA 79:7195-7199,1982.
67. Montoya, J.; Gaines, G.L.; Attardi, G. Cell 34:151-159,1983.
68. Montoya, J.; Ojala D.; Attardi, G. Nature 290:465-470,1981.
69. Moraes, C. T.; DiMauro, S.; Zeviani, M.; Lombes, A.; Shanske, S.; Miranda, A.F.; Nakase, H.; Bonilla, E.; Werneck, L.C.; Servidei, S.; Nonaka, I.; Koga, Y.; Spiro, A.J.; Brownell, K.W.; Schmidt, B.; Schotland, D.L.; Zupanc, M.; DeVivo, D.C.; Schon, E.A.; Rowland, L.P. New Eng. J. Med. 320:1293-1299,1989.
70. Munro, E.; Webb, M.; Kearsey, S.E.; Craig, I.W. Exp. Cell Res. 147:329-339,1983.
71. Myers, R.H.; Goldman, D.; Bird, E.D.; Sax D.S.; Merril, C.R.; Schoenfeld, M.; Wolf, P.A. Lancet i:208-210,1983.
72. Neckelmann, N.; Li, K.; Wade, R.P.; Shuster, R.; Wallace, D.C. Proc. Natl. Acad. Sci. 84: 7580-7584, 1987.
73. Nierlich, D.P. Mol. Cell. Biol. 2:207-209,1982.
74. Novotny, E.J., Jr.; Singh, G.; Wallace, D.C.; Dorfman, L.J.; Louis, A.; Sogg, R.L.; Steinman, L. Neurology 36:1053-1060,1986.
75. Ojala, D.; Crews, S.; Montoya, J.; Gelfand, R.; Attardi, G. J. Mol. Biol. 150:303-314,1981b.
76. Ojala, D.; Montoya, J.; Attardi, G. Nature 290:470-474,1981a.
77. Oliver, N.A.; Greenberg, B.D.; Wallace, D.C. J.Biol.Chem. 258:5834-5839,1983.
78. Oliver, N.; McCarthy, J.; Wallace, D.C. Som. Cell. Mol. Gen. 10:639-643,1984.
79. Oliver, N.A.; Wallace, D.C. Mol. Cell. Biol. 2:30-41,1982.
80. Ozawa, T.; Yoneda, M.; Tanaka, M.; Ohno, K.; Sato, W.; Suzuki, H.; Nishikimi, M.; Yamamoto, M.; Nonaka, I.; Horai, S. Biochem. Biophys. Res. Comm. 154:1240-1247,1988.
81. Paabo, S.; Gifford, J.A.; Wilson, A.C. Nucl. Acids Res. 16:9775-9787,1988.
82. Parker, W.D.Jr.; Oley, C.A.; Parks, J.K. New Eng. J. Med. 320: 1331-1333, 1989.
83. Poulton, J.; Deadman, M.E.; Gardiner, R.M. Lancet i:236-240,1989.
84. Rosing, H.; Hopkins, L.C.; Wallace, D.C.; Epstein, C.M.; Weidenheim, K. Ann. Neurol. 17:228-237,1985.
85. Rötig, A.; Colonna, M.; Blanche, S.; Fischer, A.; Le Deist, F.; Frezal, J.; Saudubray, J.; Munnich, A. Lancet ii:567-568,1988.
86. Rötig, A.; Colonna, M.; Bonnefont, J.P.; Blanche, S.; Fischer, A.; Saudubray, J.M.; Munnich, A. Lancet i:902-903,1989a.
87. Rötig, A.; Cormier, V.; Colonna, M.; Bonnefond, J.P.; Saudubray, J.M.; Frezal, J.; Munnich, A. HGM10, June 1989 New Haven, CT, USA, (1989b).
88. Saifuddin, A.N.; Marzuki, s.; Trounce, I.; Byrne, E. Lancet ii:1253-1254,1988.
89. Santachiara Benerecetti, A.S.; Scozzari, R.; Semino, O.; Torroni, A; Brega, A; Wallace, D.C. Ann. Hum. Genet. 52: 39-56, 1988.
90. Schon, E.A.; Rizzuto, R.; Moraes, C.T.; Nakase, H.; Zeviani, M.; DiMauro, S. Science 244:346-349,1989.
91. Scozzari, R.; Torroni, A.; Semino, O.; Sirugo, G.; Brega, A.; and Santachiara-Benerecetti, A.S. Am. J. Hum. Genet. 43:534-544,1988.

92. Sherman, J.; Angulo, M.; Boxer, R.; Gluck, R.A. New. Eng. J. Med. 313:186-187,1985.
93. Shoffner, J.M.; Lott, M.T.; Voljavec, A.S.; Soueidan, S.A.; Costigan, D.A.; Wallace, D.C. Proc. Natl. Acad. Sci. USA (in press).
94. Singh, G.; Lott, M.T.; Wallace, D.C. New Eng. J. Med. 320:1300-1305,1989.
95. Singh, G.; Neckelmann, N.; Wallace, D.C. Nature 329:270-272,1987.
96. Stoneking, M.; Bhatia, K.; Wilson, A.C. Cold Spring Harbor Symposia on Quantitative Biology, 51:433-439,1986.
97. Tapper, D.P.; Clayton, D.A. J. Biol. Chem. 256:5109-5115,1981.
98. Thibault, M.C.; Gelinas, Y.; Turcotte, L.; Bouchard, C.; Dionne, F.T. HGM10, June 1989, New Haven, CT, USA.
99. Vigilant, L.; Stoneking, M.; Wilson, A.C. Nucl. Acids Res. 16:5945-55, 1988.
100. Vilkki, J.; Savontaus, M.-J.; Nikoskelainen, E.K. Hum. Genet. 80:317-321,1988.
101. Vilkki, J.; Savontaus, M-L.; Nikoskelainen, E.K. HGM10, June 1989, New Haven, CT, USA.
102. Walberg, M.W.; Clayton, D.A. J. Biol. Chem. 258:1268-1275,1983.
103. Wallace, D.C. Mol. Cell. Biol. 1:697-710,1981.
104. Wallace, D.C. Microbiol. Rev. 46:208-240,1982.
105. Wallace, D.C. In Endocytobiology II. Intracellular space as oligogenetic Ecosystem, Schenk HEA and Schwemmler W, (eds.), de Gruyter, N.Y. pp.87-100,1983.
106. Wallace, D.C. In Medical and Experimental Mammalian Genetics: A Perspective, McKusick, V.A., Roderick, T.H., Mori, J. and Paul, N.W. (eds) , A.R. Liss, Inc. for the March of Dimes Foundation, N.Y., BD:OAS, 23:137-190,1987.
107. Wallace, D.C. Trends in Genetics 5:9-13,1989a.
108. Wallace, D.C.; Garrison, K.; Knowler, W.C. Amer. J. Phys. Anthrop. 68:149-155,1985.
109. Wallace, D.C.; Lott, M.T.; Voljavec, A.S.; Shoffner, J.M. HGM10, June 1989, New Haven, CT, USA, (1989b).
110. Wallace, D.C.; Singh, G.; Hopkins, L.C.; Novotny, E.J. In Achievements and Perspectives of Mitochondrial Research. Vol. II: Biogenesis, Quagliariello, E., Slater, E.C., Palmieri, F., Saccone, C. and Kroon, A.M. (eds.), Elsevier Science Publishers, Amsterdam, The Netherlands, pp. 427-436,1985b.
111. Wallace, D.C.; Singh, G.; Lott, M.T.; Hodge, J.A.; Schurr, T.G.; Lezza, A.M.; Elsas, L.J., II; Nikoskelainen, E.K. Science 242:1427-1430,1988.
112. Wallace, D.C.; Yang, J.; Ye, J.; Lott, M.T.; Oliver, N.A.; McCarthy, J. Amer. J. Hum. Genet. 38:461-481, 1986.
113. Wallace, D.C.; Ye, J.; Neckelmann, S.N.; Singh, G.; Webster, K.A.; Greenberg, B.D. Current Genetics. 12:81-90,1987.
114. Wallace, D.C.; Zheng, X.; Lott, M.T.; Shoffner, J.M.; Hodge, J.A.; Kelley, R.I.; Epstein, C.M.; Hopkins, L.C. Cell 55:601-610,1988.
115. Wong, T.W.; Clayton, D.A. Cell 45:817-825,1986.
116. Wrischnik, L.A.; Higuchi, R.G.; Stoneking, M.; Erlich, H.A.; Arnheim, N.; Wilson, A.C. Nucl. Acids Res. 15:529-543,1987.
117. Yoza, B.K.; Bogenhagen, D.F. J. Biol. Chem. 259:3909-3915,1984.
118. Zeviani, M.; Servidei, S.; Gelleva, C.; Bertini, E.; DiMauro, S.; DiDonato, S. Nature 339: 309-311, 1989.

This work was supported by National Institutes of Health Grant NS21328, a Muscular Dystrophy Foundation Clinical Reasearch Grant and a National Science Foundation Grant BNS8718775 awarded to D.C.W.